Image Processing and Jump Regression Analysis

WILEY SERIES IN PROBABILITY AND STATISTICS

Established by WALTER A. SHEWHART and SAMUEL S. WILKS

A complete list of the titles in this series appears at the end of this volume.

Image Processing and Jump Regression Analysis

Peihua Qiu

WILEY-INTERSCIENCE

A JOHN WILEY & SONS, INC., PUBLICATION

Published by John Wiley & Sons, Inc., Hoboken, New Jersey.
Published simultaneously in Canada.

Library of Congress Cataloging-in-Publication Data:

Qiu, Peihua, 1965–
Image processing and jump regression analysis / Peihua Qiu.
p. cm.
"A Wiley-Interscience publication."
Includes bibliographical references and index.
ISBN 0-471-42099-9 (cloth)
1. Image processing. 2. Regression analysis. I. Title.

TA1637.Q58 2005
006.3'7—dc22 2004053000

10 9 8 7 6 5 4 3 2 1

Contents

List of Figures

List of Tables

Preface

Regression analysis provides a major statistical tool for estimating curves and surfaces from data with noise, using various smoothing techniques. From linear regression to parametric regression to nonparametric regression, this area has experienced more than 100 years of development. Conventional nonparametric regression analysis was developed only to estimate *continuous* curves or surfaces. When a conventional nonparametric regression procedure is used for estimating a jump curve or surface, the estimator does not converge to the truth at the jump locations as the sample size increases. Therefore, new statistical tools are necessary for estimating curves and surfaces with jumps. This book introduces some recent statistical methodologies for this purpose.

Estimation of jump curves and surfaces has broad applications. One important application is image processing, since the image intensity function of a monochrome image can be regarded as a jump surface with jumps at the outlines of objects. Because of this connection, this book also introduces some image processing techniques, mainly for edge detection and image restoration, and discusses the similarities and differences between these methods and the related methods on estimating jump curves and surfaces in the statistical literature.

I started my research in nonparametric regression when I was a graduate student in China. At that time, most existing nonparametric regression methods assumed that the curves or surfaces to estimate were continuous. In my opinion, this assumption was faulty as a general rule because curves or surfaces could be discontinuous in some applications, and thus I decided to investigate this problem in my Master's thesis at Fudan University in China. After coming to the United States in 1991, I realized that this topic was closely related to image processing in computer science. I then went to the computer science departments at University of Georgia and University of Wisconsin to take courses on computer graphics and vision, and I also read hundreds of research papers in computer science journals at that time.

As I began my studies here in the United States during the early 1990s, several procedures were proposed in the statistical literature for estimating jump curves or surfaces. However, these procedures often imposed restrictive assumptions on the model, making the methods unavailable for many applications. Another limitation of the methods was that extensive computation was required. In addition, the existing procedures in the image processing literature did not have much theory to support them. A direct consequence was that, for a specific application problem, it was often difficult to choose one from dozens of existing procedures to handle the problem properly. Therefore, it was imperative to suggest some procedures that could work well in applications and have some necessary theory to support them; this became the goal of my Ph.D. thesis research at Wisconsin, and I have been working in the area since then. Part of this book summarizes my own research in this area.

This book has seven chapters. The first chapter introduces some basic concepts and terminologies in the areas of computer image processing and statistical regression analysis, along with presenting the overall scope of the book. Chapter 2 consists of two parts: the first part introduces some basic statistical concepts and terminologies, for the convenience of those readers who do not know or remember them well; and the second part introduces some conventional smoothing procedures in the statistical literature. These first two chapters constitute the prerequisite for the remaining chapters. Chapters 3–5 discuss some recent methodologies for fitting one-dimensional jump regression models, estimating the jump location curves of two-dimensional jump surfaces, and reconstructing two-dimensional jump surfaces with jumps preserved, respectively. Chapters 6 and 7 introduce some fundamental edge detection and image restoration procedures in the image processing literature. At the end of each chapter, some exercise problems are provided.

This book is intended for statisticians, computer scientists, and other researchers or general readers who are interested in curve/surface estimation, nonparametric regression, change-point estimation, computer vision and graphics, medical imaging, and other related areas.

The mathematical level required is intentionally low. Readers with some background in basic linear algebra, calculus through integration and differentiation, and an introductory level of statistics can easily understand most parts of the book. This book can be used as a primary text book for a one-semester course on nonparametric regression analysis and image processing or can be used as a supplemental text book for a course on computer vision and graphics. Some datasets used in this book can be downloaded from the following Wiley ftp site:

ftp://ftp.wiley.com/public/sci_tech_med/image_processing_jra

I thank Chooichiro Asano and Xianping Li for their help and support during the initial stage of my research on estimating jump curves or surfaces at Fudan University in China. With the precious help from Hubert Chen, Naihua Duan, Paul Switzer, and Bob Taylor, I had the chance to do my research in a better environment. It would have been impossible to have this book without their selfless support and help. Encouragement and help from Peter Hall and Steve Marron have had a great impact on my research as well. It was Peter who first told me the connection between jump curve/surface estimation and image processing. I am grateful to my Ph.D. thesis adviser Brian Yandell for his advice, encouragement, and the enormous amount of time spent on my thesis research during my graduate study at Wisconsin and for his continuing support since my graduation. Iréne Gijbels, Alexandre Lambert, and Jörg Polzehl read parts of the manuscript and provided many constructive suggestions and comments. The manuscript was used as lecture notes in my recent advanced topic course offered at the School of Statistics of University of Minnesota in the Spring of 2004; students from that class corrected a number of typos and mistakes in the manuscript. Mr. Jingran Sun kindly made Figure 6.14 used in Section 6.6. An anonymous reviewer assigned by Wiley reviewed the first five chapters and provided a very detailed review report, which much improved the presentation. I am fortunate to have had Jessica Kraker read the entire manuscript. She provided a great amount of constructive comments and suggestions.

Most of my research included in the book was carried out during my graduate study or work at Fudan University, University of Georgia, University of Wisconsin at Madison, Ohio State University, and University of Minnesota. I am indebted to all faculty, staff members, and graduate students of the related departments at these universities. Part of my research was finished during several short research visits to the Center for Mathematics and its Applications of Australian National University and to the Institut de Statistique of Université catholique de Louvain in Belgium. This book project was partially supported by a Grant-in-Aid of Research, Artistry and Scholarship at University of Minnesota, a National Security Agency grant, and a National Science Foundation grant.

Special thanks to my family for their love and constant support.

PEIHUA QIU

Minneapolis, Minnesota
November 2004

1

Introduction

Nonparametric regression analysis provides statistical tools for recovering regression curves or surfaces from noisy data. Conventional nonparametric regression procedures, however, are only appropriate for estimating continuous regression functions. When the underlying regression function has jumps, functions estimated by the conventional procedures are not statistically consistent at the jump positions, which implies that they would not converge to the true regression function at the jump positions when the data size gets larger.

The problem of estimating jump regression functions is important because the true regression functions are often discontinuous in applications. For example, the image intensity function of an image is discontinuous at the outlines of objects, and the equi-temperature surfaces in high sky or deep ocean are often discontinuous. In recent years, statistical analysis of jump regression models, namely regression models with jump regression functions, has been under rapid development. The first major objective of this book is to introduce recent methodologies of jump regression analysis in a systematic way.

Because an image can be regarded as a jump surface of the image intensity function, the edge detection and image restoration problems in image processing are closely related to the jump regression problem in statistics. Due to many differences between the two areas in terms of technical terminologies, researchers' mathematical backgrounds, etc., the two research groups have not yet communicated with each other sufficiently. Our second major objective is to introduce these two research areas in a single book. By describing their connections and differences, we hope this book can help build a bridge between the two areas.

In this chapter, we introduce some basic concepts and terminologies in the areas of computer image processing and statistical regression analysis. Connections between the two areas and the outline of the book are also briefly discussed. The remaining chapters of the book will give more detailed discussions about the materials covered in this chapter.

1.1 IMAGES AND IMAGE REPRESENTATION

Images are everywhere in our daily life. Some images need to be processed to improve their pictorial information for better human interpretation, e.g., deblurring and denoising of some satellite images. Others need to be processed for automatic machine perception, e.g., machine reading of mailing labels. These and similar examples demonstrate that image processing plays an important role in our modern society.

A monochrome image can be expressed by a bivariate function $f(x, y)$, where (x, y) denotes the spatial location in the image and the function value $f(x, y)$ is proportional to the brightness of the image at (x, y). In the computer science literature, the function $f(x, y)$ is often called the *image intensity function*.

Different coordinate systems can be used for expressing an image by its intensity function. An industry convention is the one with its origin at the upper-left corner of the image, with the x-axis going vertically toward the bottom of the image and the y-axis going horizontally from left to right, as demonstrated in Figure 1.1. Most images included in this book are produced by statistical software packages S-PLUS and R. Their default coordinate systems have the origin at the lower-left corner of the image, with the x-axis going from left to right and the y-axis going vertically upward.

For computers to handle an image, ranging from image storage in computer disks to image processing with some computer software packages, the image needs to be digitized beforehand, in both spatial location and brightness measure. In many image acquisition devices such as video cameras and scanners, a device called a *digitizer* is included, which automatically converts the acquired images into their digital forms. In this book, if there is no further specification, all images mentioned refer to monochrome digital images.

Digitization of the spatial location of an image is a process of sampling all possible locations. A conventional sampling scheme is *uniform sampling*, in which regularly spaced locations are used. Therefore, a conventional digital image can be expressed by a matrix $\{f(i, j), i = 1, 2, \ldots, n_1, j = 1, 2, \ldots, n_2\}$, where i is the index of rows and j is the index of columns, as shown below.

$$\begin{bmatrix} f(1,1) & f(1,2) & \ldots & f(1,n_2) \\ f(2,1) & f(2,2) & \ldots & f(2,n_2) \\ \vdots & \vdots & \ddots & \vdots \\ f(n_1,1) & f(n_1,2) & \ldots & f(n_1,n_2) \end{bmatrix}$$

Fig. 1.1 A conventional coordinate system for expressing an image in industry.

Each element of the matrix is called a *pixel*, an abbreviation of "picture element", of the image. The image *resolution* is related to the sample size $n_1 \times n_2$. The resolution is high if the values of n_1 and n_2 are large and low if their values are small.

Digitization of image brightness measure at each pixel is called *gray level quantization*. For the (i, j)-th pixel, the quantized value of $f(i, j)$ is conventionally assumed to be an integer number in the range $[0, L-1]$, with 0 denoting black and $L-1$ denoting white. The magnitude of $f(i, j)$ denotes the shade of the image at that position. In the literature, the value of $f(i, j)$ is often called *gray level* of the image at the (i, j)-th pixel. For convenience, the total number of gray levels L is usually an integer power of 2. For example, if $L = 2^8$, then a single pixel takes one byte, which equals 8 bits of disk space. For the same reason, n_1 and n_2 are often chosen to be integer powers of 2.

A saved image often contains noise (cf. related discussions in Sections 2.1 and 7.1). In some cases, noise is produced in the process from image acquisition and image digitization to image storage. In some other cases, noise is just part of the image, existing before image acquisition. As an example, Figure 1.2 shows a log-transformed C-band, HH-polarization, synthetic aperture radar (SAR) image of an area near Thetford forest, England. This image is discussed in some detail by Glasbey and Horgan (1995), and it can be downloaded from their web page:

http://peipa.essex.ac.uk/ipa/pix/books/glasbey-horgan/

For convenience, it is also available on the ftp site of this book. Please see the front cover of the book for its address. Clearly, the image contains much noise.

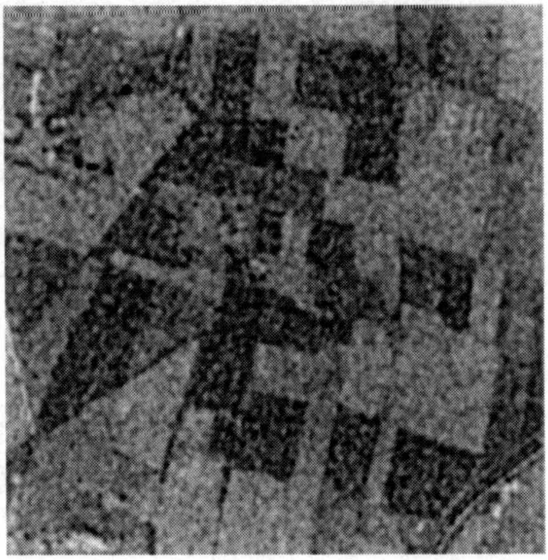

Fig. 1.2 A log-transformed C-band, HH-polarization, synthetic aperture radar image of an area near Thetford forest, England. This image contains much noise.

1.2 REGRESSION CURVES AND SURFACES WITH JUMPS

We begin by looking at an illustrative example: the small diamonds in Figure 1.3 represent the December sea-level pressures observed by a Bombay weather station in India during 1921–1992. Shea *et al.* (1994) pointed out that "a discontinuity is clearly evident around 1960. ... Some procedure should be used to adjust for the discontinuity." This discontinuity is confirmed by using the jump detection procedure suggested by Qiu and Yandell (1998), which is further discussed in Chapter 3.

The data in Figure 1.3 can be described by the following model

$$Y_i = f(x_i) + \varepsilon_i, \; i = 1, 2, \ldots, n, \tag{1.1}$$

where x_i denotes the ith value of the x variable "year", Y_i denotes the ith observation of the Y variable "sea-level pressure", $f(x_i)$ is a function of x_i which gives the mean value of Y at $x = x_i$, and ε_i is a random error term with mean 0 and variance σ^2. Model (1.1) shows that, at a given x position, Y is a random variable. Its mean value is a function of x, and its observed value equals a summation of its mean value and a random error. If readers are not familiar with terminologies such as "the mean of a random variable", Section 2.2 will provide a brief introduction to these.

Model (1.1) is a typical *regression model.* A major goal of regression analysis is to build a functional relationship between Y and x, by estimating the *regression*

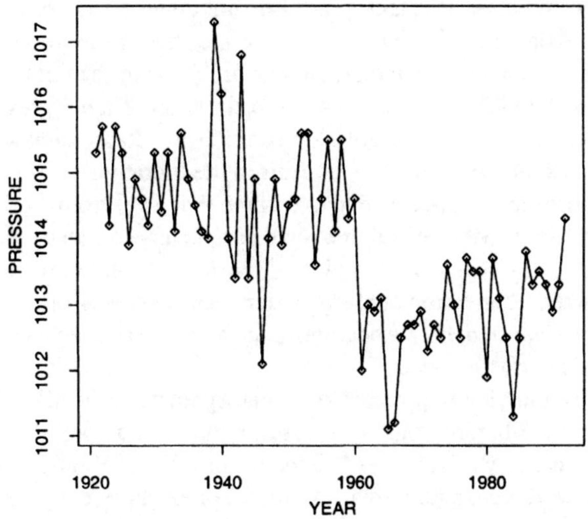

Fig. 1.3 The small diamonds represent the December sea-level pressures observed by a Bombay weather station in India during 1921–1992. They are connected by lines.

function f based on the observed data $\{(x_i, Y_i), i = 1, 2, \ldots, n\}$; then the Y variable can be predicted from the x variable based on this relationship. In statistical regression analysis, the x variable, from which the prediction is made, is often called the *explanatory variable* or *predictor*, and the Y variable, which is the variable to predict, is often called the *response variable*. The positions of the x variable at which observations of Y are obtained are called *design points*, and the space of all possible design points is called the *design space*. It should be mentioned that in some text books the explanatory variable or predictor is also called the *independent variable*, and the response variable is called the *dependent variable*. These older terms are avoided in this book because of confusion with the concepts of independence and dependence (see Section 2.2 for introduction).

In this book, all explanatory variables are assumed to be deterministic (i.e., nonrandom) if there is no further specification, which is appropriate for applications like image processing. For different applications, it might be more reasonable to treat the explanatory variables as random variables (for discussion about regression analysis in such a setup, please read text books such as Cook and Weisberg (1999)). Because the explanatory variables are assumed to be deterministic in this book, it is reasonable to assume that the design space is bounded for most applications. Without loss of generality, in this book the design space is assumed to be $[0, 1]$ in cases with only one explanatory variable involved, and $[0, 1]^p = \{(x_1, x_2, \ldots, x_p) : x_j \in [0, 1], \text{ for } j = 1, 2, \ldots, p\}$ in cases with $p > 1$ explanatory variables involved, if there is no further specification.

If the regression function f is assumed to be linear, then model (1.1) becomes

$$Y_i = \alpha + \beta g(x_i) + \varepsilon_i, \; i = 1, 2, \ldots, n, \tag{1.2}$$

where g is a known function of x, and α and β are unknown *regression coefficients*. In statistics, "a regression model is linear" often implies that the regression function is linear with respect to the unknown regression coefficients instead of the explanatory variable x. In model (1.2), it is obvious that estimation of the regression function is equivalent to estimation of the regression coefficients. Regression analysis under model (1.2) is called *linear regression analysis* in the literature.

In some applications, it is reasonable to assume that $f(x)$ from model (1.1) has a known parametric form with several unknown parameters. As an example, suppose that $f(x) = (x^\lambda - 1)/\lambda$ if $\lambda \neq 0$ and $f(x) = \log(x)$ otherwise, where λ is an unknown parameter. Regression analysis under such models is often referred to as *parametric regression analysis*. Apparently, linear regression analysis is a special case of parametric regression analysis.

In many applications, it is impossible to assume a parametric form for the regression function f. But it is often reasonable to assume that f is a continuous function in the entire design space, which implies that the mean value of Y only changes a small amount if the value of x changes a small amount. For example, suppose that a person wants to lose weight by controlling the amount of food he or she eats every day. In this example, the person's weight is a response variable and the daily intake of food is an explanatory variable. If his or her daily intake of food decreases a little bit, it may not be realistic to expect that his or her weight would decrease dramatically. Therefore, it is reasonable to assume that f is a continuous function in this example. Conventional *nonparametric regression analysis* is specifically designed for handling such cases in which f is assumed to be a continuous function without any parametric form (i.e., it is nonparametric).

In applications, f is often unknown, and the major task of nonparametric regression analysis is to estimate f from the observed data. Intuitively, a good estimator of f should be close to the data because the data carry information about f, as specified in model (1.1). It should also be smooth in conventional nonparametric regression analysis because the true regression function f is assumed to satisfy some smoothness conditions in such cases (see Section 2.1 for related discussion). However, these two goals are usually difficult to achieve simultaneously. For example, the curve obtained by connecting all data points by lines is the one that is closest to the data among all possible curves. Obviously, it is not smooth when a substantial amount of noise is involved in the observed data and thus would not be a good estimator of f in such a case. So there is a trade-off between fidelity of the estimator to the data and the smoothness of the estimator, which turns out to be a major issue in conventional nonparametric regression analysis.

In the case of Figure 1.3, the regression function has an obvious jump around the year 1960. When the true regression function has jumps in the design space and the sample size tends to infinity, it is easy to prove that the estimated regression function by a conventional nonparametric regression procedure does not converge to the true regression function at the jump positions. This implies that the conventional nonparametric regression analysis cannot handle the problem well when the true regression function has jumps. To handle such problems properly, some new statistical techniques are necessary. When the true regression function is assumed to have jumps in the design space, the corresponding regression analysis is called *jump regression analysis* (JRA) in this book.

Similar to conventional regression analysis, JRA can also be classified into parametric JRA and nonparametric JRA, depending on whether f has a parametric form or not. This book focuses on nonparametric JRA, although a special case of parametric JRA is briefly mentioned in Section 1.4 below in the piecewisely linear regression analysis discussion.

One significant application of JRA is the image processing problems mentioned in the previous section. More specifically, these image processing problems are related to two-dimensional regression analysis. A two-dimensional regression model has the following form:

$$Z_{ij} = f(x_i, y_j) + \varepsilon_{ij}, \; i = 1, 2, \ldots, n_1; \; j = 1, 2, \ldots, n_2, \tag{1.3}$$

where x and y are two explanatory variables, $\{Z_{ij},\ i = 1, 2, \ldots, n_1;\ j = 1, 2, \ldots, n_2\}$ are observations of the response variable Z observed at design points $\{(x_i, y_j),\ i = 1, 2, \ldots, n_1;\ j = 1, 2, \ldots, n_2\}$, f is the bivariate regression function, and $\{\varepsilon_{ij},\ i = 1, 2, \ldots, n_1;\ j = 1, 2, \ldots, n_2\}$ are independent and identically distributed random errors with mean 0 and variance σ^2. In the setup of digital images, x_i denotes the ith row of the image, y_j denotes its jth column, f is the image intensity function, $f(x_i, y_j)$ is the true image gray level at the (i, j)-th pixel, ε_{ij} denotes the noise at the (i, j)-th pixel, and Z_{ij} is the observed image gray level at the (i, j)-th pixel. Then, the image intensity function f has jumps at the outlines of objects. In the image processing literature, positions at which f has jumps are called *step edges*, and positions at which the first-order derivatives of f have jumps are called *roof edges* (cf. Haralick 1984). These concepts are further explained in Chapter 6.

1.3 EDGE DETECTION, IMAGE RESTORATION, AND JUMP REGRESSION ANALYSIS

Because much of the information in an image is conveyed by the edges and because our eye-brain system has evolved to extract edges by preprocessing beginning right at the retina, edges are an important part of the entire image. In various applications, outlines of some objects (e.g., the outline of a tumor in medical imaging) need to be enhanced. In others, the detail of an image can be ignored, and all we need to study is the edge information (e.g., in some applications of object recognition). In such cases, the major challenge is to detect edges properly based on the observed image, which is the goal of *edge detection* in image processing. If edges are the only concern in an application, then only the edge information needs to be saved to a storage medium, and the remaining part of the image can be completely discarded. Since much storage space can be saved in this way, edge detection can also be used for *data compression*.

For noisy images, it is often important to remove the noise (i.e., denoising) for better human and machine perception of the true images. Most denoising procedures involve data smoothing, which is a process of averaging neighboring pixels. Due to the importance of edges in human and machine perception of the entire image, we need to preserve them when smoothing the data. Thus *edge-preserving image restoration* is another important research topic in image processing. With the use of the restored image, the image resolution can easily be changed; this is often another purpose of edge-preserving image restoration. In the literature, some image restoration procedures can be used for deblurring the observed image or for reconstructing image objects from several of their projections. These procedures are not the focus of this book because they are quite different from most procedures in JRA.

As mentioned in the previous section, an image can be described by a two-dimensional regression model. Thus, edge detection and edge-preserving image restoration in image processing are essentially the same problems as jump detection and jump-preserving surface estimation in statistical regression analysis, although two different sets of terminologies are used in the two research areas.

There are some differences between an image model and a jump regression model. As mentioned in Section 1.2, image pixels can be regarded as design points of a jump regression model; the pixels are regularly spaced in the design space if the "uniform sampling" scheme is used in image digitization. But design points of a jump regression model usually do not have this regularity; in this sense, the jump regression model is more general. In some applications, such as estimation of the equi-temperature surfaces in high sky or deep ocean, the design points are often irregularly spaced. The jump regression model is appropriate for these problems, but most image processing methods in the computer science literature cannot be applied directly to such problems because these methods have made use of the properties of regularly spaced pixels.

Another major difference between an image model and a jump regression model is that gray levels of a digital image are discrete values, whereas the regression function f can take any values in an interval. In some image applications, the number of gray levels of an image is small (e.g., black-and-white images). Many image processing methods in the literature have used the discreteness of the gray levels. Consequently, they are inappropriate for handling some jump surface estimation problems such as the equi-temperature surface estimation problem mentioned above. Because most JRA procedures introduced in this book treat f as a numerical function that can take any values in an interval, they do not share this limitation.

1.4 STATISTICAL PROCESS CONTROL AND SOME OTHER RELATED TOPICS

In the statistical literature, there are several other research areas that mainly deal with jumps and discontinuities. Methodologies in these areas might be helpful for further development of JRA and image processing. We briefly introduce four such research areas: piecewisely linear regression analysis, change-point estimation for a sequence of random variables or a time series, change-point estimation in survival analysis, and shift detection in statistical process control. These research areas will not be discussed in detail in the following chapters, because their major goals are quite different from those of this book. Interested readers can read the references provided here.

Piecewisely linear regression analysis and other parametric jump regression analyses have been discussed in the literature for several decades (cf., e.g., Brown *et al.* 1975, Hinkley 1969, 1971, Kim 1993, Kim and Siegmund 1989, Quandt 1958, 1960, Worsley 1983a). In the regression model (1.1), the regression function f is assumed to be a piecewisely linear function, and the positions at which f switches from one line to the other are called *change-points*. The number and locations of the change-points are often treated as model parameters along with the regression coefficients. All model parameters are then estimated by parameter estimation procedures, such as the least squares estimation procedure and the maximum likelihood estimation procedure, which are briefly introduced in Subsection 2.2.4.

Next, we consider change-point estimation for a sequence of random variables $X_1, X_2, \ldots, X_n$. The distribution of the first r variables is F, where $1 \leq r \leq n-1$, the distribution of the remaining $n-r$ random variables is G, and $F \neq G$. Then r is called a change-point of the sequence. The main objective of the research area on change-point detection for a sequence of random variables is to estimate the value of r and the distributions F and G (cf., e.g., Gombay 2003, Hinkley 1970, Smith 1975, Worsley 1983b, 1986, Yao 1987). Some change-point detection procedures allow for several change-points in the sequence and can estimate all unknown parameters, including the number and the values of the change-points, simultaneously (e.g., Aly *et al.* 2003, Fu and Curnow 1990, Hawkins 2001, Sullivan 2002). In some applications, e.g., problems involving economic indices, the sequence of random variables is actually a time series. That is, the index of the random variables is time or time-related; neighboring random variables are correlated, and the correlation structure is assumed to follow one of various time series models. Parameters in these models may have shifts at the change-points, which often reflect the impact of some abrupt events on the related system (cf. e.g., Kumar and Wu 2001, Picard 1985, Shiohama *et al.* 2003).

Change-point estimation can also be based on the survival function $S(t)$, which is often used in statistics for describing the chance that a patient can survive a disease after a given time t. With some internal or external incentives, e.g., introduction of a new medical treatment at a specific time, this function may have jumps at some time points. Because the jump positions and jump magnitudes are often related to the effect of the medical treatments, it is important to estimate them properly for evaluating the medical treatments. For more discussion about change-point estimation in survival analysis, read Antoniadis *et al.* (2000), Müller and Wang (1990), Sen (1994), and the references cited there.

Statistical process control (SPC) consists of methods for understanding, monitoring, and improving process performance, e.g., performance of a production line of a soft-drink, over time (Woodall 2000). When the process is *in-control*, the distribution of the process measurements equals a specific distribution. If the measurement distribution changes after an unknown time point, then the process becomes *out-of-control*. The major objective of SPC procedures is to detect such a change as soon as possible so that the process can be stopped and the causes of the shift can be checked out in a timely fashion. In most SPC applications, we are concerned about shifts in the mean or variance parameters of the measurement distribution. Commonly used SPC procedures include the Shewhart charts (e.g., Göb *et al.* 1999, Hunter 1989, 2002), the cumulative sum (cusum) procedures (e.g., Gan 1993, 1994, Hawkins 1987, Johnson and Bagshaw 1974, Page 1954, 1961, 1963, Qiu and Hawkins 2001, 2003, Reynolds *et al.* 1990, Srivastava and Wu 1993, Woodall 1984, 1986, Yashchin 1992, 1993), and the exponentially weighted moving average (EWMA) procedures (Gan 1998, Jones 2002, Stoumbos and Sullivan 2002). For an overview of SPC, read text books such as Montgomery (1996) and Hawkins and Olwell (1998).

1.5 ORGANIZATION OF THE BOOK

This book has seven chapters. Chapter 2 first introduces some basic statistical concepts and terminologies for readers who know little about statistics, to help them understand the statistical jargon used in the remaining parts of the book. Further on in this chapter, we introduce some conventional nonparametric regression procedures on which most JRA procedures are based, including kernel smoothing procedures, local polynomial kernel smoothing procedures, spline smoothing procedures, and function estimation procedures based on wavelet transformations. Chapters 3–5 introduce some recent methodologies for fitting one-dimensional jump regression models, estimating the jump location curves of two-dimensional jump surfaces, and reconstructing the jump surfaces with jumps preserved, respectively. Chapters 6 and 7 introduce some fundamental edge detection and image restoration procedures proposed in the image processing literature. Connections and differences between these image processing procedures and the procedures related to two-dimensional JRA are also discussed.

In each chapter of the book, we focus mainly on the basic ideas of the related methods and procedures. Their major statistical properties are discussed. The mathematical proofs are not included, although some related references are provided in case some readers are interested in reading the proofs. Materials are written in a way that readers with some background in basic linear algebra, calculus through integration and differentiation, and an introductory level of statistics can easily understand them. For each specific topic, we select some representative procedures to introduce in some detail. Other procedures are discussed briefly, and the related references are provided. In most chapters, one section (usually the final section) discusses the strengths and limitations of the procedures introduced in the chapter and notes some related open research problems.

Problems

1.1 Figure 1.1 shows a conventional coordinate system used in industry for expressing an image. In Section 1.1, the default coordinate system in statistical software packages S-PLUS and R is also introduced. Please explain how we can display an image produced in industry in either S-PLUS or R by using their default coordinate system without changing the appearance of the image.

1.2 Suppose that a digital image has 512 rows and 512 columns of pixels. At each pixel, the range of gray levels is $[0, 255]$. To save this image to a computer disk, what is the minimum storage size required for the disk?

1.3 Can a digital image be described properly by the one-dimensional regression model (1.1)? Explain your answer.

1.4 In Section 1.2, it is mentioned that the estimated regression function by a conventional nonparametric regression procedure does not converge to the true regression function at its jump positions. Explain this statement in terms of image restoration.

1.5 Summarize the major connections and differences between an image model and a two-dimensional jump regression model.

2

Basic Statistical Concepts and Conventional Smoothing Techniques

2.1 INTRODUCTION

Observations of a specific characteristic of interest of the members in a population usually have *noise* involved. There are many different sources of noise. For example, when measuring a person's height, measurement error is almost inevitable. Another common source of noise occurs during data recording and data entry, where some digits of the observations may be altered carelessly. If the recording mistakes are made on the leading digits, they might be found and then fixed during the quality control stage of data entry or during data analysis. However, if the mistakes are made on the nonleading digits, it is often difficult to sort them out. Measurement error and mistakes made during data recording and data entry are not part of the original characteristic that we are interested in studying. Thus, in statistics, the issue of noise must be addressed.

When we study the relationship between height and weight, intuition tells us that a functional relationship with an upward trend should exist. But if we present the data of observed (height, weight) pairs in a scatter plot, the data points will never fall exactly on a curve with increasing trend, no matter how precisely each observation is measured or how accurately the data are recorded and entered. A common pattern in such data would be observations clustering around a curve with increasing trend. Some data points are above the curve, whereas others are below the curve; a reason for this phenomenon is that there are many different factors affecting both height and weight. Some of these factors are known and can be measured. Some of them are known but difficult to measure. Others are even unknown. In addition, factors that are known and convenient to measure may be excluded from a study because they are not our major concern. The factors that are related to the data under study but are not included in the data for various reasons are usually the major source of variability in the data. In statistics, variability in the data caused by such factors is also treated as noise.

In the regression model (1.1), f denotes the true functional relationship between the explanatory variable x and the response variable Y, which is simply called the *signal* in some scientific communities, and ε denotes the noise. Signal and noise are assumed to be additive, so each observation of Y is a combination of the signal part and the noise part. Because both parts cannot be observed directly, some statistical tools are needed to remove the noise and consequently to recover the signal from the noisy data. Regression analysis is a major statistical tool for this purpose.

As discussed in Chapter 1, there are several different regression models. In this chapter, we focus on the conventional nonparametric regression model, in which f is assumed to be a "smooth" function without any parametric form.

When the regression function f has kth-order derivative at each point in the design space, where $k \geq 0$ is an integer, it is called *kth-order smooth* in this book. By this definition, when $k_1 \geq k_2$ are two nonnegative integers, a k_1th-order smooth function must also be k_2th-order smooth. A zero-th-order smooth function is continuous in the entire design space, but its derivative functions may not be continuous in the design space or may not even exist at some points. Sometimes, we simply call a zero-th-order or higher-order smooth function a smooth function. Compared with the noisy data, f looks much smoother. Thus the process of recovering it from the noisy data is often called *data smoothing* in the literature.

During the past several decades, a number of nonparametric data smoothing techniques have been proposed in the literature for estimating continuous nonparametric regression functions. These techniques include kernel smoothing procedures, local polynomial kernel smoothing procedures, spline smoothing procedures, discrete wavelet transformation and thresholding procedures, and many others. To better understand these and other procedures discussed in the book, some basic statistical concepts and terminologies are introduced in Section 2.2. The remaining part of this chapter discusses some of the conventional data smoothing procedures mentioned above.

2.2 SOME BASIC STATISTICAL CONCEPTS AND TERMINOLOGIES

This section is only for the convenience of those readers who do not know or remember some of the basic statistical concepts and terminologies well. It can be skipped by readers with background in introductory-level statistics. The introduction here is kept to a minimum. For more complete discussion about statistical theory and inference, see, for example, Lehmann (1983, 1986) and Casella and Berger (1990).

2.2.1 Populations, samples, and distributions

First, we would like to mention that any statistical statement is only applicable to a specific *population*, which is the entire collection of members or objects about which information is desired. For example, if we are interested in knowing the approval rate of the current president's job performance, then the collection of all citizens of voting age in this country is our population.

In any particular application, usually we are only interested in one or several specific characteristics of the members in the population, e.g., the approval status of each individual in the above example. These characteristics are often called variables because their values vary among different members in the population. As another example, suppose that we are interested in knowing the numerical relationship between the height and weight of all current college students in this country. Then the collection of all current college students in this country is our population, and height and weight are the two variables of interest.

Let us focus on single-variable cases first. Suppose that a member is randomly selected from a population. After the selection, the variable value of the selected member becomes known. But before the selection, the variable value of the member to be selected could be any one in the population. In that sense, the variable is random and is therefore called a *random variable*. In the case when several variables are involved, then the random variable is multivariate, or it is a random vector.

For a given population, it is often of interest to know how all the values in the population are distributed or, equivalently, how all the possible values of the related random variable are distributed. This distribution is called the *population distribution*. By the connection between a population and a random variable, the population distribution is the same as the distribution of the related random variable.

We now consider how to describe the distribution of a random variable. If the random variable is categorical in the sense that all its possible values belong to several categories, then a table listing all the categories and the corresponding proportions of the categories in the population is sufficient for describing the population distribution. This table is often called a *probability distribution function* or *probability mass function*. If the random variable is a univariate numerical variable, then its distribution can be described by the following *cumulative distribution function* (cdf):

$$F(x) = P(X \leq x), \quad \text{for } x \in R,$$

where X denotes the random variable of interest and $P(X \leq x)$ is the probability of the event that X is less than or equal to a given value x on the number line R. In the case that the cdf $F(x)$ is absolutely continuous, in the sense that there is a nonnegative, real-valued, measurable function f on R such that

$$F(x) = \int_{-\infty}^{x} f(u)\, du, \quad \text{for } x \in R,$$

the random variable X is said to be *absolutely continuous*, and f is called its *probability density function* (pdf). The corresponding curve of a pdf is called a (probability) density curve. Thus, for an absolutely continuous random variable X, its pdf f can be computed easily from its cdf F by the relationship

$$f(x) = F'(x), \quad \text{for } x \in R.$$

We next consider some properties of probability density functions. A pdf must be a nonnegative integrable function, and the area under its entire curve is one. These conditions are also sufficient for a measurable function to be a pdf. If X has a pdf f, then the area under the density curve and above an interval $[a, b]$ equals $P(a \leq X \leq b)$, for any $-\infty \leq a \leq b \leq \infty$. That is, areas underneath the density curve give probabilities for the random variable X. The center of all the possible values of X can be measured by

$$\mu_X = \int_{-\infty}^{\infty} u f(u)\, du.$$

In the literature, μ_X is often called the *mean* of X or its *expected value*. Another commonly used notation for μ_X is $\mathrm{E}(X)$, where E is the first letter of "expected value". The spread of all possible values of X can be measured by

$$\sigma_X^2 = \int_{-\infty}^{\infty} (u - \mu_X)^2 f(u)\, du,$$

where σ_X^2 is called the *variance* of X. Sometimes we also write σ_X^2 as $\mathrm{Var}(X)$, where Var denotes variance. Its square root σ_X is called the *standard deviation* of X. Because σ_X has the same unit as X, it is often more convenient for measuring the spread, but, mathematically, σ_X^2 is easier to handle.

A very important class of distributions are the *normal* (or *Gaussian*) *distributions*. When an absolutely continuous random variable X has the following pdf:

$$f(x) = \frac{1}{\sqrt{2\pi}\sigma} \exp\left[-\frac{(x-\mu)^2}{2\sigma^2}\right], \tag{2.1}$$

where μ and σ are two parameters, then its distribution is called a normal distribution. It can be easily checked that if X has a normal distribution with parameters μ and σ as defined in equation (2.1), then $\mu_X = \mu$ and $\sigma_X^2 = \sigma^2$. Therefore, a normal distribution is uniquely determined by its mean and variance. We use the conventional notation $X \sim N(\mu, \sigma^2)$ to denote that X is a normally distributed random variable with mean μ and variance σ^2. The normal distribution is important in statistics because of two major reasons. One is that distributions of many continuous variables in applications, such as the height or weight of all people in this country, can be described reasonably well by normal distributions. The second major reason is that much statistical theory is developed with the use of the normal distribution.

Some properties of the normal distribution are as follows. The density curve of a normal distribution is bell-shaped, it is symmetric about the mean μ, and its spread is controlled by the standard deviation σ. Therefore, μ is a *location parameter* and σ is a *scale parameter*. If $X \sim N(\mu, \sigma^2)$, then the random variable defined by

$$Z = \frac{X - \mu}{\sigma}$$

has a normal distribution with mean zero and variance one; this specific normal distribution is called the *standard normal distribution*. The density curve of the standard normal distribution is displayed in Figure 2.1.

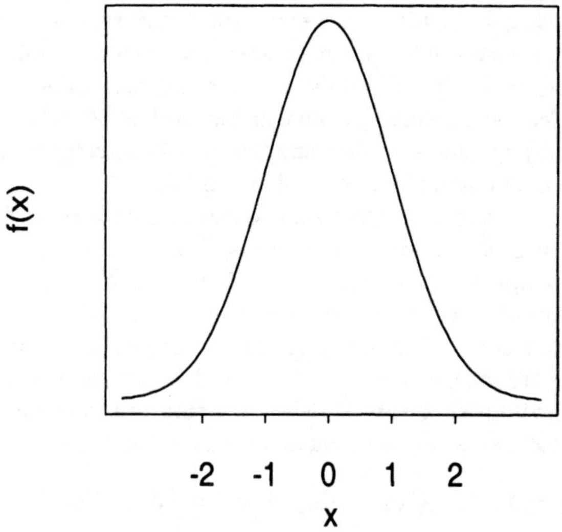

Fig. 2.1 Probability density curve of the standard normal distribution.

In applications, it is often of interest to know the population distribution. In the approval rate of the current president example above, we are interested in knowing the proportion of approvals in the entire population. Because the individuals of the population in this example take only two possible values, either approval or disapproval, the population distribution is uniquely determined by the approval rate in the population, or the *population proportion* of approvals. More generally, when a population distribution has a parametric form such as the one defined in equation (2.1) for a normal distribution, the population parameters appearing in the parametric form uniquely determine the entire population distribution. Therefore, it suffices to know these parameters in order to know the population distribution. In some instances, our major interest is in one or more population parameters instead of the entire population distribution. For example, it is often sufficient to know the average exam score of the students in a class in order to have a rough idea about the overall performance of these students on the exam.

However, to know the population distribution or its parameters, we would need to know variable values for all members in the population. In many applications, the related population is large. It is time-consuming and sometimes even impossible to observe the variable for each member in the population. To overcome this difficulty, the idea of sampling the population enters the picture.

A *sample* of the population is a subset of the population, selected in some prescribed manner for study. After a sample is obtained, the population distribution or its parameters can be *estimated* based on the sample. To have accurate estimation, the sample should represent the population well. In the literature, many different sampling techniques have been proposed to handle different cases. Interested readers can read Cochran (1977) or other text books on statistical sampling for a systematic discussion about this topic. In this book, if there is no further specification, we assume that all samples are *simple random samples*. A simple random sample of size n is a sample that consists of n selected members, which is generated in a way such that the result of one selection has nothing to do with the result of any other selection, and every member in the population has the same chance to be selected in each selection. Such samples turn out to have nice statistical properties.

One property of the simple random sample can be described as follows. Suppose that the cdf of a population distribution is F and that $\{X_1, X_2, \ldots, X_n\}$ is a simple random sample from the population. Then $\{X_1, X_2, \ldots, X_n\}$ can be regarded as a sequence of independent and identically distributed (i.i.d.) random variables with a common cdf F. Intuitively speaking, a sequence of random variables $\{X_1, X_2, \ldots, X_n\}$ are independent of each other if the values of any subset of the sequence provide no information about the values of the remaining random variables in the sequence. Mathematically, they are *independent* if and only if

$$P(X_1 \leq x_1, X_2 \leq x_2, \ldots, X_n \leq x_n) = \Pi_{i=1}^{n} P(X_i \leq x_i)$$

holds for any $x_1, x_2, \ldots, x_n$ on the number line R. The assumption of independent random variables is fundamental in many statistical theories.

2.2.2 Point estimation of population parameters

For statisticians, the sample consists of the data that provide information about the population. After a sample is obtained, the next question is how to estimate the population distribution or its parameters from the sample. Our discussion here focuses on estimating population parameters. For nonparametric estimation of population pdf, please read Härdle (1990) and the references cited there.

Any quantity calculated from the sample and uniquely determined by the sample is called a *statistic*. For example, the *sample mean*

$$\overline{X} = \frac{1}{n} \sum_{i=1}^{n} X_i$$

and the *sample variance*

$$s^2 = \frac{1}{n-1} \sum_{i=1}^{n} (X_i - \overline{X})^2$$

are two statistics. The square root of the sample variance, s, is another statistic which is called the *sample standard deviation*. The sample mean is a measurement of the center of the data, whereas the sample variance or the sample standard deviation is a measurement of the spread of the data. Similar to the population variance and population standard deviation, the sample standard deviation is more convenient to measure the spread of the data because it has the same unit as the observations and the sample variance is easier to handle mathematically.

If there are some extremely large or extremely small values in the data, which are often called *outliers*, then they would affect the sample mean and the sample variance quite dramatically. Usually, the sample mean is pulled up (down) by the extremely large (small) values in the data, and the sample variance gets larger if outliers exist. When outliers exist, the sample mean and sample variance might not be appropriate for measuring the center and spread of the data. Instead, people often use the *sample median* for measuring the center of the data and the *sample interquartile range* for measuring the data spread in such cases. To calculate these statistics, first suppose that observations in the sample $\{X_1, X_2, \ldots, X_n\}$ are ordered from smallest to largest as follows: $X_{(1)} \leq X_{(2)} \leq \cdots \leq X_{(n)}$. Then $X_{(1)}$ is the first *order statistic*, $X_{(2)}$ is the second order statistic, and so on. Roughly speaking, the sample median is defined by the observation at the middle position of the ordered data. So about half of the observations in the data are smaller than the sample median, and the other half are larger than the sample median. More specifically, when n is odd, the sample median is defined by $X_{((n+1)/2)}$; it is defined by $[X_{(n/2)} + X_{(n/2+1)}]/2$ when n is even. Obviously, the sample median is not affected by a small amount of outliers in the data. Similarly, the sample first quartile is defined by the observation such that about a quarter of the data points are smaller than that observation and another three quarters of the data points are larger than it. The sample third quartile is defined the other way around. The sample interquartile range is defined by the distance from the sample first quartile to the sample third quartile, and it is the range of the middle half of the ordered observations.

To estimate a specific population parameter θ (e.g., θ is the population mean or standard deviation), an appropriate statistic should be chosen as an estimator. It has become a convention in the literature to put a hat above the parameter to denote its estimator. So $\widehat{\theta}$ is an estimator of θ. Although $\widehat{\theta}$ is a function of the sample, this is often not explicit in notation, for simplicity. To estimate θ by $\widehat{\theta}$, we use a single-valued statistic calculated from the sample $\widehat{\theta}$ for estimating a single-valued parameter of the population θ. This parameter estimation method is therefore called *point estimation* in the literature, and the estimator is often called a *point estimator*. For example, $\overline{X}$ is a point estimator of μ_X and s^2 is a point estimator of σ_X^2.

Because the sample is random, any point estimator of a parameter is also random. Thus it has its own distribution, which is called a *sampling distribution*. To assess the accuracy of a point estimator, we need to study its sampling distribution, especially its mean and variance. For the mean of the sampling distribution of $\widehat{\theta}$, $\mu_{\widehat{\theta}}$, if

$$\mu_{\widehat{\theta}} = \theta$$

holds for all values of θ, then $\widehat{\theta}$ is called an *unbiased* estimator of θ. Otherwise, the estimator is biased and the bias is defined by $\text{Bias}(\widehat{\theta}, \theta) = \mu_{\widehat{\theta}} - \theta$. On average, an unbiased estimator equals the parameter to estimate. Among all unbiased estimators, the one with the smallest variance is obviously the best unbiased estimator. This estimator is often called the *minimum variance unbiased estimator* (MVUE).

Another commonly used criterion for choosing point estimators is the following *mean squared error* (MSE):

$$\text{MSE}(\widehat{\theta}, \theta) = \text{E}(\widehat{\theta} - \theta)^2 = \sigma^2_{\widehat{\theta}} + \text{Bias}^2(\widehat{\theta}, \theta).$$

The MSE measures the averaged, squared distance between the point estimator $\widehat{\theta}$ and the parameter θ. By this criterion, the best point estimator has the smallest MSE value among all possible point estimators. Obviously, the best point estimator may not be unbiased, and the MVUE may not be the best point estimator either; the MVUE has the smallest MSE among all *unbiased* point estimators only, instead of among all point estimators.

Although it is not explicit in notation, both $\text{Bias}(\widehat{\theta}, \theta)$ and $\text{MSE}(\widehat{\theta}, \theta)$ depend on the sample size n. Generally speaking, when n is larger, the sample carries more information about the population, so it is natural to expect that both $\text{Bias}(\widehat{\theta}, \theta)$ and $\text{MSE}(\widehat{\theta}, \theta)$ would get smaller. If an estimator $\widehat{\theta}$ of θ is biased but the bias converges to zero as the sample size increases, then it is an *asymptotically unbiased estimator*. If $\widehat{\theta}$ satisfies the condition that

$$\lim_{n \to \infty} \text{MSE}(\widehat{\theta}, \theta) = 0,$$

then we say that $\widehat{\theta}$ is *L_2 consistent*.

"Consistency" is a kind of large-sample property of a point estimator. There are several different versions of consistency in the literature. If the cdf of $\widehat{\theta}$ converges to the cdf of the constant θ at all continuity points of the latter cdf, then $\widehat{\theta}$ is said to be *consistent in distribution*. If for any constant $\epsilon > 0$,

$$\lim_{n \to \infty} P(|\widehat{\theta} - \theta| > \epsilon) = 0,$$

then $\widehat{\theta}$ is said to be *consistent in probability*. Another commonly used consistency is defined as follows: if

$$P(\lim_{n \to \infty} \widehat{\theta} = \theta) = 1,$$

then $\widehat{\theta}$ is said to be *almost surely (a.s.) consistent*. Based on some routine mathematical manipulations, it is easy to check the following relations among the four types of consistency mentioned above.

- If $\widehat{\theta}$ is consistent in probability, then it must also be consistent in distribution.
- If $\widehat{\theta}$ is L_2 consistent or a.s. consistent, then it must also be consistent in probability.
- There are some estimators that are L_2 consistent but not a.s. consistent, and vice versa.

For each type of consistency mentioned above, there is a convergence rate associated with it, which tells us how fast the related convergence is. For example, if

$n^\nu \text{MSE}(\widehat{\theta}, \theta) = O(1)$ for some positive constant ν, then we say that $\widehat{\theta}$ is L_2 consistent with convergence rate $O(n^{-\nu})$. Here the big O notation $a_n = O(b_n)$ has been used. Its formal definition is that there exist two positive constants A and B such that $A \leq a_n/b_n \leq B$. Thus, if $a_n = O(b_n)$ and b_n converges to 0 as n tends to infinity, then a_n also converges to 0 with a *same* convergence rate. Later in the book, the small o notation will also be used. By definition, the expression $a_n = o(b_n)$ means that $\lim_{n\to\infty} a_n/b_n = 0$. So if $a_n = o(b_n)$ and b_n converges to 0 as n tends to infinity, then a_n also converges to 0 with a *faster* rate. For other types of consistency, the convergence rate can be discussed similarly. More discussion about large-sample properties of point estimators can be found in text books such as Ash (1972) and Chung (1974).

2.2.3 Confidence intervals and hypothesis testing

Besides point estimation discussed in the previous subsection, there are two other methods of inference about a population parameter. The first method is to use confidence intervals, in which a population parameter θ is estimated by an interval. Suppose that a point estimator $\widehat{\theta}$ of θ has a normal distribution with mean θ and variance $\sigma^2_{\widehat{\theta}}$. Then

$$P\left(\widehat{\theta} - Z_{1-\alpha/2}\sigma_{\widehat{\theta}} < \theta < \widehat{\theta} + Z_{1-\alpha/2}\sigma_{\widehat{\theta}}\right) = 1 - \alpha,$$

where α is a given number between 0 and 1, and $Z_{1-\alpha/2}$ is the $(1-\alpha/2)$-th *quantile* of the standard normal distribution (cf. Figure 2.1), defined by $P(Z \leq Z_{1-\alpha/2}) = 1 - \alpha/2$, where Z is a random variable with the standard normal distribution (in statistics, it is a convention to use Z to denote a random variable with standard normal distribution). That is, the random interval

$$\left(\widehat{\theta} - Z_{1-\alpha/2}\sigma_{\widehat{\theta}},\ \widehat{\theta} + Z_{1-\alpha/2}\sigma_{\widehat{\theta}}\right)$$

has $100(1-\alpha)\%$ chance to cover the true value of θ. This interval is called a $100(1-\alpha)\%$ *confidence interval* for θ, and the number $100(1-\alpha)\%$ is called the *confidence level*. Usually, the standard deviation $\sigma_{\widehat{\theta}}$ has some unknown population parameters, such as the population standard deviation σ, involved. So the above confidence interval formula can be used only after these parameters are replaced by their point estimators. In other words, the standard deviation $\sigma_{\widehat{\theta}}$ needs to be replaced by the *standard error* $\widehat{\sigma}_{\widehat{\theta}}$, in which the unknown parameters have been replaced by their point estimators.

In applications, the exact distribution of $\widehat{\theta}$ is often unknown for a fixed sample size. If its *asymptotic distribution*, which is the limit distribution of $\widehat{\theta}$ when n tends to infinity, can be derived, then the confidence interval for θ can be constructed based on this asymptotic distribution. Of course, in such cases it is only asymptotically true that the confidence interval has $100(1-\alpha)\%$ chance to cover the true value of θ.

In many applications, we are interested in knowing the population mean μ. A good point estimator of μ is the sample mean. For the sample mean $\overline{X}$ of a simple random sample $\{X_1, X_2, \ldots, X_n\}$, if the population has a normal distribution $N(\mu, \sigma^2)$, then its sampling distribution is $N(\mu, \sigma^2/n)$. If the population distribution is not normal, then the asymptotic sampling distribution of $\overline{X}$ is still normal under some regularity conditions. More specifically, the limit distribution of $\sqrt{n}(\overline{X} - \mu)/\sigma$ is $N(0,1)$ in the latter case. Because this result plays a central role in statistical theory and inference, it is called the *central limit theorem*. Therefore, by the central limit theorem, when the sample size n is large, the interval $(\overline{X} - Z_{1-\alpha/2}\frac{s}{\sqrt{n}}, \overline{X} + Z_{1-\alpha/2}\frac{s}{\sqrt{n}})$ is an approximate $100(1-\alpha)\%$ confidence interval for the population mean μ.

The other method of inference about a population parameter involves testing hypotheses. In our daily life, we often make some statement or *hypothesis* about a population parameter. For example, some reports claim that smoking would increase the chance of lung cancer under a general circumstance. In this example, all smokers constitute the population. Suppose that the prevalence rate of lung cancer in this population is an unknown parameter π and that the prevalence rate of the same disease among all nonsmokers is known to be π_0. Then the above statement basically says that $\pi > \pi_0$. A major goal of many research projects is to collect data for verifying such a hypothesis. If the hypothesis is supported by data from one or more repeated experiments, then it becomes a new theory. In statistics, the hypothesis that we want to validate is called an *alternative hypothesis*. Usually, an alternative hypothesis represents a potential new theory, new method, new discovery, etc. It is a competitor to a so-called *null hypothesis*, which often represents existing theory, existing method, existing knowledge, etc. The null hypothesis is initially assumed to be true while we try to verify the validity of the alternative hypothesis.

By convention, the null hypothesis is denoted by H_0, and it always takes the equality form

$$H_0 : \theta = \theta_0,$$

where θ_0 is the hypothesized value of the population parameter θ. The alternative hypothesis is denoted by H_1 or H_a, and it can take one of the following three forms:

$$\begin{aligned} H_1 : \theta &> \theta_0 \\ H_1 : \theta &< \theta_0 \\ H_1 : \theta &\neq \theta_0. \end{aligned}$$

The first form is called right-sided or right-tailed, the second one is called left-sided or left-tailed, and the last one is called two-sided or two-tailed.

Now, we need some criterion to decide whether the alternative hypothesis is valid, based on the observed data. Suppose that $\{X_1, X_2, \ldots, X_n\}$ is a simple random sample from the population of interest and that $\widehat{\theta}$ is a point estimator of θ. As an example of constructing hypothesis testing procedures, let us assume that $\widehat{\theta} \sim N(\theta, \sigma^2_{\widehat{\theta}})$ and that $\sigma^2_{\widehat{\theta}}$ does not include any unknown parameters. Then, when H_0 is true, the statistic

$$T = \frac{\widehat{\theta} - \theta_0}{\sigma_{\widehat{\theta}}} \sim N(0,1).$$

Let us assume that the alternative hypothesis is right-sided and the observed value of T is T^*. Then the probability

$$P_{H_0}(T \geq T^*)$$

tells us the chance that we can observe T^* or values of T that are more inconsistent with H_0 when H_0 is assumed to be true, where P_{H_0} denotes the probability under H_0. The probability $P_{H_0}(T \geq T^*)$ is called the *p-value* and the statistic T is called a *test statistic*. The p-value is defined by $P_{H_0}(T \leq T^*)$ when H_1 is left-sided, and by $P_{H_0}(|T| \geq |T^*|)$ when H_1 is two-sided.

From the definition of p-value, the data provide more evidence against H_0 if the p-value is smaller. Then the question becomes: how small is small? To answer this question, we usually compare the p-value to a *significance level* α, which is given beforehand. If the p-value is smaller than or equal to α, then we reject H_0 and conclude that the data provide significant evidence to support H_1. Otherwise, we fail to reject H_0 and conclude that there is no significant evidence in the data to support H_1. This process of decision making is called *hypothesis testing*.

For a given significance level α, an alternative approach to testing hypotheses is to compare the observed value T^* of the test statistic T with its *α-critical value*. In the case considered above, when $T = \frac{\widehat{\theta}-\theta_0}{\sigma_{\widehat{\theta}}} \sim N(0,1)$, its α-critical value is defined by $Z_{1-\alpha}$ if the alternative hypothesis H_1 is right-sided, by Z_α if H_1 is left-sided, and by $Z_{1-\alpha/2}$ if H_1 is two-sided. The null hypothesis H_0 is rejected when $T^* > Z_{1-\alpha}$, $T^* < Z_\alpha$, and $|T^*| > Z_{1-\alpha/2}$, respectively, for the three types of H_1 considered above. It is easy to see that decisions made by the two approaches are actually equivalent to each other. In applications, the p-value approach is often preferred because it provides decisions regarding whether H_0 needs to be rejected at a given significance level and tells the strength of the evidence in the sample against H_0.

Mistakes are inevitable in our decision making using any hypothesis-testing procedure. There are two types of mistakes, or, more conventionally, two types of errors that we can make. *Type I error* is the case in which H_0 is rejected when it is actually true. *Type II error* is the case in which H_0 fails to be rejected when it is actually false. The probabilities of type I and type II errors are denoted by α and β, respectively. Note that α is used to denote both the significance level and the probability of type I error because these two quantities are the same in many situations.

Intuitively, an ideal hypothesis-testing procedure should have both α and β small. However, in reality, if α is made small, then β will be large, and vice versa. To handle this situation, a statistician's strategy is to control α at some fixed level and then to let β be as small as possible. By this strategy, H_0 is protected to some degree because the chance to reject it when it is actually true cannot exceed a fixed level; this is reasonable in the sense that H_0 often represents existing methods or knowledge and has been justified in the past. Selection of α usually depends on the consequence of type I error. If the consequence is serious, then α should be chosen small. Otherwise, it could be chosen relatively large. Commonly used α values include 0.1, 0.05, 0.01, 0.005, and 0.001. A default α value adopted by most scientific communities is 0.05.

From the above discussion, it can be seen that a major step for solving a hypothesis-testing problem is to find an appropriate test statistic, which should have the property that its distribution under H_0, or its *null distribution*, is known or can be computed. For a given hypothesis-testing problem, different testing procedures are possible. In statistics, we usually only compare procedures with type I error probabilities all below a given level. Among those procedures, the one with the smallest type II error probability β, or equivalently the largest *power* defined by $1 - \beta$, is the best.

2.2.4 Maximum likelihood estimation and least squares estimation

Assume that a population distribution has a pdf which has the parametric form $f(x; \theta_1, \ldots, \theta_r)$, where $\theta_1, \ldots, \theta_r$ are r unknown population parameters. To estimate these population parameters based on a simple random sample $\{X_1, X_2, \ldots, X_n\}$, a standard estimation procedure, called the *maximum likelihood estimation* procedure, is based on the following *likelihood function*:

$$L(\theta_1, \ldots, \theta_r; X_1, X_2, \ldots, X_n) = \Pi_{i=1}^n f(X_i; \theta_1, \ldots, \theta_r).$$

The likelihood function $L(\theta_1, \ldots, \theta_r; X_1, X_2, \ldots, X_n)$ is treated as a function of the unknown parameters $\theta_1, \ldots, \theta_r$ only. The sample $\{X_1, X_2, \ldots, X_n\}$ is assumed to be given. From the discussion about the pdf in Subsection 2.2.1, the value of $f(x; \theta_1, \ldots, \theta_r)\Delta x$ is roughly equal to $P(X \in [x, x+\Delta x])$, where Δx is a small positive number. So the likelihood function is proportional to the likelihood that the observations in the sample take values around the observed sample $\{X_1, X_2, \ldots, X_n\}$.

The *maximum likelihood estimators* (MLEs) $\widehat{\theta}_1, \ldots, \widehat{\theta}_r$ of $\theta_1, \ldots, \theta_r$ are defined by the maximizers of the likelihood function $L(\theta_1, \ldots, \theta_r; X_1, X_2, \ldots, X_n)$. So the likelihood that the observations in the sample take values around $\{X_1, X_2, \ldots, X_n\}$ reaches the maximum when the parameters equal their MLEs, which is reasonable because the sample $\{X_1, X_2, \ldots, X_n\}$ is assumed to have been observed before parameter estimation.

It is often more convenient to work with the logarithm of the likelihood function because the likelihood function has an exponential form in many cases. Then the MLEs of $\theta_1, \ldots, \theta_r$ are the solutions to $\tilde{\theta}_1, \ldots, \tilde{\theta}_r$ of the following maximization problem:

$$\max_{\tilde{\theta}_1, \ldots, \tilde{\theta}_r \in R} \sum_{i=1}^n \log \left[f(X_i; \tilde{\theta}_1, \ldots, \tilde{\theta}_r) \right].$$

For example, when the population has a normal distribution $N(\mu, \sigma^2)$, it is easy to check that the MLEs of μ and σ^2 are $\overline{X}$ and $\frac{n-1}{n} s^2$, respectively, by using the logarithm of the likelihood function.

Next, let us discuss another popular parameter estimation procedure applicable in a regression setting. Suppose that the data $\{(x_i, Y_i), i = 1, 2, \ldots, n\}$ are generated from the linear regression model

$$Y_i = \alpha + \beta x_i + \varepsilon_i, \; i = 1, 2, \ldots, n,$$

where α and β are two unknown regression coefficients. A widely used criterion for measuring the goodness-of-fit of a candidate estimator $a + bx$ of the true linear regression function $\alpha + \beta x$ is the *residual sum of squares* (RSS), defined by

$$\text{RSS}(a, b) = \sum_{i=1}^{n} [Y_i - (a + bx_i)]^2.$$

The *least squares* (LS) estimators of α and β are defined to be the minimizers of $\text{RSS}(a, b)$. They can be calculated by the following formulas:

$$\begin{aligned} \widehat{\beta} &= \frac{\sum_{i=1}^{n}(x_i - \overline{x})(Y_i - \overline{Y})}{\sum_{i=1}^{n}(x_i - \overline{x})^2} \quad \text{and} \\ \widehat{\alpha} &= \overline{Y} - \widehat{\beta}\overline{x}, \end{aligned}$$

where $\overline{x}$ and $\overline{Y}$ are the sample means of x and Y values, respectively. If we assume that $\{\varepsilon_i, i = 1, 2, \ldots, n\}$ are i.i.d. and $\varepsilon_1 \sim N(0, \sigma^2)$, then $Y_i \sim N(\alpha + \beta x_i, \sigma^2)$ for $i = 1, 2, \ldots, n$. It can be checked that the LS estimators of α and β are the same as their MLEs in such a case.

2.3 NADARAYA-WATSON AND OTHER KERNEL SMOOTHING TECHNIQUES

2.3.1 Univariate kernel estimators

In the one-dimensional nonparametric regression model (1.1), suppose we want to estimate the value of f at any given point $x \in [0, 1]$ based on i.i.d. observations $\{(x_i, Y_i), i = 1, 2, \ldots, n\}$, where we assume that the design space is the interval $[0, 1]$ and the nonparametric regression function f is continuous in the entire design interval. Intuitively, observations farther away from x provide less information about $f(x)$. Therefore, one natural idea is to simply average the observations in a neighborhood $[x - h_n/2, x + h_n/2]$ of the point x and then to use this average as an estimator of $f(x)$, where h_n is the width of the neighborhood and is often called the *bandwidth* or *window width* in the literature. The bandwidth is usually chosen relatively small, especially when the sample size n is large. From this idea, $f(x)$ can be estimated by the following simple average (SA):

$$\widehat{f}_{SA}(x) = \frac{1}{N_n(x)} \sum_{x_i \in [x - h_n/2, x + h_n/2]} Y_i,$$

where $N_n(x)$ denotes the number of design points in $[x - h_n/2, x + h_n/2]$.

The estimator $\widehat{f}_{SA}(x)$ is not well defined when the point x is beyond the interior region $[h_n/2, 1 - h_n/2]$ of the design interval. This is the so-called "boundary problem" in the literature and will be further discussed later. For the time being, let us assume that x is in the region $[h_n/2, 1 - h_n/2]$.

Based on model (1.1),

$$\widehat{f}_{SA}(x) = \frac{1}{N_n(x)} \sum_{x_i \in [x-h_n/2, x+h_n/2]} f(x_i) + \frac{1}{N_n(x)} \sum_{x_i \in [x-h_n/2, x+h_n/2]} \varepsilon_i. \quad (2.2)$$

This should estimate $f(x)$ well because of the following two facts.

- All values of $f(t)$ for $t \in [x-h_n/2, x+h_n/2]$ should be close to $f(x)$ because f is continuous in the design interval and h_n is small. Consequently, the first term on the right-hand side of (2.2) is close to $f(x)$.

- The second term on the right-hand side of (2.2) is close to zero as long as there are enough terms in the summation, which is guaranteed by the central limit theorem mentioned in the previous section. Positive errors and negative errors are canceled out, and thus, on average, the summation of many i.i.d. error terms is close to zero.

The above function estimation procedure, which results in the estimator $\widehat{f}_{SA}(x)$, is a simple example of data smoothing. Almost all data smoothing procedures in the literature involve averaging observations for estimating $f(x)$. Procedures that average all observations in the design interval are referred to as *global smoothing* procedures in the literature. The smoothing spline procedure discussed in Section 2.5 of this chapter is an example of global smoothing. Other procedures only average observations in a neighborhood of x, and are referred to as *local smoothing* procedures. All procedures discussed in this section and the next section belong to the latter category. Such local smoothing procedures are commonly analyzed based on the two facts mentioned above about the two terms on the right-hand side of equation (2.2).

By using $\widehat{f}_{SA}(x)$ for estimating $f(x)$, all observations outside the neighborhood $[x - h_n/2, x + h_n/2]$ are ignored completely, and all observations inside the neighborhood are treated equally. A natural generalization of $\widehat{f}_{SA}(x)$, which does not treat all observations in the neighborhood equally, is

$$\widehat{f}_{NW}(x) = \frac{\sum_{i=1}^{n} Y_i K\left(\frac{x_i - x}{h_n}\right)}{\sum_{i=1}^{n} K\left(\frac{x_i - x}{h_n}\right)}, \quad (2.3)$$

where K is a function which has support $[-1/2, 1/2]$ and which satisfies the condition that $\int_{-1/2}^{1/2} K(u)\, du = 1$. In the literature, K is often called a *kernel* function.

The estimator $\widehat{f}_{NW}(x)$ is simply a weighted average of the observations in the neighborhood $[x - h_n/2, x + h_n/2]$, with weights controlled by the kernel function. Usually, the kernel function is chosen to be a smooth density function that is symmetric about 0 and nondecreasing on $[-1/2, 0]$. Commonly used kernel functions include the uniform kernel function $K(u) = I(-1/2 \le u \le 1/2)$, the Epanechnikov kernel function $K(u) = \frac{12}{11}(1 - u^2)I(-1/2 \le u \le 1/2)$, and the Gaussian kernel function $K(u) = 0.6171 \exp(-u^2/2)I(-1/2 \le u \le 1/2)$, where $I(a)$ is an indicator function defined by $I(a) = 1$ if a is "true" and 0 otherwise. With the use of the

Epanechnikov kernel function or the Gaussian kernel function, observations closer to x receive more weight in defining $\widehat{f}_{NW}(x)$. Obviously, the estimator $\widehat{f}_{SA}(x)$ is a special case of $\widehat{f}_{NW}(x)$ when the uniform kernel function is used.

The kernel estimator $\widehat{f}_{NW}(x)$ was first suggested by Nadaraya (1964) and Watson (1964). Therefore, it is often called a Nadaraya-Watson (NW) kernel estimator in the literature, as identified by the subscript of $\widehat{f}_{NW}(x)$.

In the literature, there are several other kernel estimators. For example, Priestley and Chao (PC, 1972) suggested the following kernel estimator

$$\widehat{f}_{PC}(x) = \frac{1}{h_n}\sum_{i=1}^{n} Y_i(x_i - x_{i-1})K\left(\frac{x_i - x}{h_n}\right), \tag{2.4}$$

where $x_0 = 0$. The kernel estimator suggested by Gasser and Müller (GM, 1979) is

$$\widehat{f}_{GM}(x) = \frac{1}{h_n}\sum_{i=1}^{n} Y_i \int_{s_{i-1}}^{s_i} K\left(\frac{u - x}{h_n}\right) du,$$

where $s_i = (x_i + x_{i+1})/2, x_0 = 0$ and $x_{n+1} = 1$. Cheng and Lin (1981) proposed a similar estimator to $\widehat{f}_{GM}(x)$ and proved that their estimator and several other kernel estimators were asymptotically equivalent.

2.3.2 Some statistical properties of kernel estimators

For the Nadaraya-Watson kernel estimator, if the following conditions hold

- the kernel function satisfies the condition that $\int_{-1/2}^{1/2} uk(u)\, du = 0$,
- the design points have some homogeneity on $[0, 1]$, and
- f has continuous second-order derivatives in a neighborhood of a given point $x \in (0, 1)$,

then, under some other regularity conditions,

$$\text{Bias}\left[\widehat{f}_{NW}(x), f(x)\right] \sim h_n^2 K_{21} f''(x)/2,$$

$$\text{Var}\left[\widehat{f}_{NW}(x)\right] \sim (nh_n)^{-1} K_{02}\sigma^2,$$

and consequently,

$$\text{MSE}\left[\widehat{f}_{NW}(x), f(x)\right] \sim h_n^4 K_{21}^2 \left[f''(x)\right]^2/4 + (nh_n)^{-1} K_{02}\sigma^2. \tag{2.5}$$

Note that "$\sim$" denotes asymptotic equivalence, $K_{21} = \int_{-1/2}^{1/2} u^2 K(u)\, du$, and $K_{02} = \int_{-1/2}^{1/2} K^2(u)\, du$. Therefore, as long as the bandwidth h_n is chosen such that

$$\lim_{n\to\infty} h_n = 0; \quad \lim_{n\to\infty} 1/(nh_n) = 0,$$

$\widehat{f}_{NW}(x)$ is asymptotically unbiased, and its variance tends to zero when n tends to infinity. Equation (2.5) shows that $\widehat{f}_{NW}(x)$ is L_2 consistent with optimal convergence rate $O(n^{-4/5})$ when $h_n \sim n^{-1/5}$.

When the given point x is in the boundary regions $[0, h_n/2)$ and $(1 - h_n/2, 1]$, $\widehat{f}_{NW}(x)$ can still be defined by equation (2.3), but part of the neighborhood $[x - h_n/2, x + h_n/2]$ is beyond the design interval $[0, 1]$. In such cases, the number of observations used in defining $\widehat{f}_{NW}(x)$ is usually less than the number of observations used in defining the same estimator when x is located in the interior region $[h_n/2, 1 - h_n/2]$. For instance, suppose that $x = \tau h_n/2$ (so x is in the boundary region $[0, h_n/2)$), where $0 \leq \tau < 1$ is a constant. Then, it can be checked that

$$\text{Bias}\left[\widehat{f}_{NW}(x), f(x)\right] \sim h_n K_\tau^{(1)} f'(x)$$

$$\text{Var}\left[\widehat{f}_{NW}(x)\right] \sim (nh_n)^{-1} K_\tau^{(2)} \sigma^2,$$

where $K_\tau^{(1)}$ and $K_\tau^{(2)}$ are two constants that depend on the kernel function K and the constant τ. Similar results can be derived in the case when x is in the boundary region $(1 - h_n/2, 1]$.

From the above equations, it can be seen that the bias of the kernel estimator $\widehat{f}_{NW}(x)$ when x is in boundary regions is larger than its bias when x is in the interior region. Consequently, the optimal L_2 convergence rate of $\widehat{f}_{NW}(x)$ when x is in the boundary regions is only $O(n^{-2/3})$ for $h_n \sim n^{-1/3}$, which is much slower than its optimal rate $O(n^{-4/5})$ when x is in the interior region. This is the notorious "boundary problem" of the kernel estimators.

To deal with the boundary problem, various boundary kernel functions have been proposed in the literature. Interested readers can read Gasser and Müller (1979, 1984) and Rice (1984). One simple way to overcome the boundary problem is to use the local linear kernel smoothing procedure, which is discussed in Section 2.4.

Next, we briefly discuss some properties of the Priestley-Chao kernel estimator $\widehat{f}_{PC}$. Cheng and Lin (1981) proved the almost sure consistency of their kernel estimator. By a similar proof, Qiu (1994) gave the following result about the almost sure consistency of $\widehat{f}_{PC}$.

Theorem 2.1 *Let ν be a positive number and β_n be a series of numbers that satisfy $\lim_{n\to\infty} \beta_n = \infty$. In model (1.1), suppose that f is Lipschitz (α) continuous with $\alpha > 0$, $E|\varepsilon_1|^\rho < \infty$ with $\rho \geq 2$, and $\max_{1\leq i\leq n} |(x_i - x_{i-1}) - \frac{1}{n}| = O(n^{-1-\lambda})$ with $\lambda > 0$. The kernel function K is assumed to be a nonnegative, bounded density function which is Lipschitz (β) continuous with $\beta > 0$. The bandwidth h_n is assumed to satisfy the conditions that (1) $h_n = o(1)$, (2) $1/(nh_n) = o(1)$, (3) $\frac{n^\nu}{\beta_n \log n}[h_n^\alpha + \frac{1}{n^\lambda h_n} + \frac{1}{n^\beta h_n^{\beta+1}}] = o(1)$, (4) $\frac{n^{2\nu}}{n\beta_n h_n} = O(1)$, and (5) $\frac{n^{\nu+\frac{1}{\rho}-1}}{h_n \beta_n \log n} = o(1)$. Then, for any $0 < a \leq b < 1$, we have*

$$\lim_{n\to\infty} \frac{n^\nu}{\beta_n \log n} \|\widehat{f}_{PC} - f\|_{[a,b]} = 0, \ a.s.$$

where $\|f\|_\Omega = \max_{x\in\Omega} |f(x)|$.

In Theorem 2.1, if $\alpha = \beta = 1$, $\rho \geq 3$, $\lambda \geq 2/3$, and $h_n \sim n^{-1/3}$, then the conditions on h_n are all satisfied, and, under the other conditions, we have

$$\|\widehat{f}_{PC} - f\|_{[a,b]} = O(n^{-1/3}),\ a.s. \tag{2.6}$$

Since $\lim_{n\to\infty} \frac{1}{nh_n}\sum_{i=1}^{n} K[(x_i - x)/h_n] = 1$ under the conditions given in Theorem 2.1, the Nadaraya-Watson kernel estimator $\widehat{f}_{NW}$ should also have the property stated in Theorem 2.1.

2.3.3 Multivariate kernel estimators

When there are p explanatory variables $x_1, x_2, \ldots, x_p$, where $p > 1$ is an integer, the corresponding regression problem is multivariate or, more specifically, p-dimensional. Suppose that the p-dimensional regression model is

$$Y_i = f(x_{1i}, x_{2i}, \ldots, x_{pi}) + \varepsilon_i,\ i = 1, 2, \ldots, n, \tag{2.7}$$

where $\{(x_{1i}, x_{2i}, \ldots, x_{pi}), i = 1, 2, \ldots, n\}$ are the design points in a p-dimensional design space, $\{Y_i, i = 1, 2, \ldots, n\}$ are observations of the response variable Y, f is a p-dimensional regression function, and $\{\varepsilon_i, i = 1, 2, \ldots, n\}$ are i.i.d. random errors. Then, kernel estimators of f in p-dimensional cases can be defined similarly to those in the case with $p = 1$. For example, let K be a p-dimensional kernel function satisfying some regularity conditions. Then the Nadaraya-Watson kernel estimator can be defined by

$$\widehat{f}_{NW}(x_1, x_2, \ldots, x_p) = \frac{\sum_{i=1}^{n} Y_i K\left(\frac{x_{1i}-x_1}{h_{1n}}, \ldots, \frac{x_{pi}-x_p}{h_{pn}}\right)}{\sum_{i=1}^{n} K\left(\frac{x_{1i}-x_1}{h_{1n}}, \ldots, \frac{x_{pi}-x_p}{h_{pn}}\right)},$$

where $h_{1n}, h_{2n}, \ldots, h_{pn}$ are p bandwidths and $(x_1, x_2, \ldots, x_p)$ is a given point in the design space.

In applications, the p-dimensional kernel function K is often chosen to be a product of p univariate kernel functions. When p is large, a p-dimensional regression function f is in general difficult to estimate, due to the *curse of dimensionality*. To illustrate, suppose that we have n =1,000 data points. This data might be large enough to estimate the regression function well when $p = 1$. But 1,000 data points could be extremely sparse in the case when $p = 5$. Consequently, f is difficult to estimate in the latter case. More discussion about multivariate kernel estimation can be found in Müller (1988, Chapter 6).

2.4 LOCAL POLYNOMIAL KERNEL SMOOTHING TECHNIQUES

2.4.1 Univariate local polynomial kernel estimators

It can be easily checked that the Nadaraya-Watson kernel estimator $\widehat{f}_{NW}(x)$ is the solution to a of the following minimization problem:

$$\min_{a \in R} \sum_{i=1}^{n} (Y_i - a)^2 K\left(\frac{x_i - x}{h_n}\right). \tag{2.8}$$

Therefore, $\widehat{f}_{NW}(x)$ has the property that, among all constants, its weighted distance to the observations in the neighborhood $[x - h_n/2, x + h_n/2]$ is the smallest. This is illustrated by Figure 2.2(a), in which the solid curve at bottom denotes the weights $K[(x_i - x)/h_n]$ and the dashed horizontal line denotes $\widehat{f}_{NW}(x)$.

A natural generalization of (2.8) is the following minimization problem:

$$\min_{a, b_1, \ldots, b_m \in R} \sum_{i=1}^{n} \left[Y_i - (a + b_1(x_i - x) + \ldots + b_m(x_i - x)^m)\right]^2 K\left(\frac{x_i - x}{h_n}\right), \tag{2.9}$$

where m is a positive integer. Equation (2.9) is used to search for a polynomial function of order m whose weighted distance to the data in the neighborhood $[x - h_n/2, x + h_n/2]$ reaches the minimum. Then, the solution to a in (2.9) can be defined as an estimator of $f(x)$, and is called the mth-order *local polynomial kernel estimator*. As a by-product, the solution to b_j in (2.9) can be used as an estimator of $j!f^{(j)}(x)$ for $j = 1, 2, \ldots, m$, where $f^{(j)}(x)$ is the jth-order derivative of f at x.

Obviously, (2.8) is a special case of (2.9) when $m = 0$. That is, the Nadaraya-Watson kernel estimator is the zero-th-order local polynomial kernel estimator or the *local constant kernel estimator*.

In applications, the most commonly used local polynomial kernel estimator is the *local linear kernel (LK) estimator*, which is the solution to a in (2.9) when $m = 1$, and is denoted by $\widehat{f}_{LK}(x)$. This estimator is illustrated in Figure 2.2(b) by the dashed line. By some routine algebra, when $m = 1$, the solutions to a and b_1 of (2.9) have the following expressions:

$$\begin{aligned} \widehat{a}(x) &= \sum_{i=1}^{n} \frac{w_2 - w_1(x_i - x)}{w_0 w_2 - w_1^2} Y_i K\left(\frac{x_i - x}{h_n}\right) \\ \widehat{b}_1(x) &= \sum_{i=1}^{n} \frac{w_0(x_i - x) - w_1}{w_0 w_2 - w_1^2} Y_i K\left(\frac{x_i - x}{h_n}\right), \end{aligned} \tag{2.10}$$

where $w_j = \sum_{i=1}^{n} (x_i - x)^j K(\frac{x_i - x}{h_n})$ for $j = 0, 1$ and 2. Therefore, $\widehat{f}_{LK}(x) = \widehat{a}(x)$ and $\widehat{f}'(x) = \widehat{b}_1(x)$.

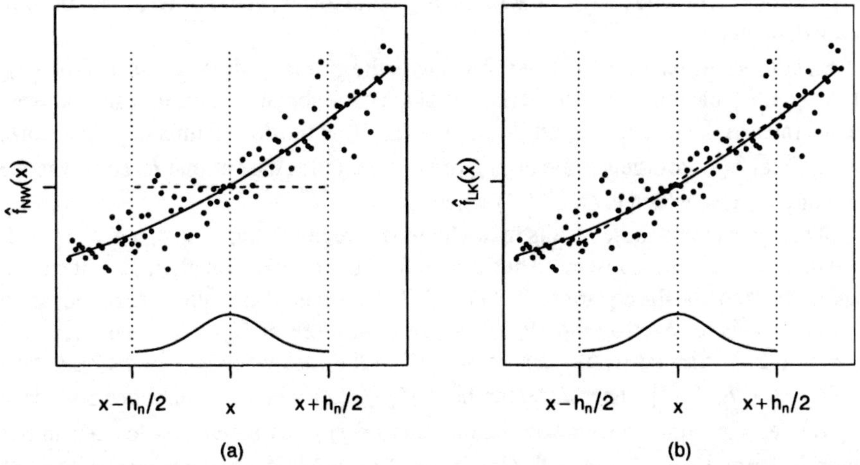

Fig. 2.2 (a): The Nadaraya-Watson (NW) kernel estimator of $f(x)$ (dashed horizontal line) has the property that, among all constants, its weighted distance to observations in the neighborhood $[x - h_n/2, x + h_n/2]$ is the smallest. (b): The local linear kernel (LK) estimator equals the value of the dashed line at x. The dashed line has the property that its weighted distance to observations in the neighborhood $[x - h_n/2, x + h_n/2]$ is the smallest among all possible lines. Small circles in both plots denote observations. Solid curve going through the data denotes the true regression function. Solid curve at bottom denotes the weights that are controlled by the kernel function.

2.4.2 Some statistical properties

One benefit of using the local polynomial kernel estimator instead of the local constant kernel estimator is that the first m derivatives of f should have little impact on the performance of the mth-order local polynomial kernel estimator; by the definition, most of the effects of the derivatives has been accommodated in the estimation by fitting a local polynomial. The local polynomial kernel estimator has some other good properties, which are discussed below.

Under some regularity conditions, it can be checked that, when $x \in (0,1)$,

$$\text{Bias}\left[\widehat{f}_{LK}(x), f(x)\right] \sim h_n^2 f''(x)\frac{K_{21}^2 - K_{11}K_{31}}{2(K_{01}K_{21} - K_{11}^2)} \tag{2.11}$$

and

$$\text{Var}\left[\widehat{f}_{LK}(x)\right] \sim (nh_n)^{-1}C(K)\sigma^2, \tag{2.12}$$

where $K_{j1} = \int_{-1/2}^{1/2} u^j K(u)du$ for $j = 0, 1, 2, 3$, and $C(K) = \frac{1}{(K_{01}K_{21} - K_{11})^2} \int_{-1/2}^{1/2} (K_{21} - K_{11}u)^2 K^2(u)du$. If the kernel function K is chosen to be a symmetric density function, as discussed in the previous section, then $K_{01} = 1$, $K_{11} = 0$, $C(K) = K_{02} = \int_{-1/2}^{1/2} K^2(u)du$, and

$$\text{Bias}\left[\widehat{f}_{LK}(x), f(x)\right] \sim h_n^2 K_{21} f''(x)/2, \ \text{Var}\left[\widehat{f}_{LK}(x)\right] \sim (nh_n)^{-1}K_{02}\sigma^2.$$

Expressions (2.11) and (2.12) are written in general forms for the convenience of later use in Chapter 3.

By comparing results (2.11) and (2.12) with the corresponding results of $\widehat{f}_{NW}(x)$, it can be seen that the biases and variances of the two estimators have the same convergence rates in the interior region $[h_n/2, 1 - h_n/2]$ of the design interval. Therefore, the optimal L_2 convergence rate of $\widehat{f}_{LK}(x)$ is the same as the optimal L_2 convergence rate of $\widehat{f}_{NW}(x)$, which is $O(n^{-4/5})$ when $h_n \sim n^{-1/5}$.

We now examine these results in the boundary regions. Suppose that $x = \tau h_n/2$, where $0 \leq \tau < 1$ is a constant. Then x is in the boundary region $[0, h_n/2)$. It can be easily checked that the equations (2.11) and (2.12) still hold after the integration range $[-1/2, 1/2]$ is replaced by $[-\tau/2, 1/2]$ in the definitions of K_{j1}, K_{02}, and $C(K)$ for $j = 0, 1, 2, 3$. Similar results can be derived in the case when x is in the boundary region $(1 - h_n/2, 1]$. Therefore, the bias of $\widehat{f}_{LK}(x)$ when x is in either boundary region has the same convergence rate as that of $\widehat{f}_{LK}(x)$ when x is located in the interior region $[h_n/2, 1 - h_n/2]$. Consequently, the L_2 convergence rate of $\widehat{f}_{LK}(x)$ remains the same no matter whether x is in the interior region or not.

This *automatic boundary carpentry* property of the local linear kernel estimator is illustrated in Figure 2.3(b) (see also Fan and Gijbels 1996; Hastie and Loader 1993). In the plot, x is in the boundary region $[0, h_n/2)$. It can be seen that this does not cause large bias of $\widehat{f}_{LK}(x)$. As a comparison, the Nadaraya-Watson and other kernel estimators suffer the boundary problem, as demonstrated by Figure 2.3(a). From the plot, we can see that the bias of $\widehat{f}_{NW}(x)$ is quite large when x is in a boundary region. This kind of bias is mainly caused by the large slope of f in the boundary region.

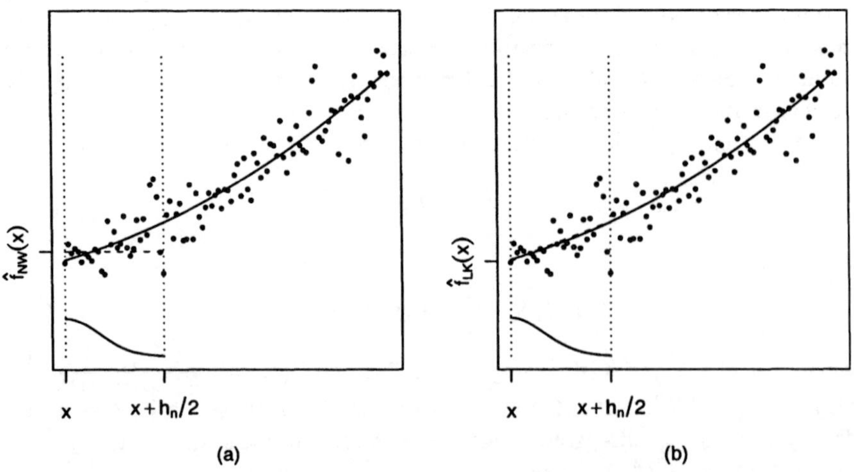

Fig. 2.3 Behavior of the Nadaraya-Watson (NW) kernel estimator [plot (a)] and the local linear (LK) kernel estimator [plot (b)] of $f(x)$ when x is located in a boundary region. Small circles, dashed lines, and solid curves have the same meanings as those in Figure 2.2.

The performance of the Nadaraya-Watson kernel estimator is also affected by the distribution of the design points in the design space, as demonstrated by Figure 2.4(a). In the plot, the design points are distributed unevenly in the neighborhood $[x - h_n/2, x + h_n/2]$, which results in relatively large bias of $\widehat{f}_{NW}(x)$. As a comparison, the local linear kernel estimator $\widehat{f}_{LK}(x)$ is almost immune to the distribution of design points (cf. Figure 2.4(b)), which is another of its strengths compared to the kernel estimators.

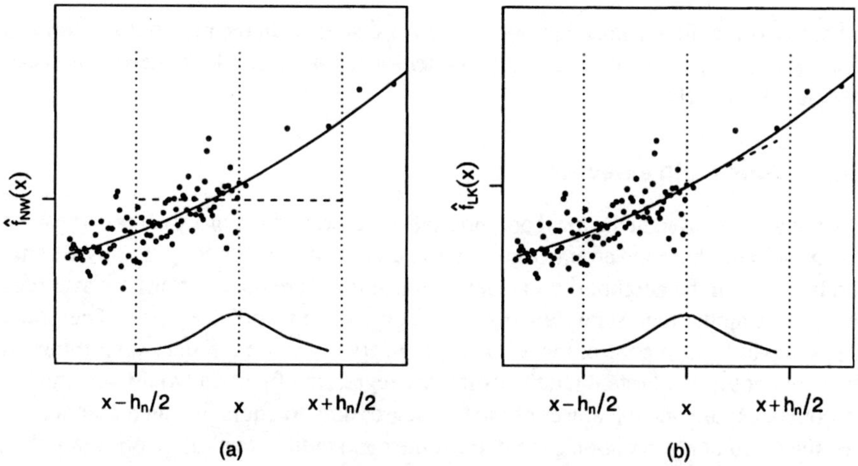

Fig. 2.4 Behavior of the Nadaraya-Watson (NW) kernel estimator [plot (a)] and the local linear kernel (LK) estimator [plot (b)] of $f(x)$ when the design points are distributed unevenly in the design space. Small circles, dashed lines, and solid curves have the same meanings as those in Figure 2.2.

We next consider the choice of the power of the polynomial for local polynomial kernel estimation. By some routine algebra, it can be shown, under some regularity conditions, that (i) the variance of a $(2k + 1)$-th-order local polynomial kernel estimator is asymptotically the same as the variance of a $(2k)$-th-order local polynomial kernel estimator, (ii) the variance of the $(2k + 1)$-th-order local polynomial kernel estimator increases with k, and (iii) the bias of a higher order local polynomial kernel estimator is smaller than the bias of a lower order estimator. By these properties, the order of a local polynomial kernel estimator should be chosen odd because a $(2k+1)$-th-order estimator is generally better than a $(2k)$-th-order estimator. In most cases, the local linear kernel estimator should be good enough because the bias reduction is limited by using a higher order estimator and because the variability of a higher order estimator is often much larger.

2.4.3 Multivariate local polynomial kernel estimators

The local polynomial kernel smoothing techniques can be used for estimating multivariate regression functions. Suppose that $\{(x_{1i}, x_{2i}, \ldots, x_{pi}, Y_i), i = 1, 2, \ldots, n\}$

are n observations generated from the p-dimensional model (2.7). Then, the local linear kernel estimator of the p-dimensional regression function f is defined by the solution to a of the following minimization problem:

$$\min_{a,b_1,\ldots,b_p\in R} \sum_{i=1}^{n} [Y_i - (a + b_1(x_{1i} - x_1) + \ldots + b_p(x_{pi} - x_p))]^2 \cdot K\left(\frac{x_{1i} - x_1}{h_{1n}}, \ldots, \frac{x_{pi} - x_p}{h_{pn}}\right),$$

where K is a p-dimensional kernel function as discussed in the previous section and $h_{1n}, h_{2n}, \ldots, h_{pn}$ are the bandwidths. Other local polynomial kernel estimators can be defined similarly.

2.4.4 Bandwidth selection

The bandwidths used in the local polynomial kernel smoothing procedures control the size of the neighborhood around a given point used for data smoothing. If they are chosen larger, then the neighborhood is larger, meaning more observations are averaged for estimating the regression function at the given point, and vice versa. Therefore, the bandwidths also control the amount of observations used in data smoothing. If they are chosen too large, then the estimated regression function would become too smooth to capture local features of the true regression function, in which case we say that the estimator *oversmooths* the data. If the bandwidths are chosen too small, then the estimated regression function would capture too many details of the data, some of which are just noise, in which case the estimator *undersmooths* the data.

There are several existing procedures in the literature for choosing the bandwidths properly. We next introduce three of them: the cross-validation (CV) procedure, the Mallow's C_p criterion, and the plug-in algorithm. For simplicity, they are introduced only in the case of one-dimensional local linear kernel smoothing. Actually, these procedures are quite general and can be used for selecting bandwidths and other parameters used in most data smoothing procedures.

For bandwidth selection of the local linear kernel estimator $\widehat{f}_{LK}$, a natural criterion is the following residual mean squares (RMS):

$$\text{RMS}(h_n) = \frac{1}{n}\sum_{i=1}^{n}\left[Y_i - \widehat{f}_{LK}(x_i)\right]^2. \tag{2.13}$$

Since $\text{RMS}(h_n)$ measures the distance between the data and the estimator, the bandwidth h_n can be chosen by minimizing $\text{RMS}(h_n)$. However, it can be easily checked that the best bandwidth is $h_n = 0$ by this criterion because the corresponding estimator connects all the data points in such a case and thus sets $\text{RMS}(h_n) = 0$. Obviously, this result is not what we want because no data smoothing is actually involved.

We can amend this criterion to allow for data smoothing as follows. Because $\widehat{f}_{LK}(x_i)$ is constructed by all observations in the neighborhood $[x_i - h_n/2, x_i + h_n/2]$, including Y_i, the residual $Y_i - \widehat{f}_{LK}(x_i)$ tends to be small for measuring the distance between the estimator and the data. This is a major reason why the criterion (2.13) is inappropriate for choosing h_n. As a remedy, let $\widehat{f}_{LK,-i}(x_i)$ be the local linear kernel estimator of $f(x_i)$ constructed from all observations in $[x_i - h_n/2, x_i + h_n/2]$ except Y_i, for $i = 1, 2, \cdots, n$. Then, a similar criterion to equation (2.13) is the following CV score:

$$CV(h_n) = \frac{1}{n}\sum_{i=1}^{n}\left[Y_i - \widehat{f}_{LK,-i}(x_i)\right]^2. \tag{2.14}$$

The optimal value of h_n can be estimated by the minimizer of the CV score. This method for choosing the bandwidths or other procedure parameters is called the *CV procedure* in the literature (Allen 1974). More discussion about this procedure can be found in the next section when we discuss the smoothing spline procedure.

We next discuss the Mallows' C_p criterion. Let $\mathbf{Y} = (Y_1, Y_2, \ldots, Y_n)'$ and $(\widehat{f}_{LK}(x_1), \widehat{f}_{LK}(x_2), \ldots, \widehat{f}_{LK}(x_n))' = \mathbf{HY}$. Then $\mathbf{H}$ is called the *hat matrix* in regression analysis. The Mallows' C_p criterion (Mallows 1973) is defined by

$$C_p(h_n) = \frac{1}{\sigma^2}\|(I_{n\times n} - \mathbf{H})\mathbf{Y}\|^2 - n + 2\,\mathrm{tr}(\mathbf{H}), \tag{2.15}$$

where $I_{n\times n}$ is the $n \times n$ identity matrix, $\|\cdot\|$ is the Euclidean norm, and $\mathrm{tr}(\mathbf{H})$ is the trace of the matrix $\mathbf{H}$ (i.e., the summation of all diagonal elements of $\mathbf{H}$). By this criterion, the optimal value of h_n is estimated by the minimizer of $C_p(h_n)$.

The plug-in algorithm is based on the MSE criterion. From expressions (2.11) and (2.12), the MSE of the local linear kernel estimator $\widehat{f}_{LK}(x)$ is

$$\mathrm{MSE}\left(\widehat{f}_{LK}(x), f(x)\right) \sim \left[h_n^2 f''(x)\frac{K_{21}^2 - K_{11}K_{31}}{2(K_{01}K_{21} - K_{11}^2)}\right]^2 + (nh_n)^{-1}C(K)\sigma^2. \tag{2.16}$$

By criterion (2.16), the optimal value of h_n is asymptotically equal to

$$h_{n,opt} = \left(\frac{C_1(K)\sigma^2}{n[f''(x)]^2}\right)^{1/5}, \tag{2.17}$$

where $C_1(K) = C(K)(K_{01}K_{21} - K_{11}^2)^2/(K_{21}^2 - K_{11}K_{21})^2$. Because the quantity $f''(x)$ is unknown, formula (2.17) cannot be used directly for choosing h_n. One way out is the following iterative algorithm: (1) we assign an initial value for h_n and estimate $f''(x)$ by a local polynomial kernel estimator based on this initial bandwidth, (2) the value of h_n is updated by (2.17) after $f''(x)$ is replaced by its estimator, (3) we go back to step (1) with the updated bandwidth obtained from the previous step, and (4) steps (1)–(3) continue until some convergence criterion is satisfied. This *plug-in* bandwidth selection procedure has several different versions. For more discussion, please read Gasser *et al.* (1991), Härdle *et al.* (1992), Loader (1999), Ruppert *et al.* (1995), and the references cited therein.

2.5 SPLINE SMOOTHING PROCEDURES

2.5.1 Univariate smoothing spline estimation

In mathematics, the Weierstrass approximation theorem says that any continuous function defined on a closed interval can be approximated well by a polynomial function (e.g., Stoll 2001, Section 8.6). By this result, the continuous regression function f in model (1.1) can also be approximated well by a polynomial function. *Polynomial regression* analysis provides us methodologies for this purpose. In the literature, however, it has been found that the estimated regression function by polynomial approximation is often unstable and has artificial patterns when monomials are used as basis functions. This is especially true when the true regression function behaves quite differently in one part of the design space from another. This situation can be improved a little by using orthogonal polynomial basis functions such as the Chebyshev's polynomials (Seber 1977, Chapter 8). Unfortunately, the problems are not essentially solved.

In numerical analysis, a more suitable tool for approximating a continuous function is the *(natural) polynomial spline function* (see e.g., Greville 1968, Reinsch 1967, Schoenberg 1964), first suggested for interpolating deterministic data in mathematics. Suppose that the design space is $[0, 1]$ and $0 < t_1 < t_2 < \ldots < t_s < 1$ are s given points in the design space. Then, for a given positive integer m, a function g on $[0, 1]$ is called a (natural) polynomial spline function of order $2m - 1$ with *knots* $0 < t_1 < t_2 < \ldots < t_s < 1$ if it satisfies the following conditions.

- In each interval $[t_{j-1}, t_j]$ for $j = 2, \ldots, s$, g is a polynomial function of order $2m - 1$.
- In intervals $[0, t_1)$ and $(t_s, 1]$, g is a polynomial function of order $m - 1$.
- g has continuous $(2m - 2)$-th-order derivative at all knots.

Therefore, a polynomial spline function of order $2m - 1$ is a piecewise polynomial function connected at its knots in a way such that it has continuous $(2m - 2)$-th-order derivative in the entire design space. From the above definition, it can be seen that the polynomial spline function is more flexible than the regular polynomial function because of its nature as a piecewise polynomial. In addition, in each interval $[t_{j-1}, t_j]$ for $j = 1, 2, \ldots, s + 1$, where $t_0 = 0$ and $t_{s+1} = 1$, it is a polynomial function and has all nice properties of polynomial functions.

In statistics, the polynomial spline function is used for estimating the regression function f in model (1.1). The response variable Y is assumed to be random, and the observations $\{Y_i, i = 1, 2, \ldots, n\}$ have random errors involved. Because the data carry information about the underlying regression function f and they are random, the estimator of f should be close to the data but does not need to go through all the data points. For a possible estimator g, one convenient measurement of the fidelity to the data is the residual sum of squares $\sum_{i=1}^{n} [Y_i - g(x_i)]^2$. The residual sum of squares of g should be small but not necessarily zero. On the other hand, an ideal estimator should be smooth because the true regression function is assumed to be smooth. One measurement of the smoothness of the function g in design space $[0,1]$ is $\int_0^1 [g^{(m)}(u)]^2 du$, where $g^{(m)}$ denotes the mth-order derivative of g. When $m = 2$, $\int_0^1 [g''(u)]^2 du$ measures the curvature of g.

A trade-off between the smoothness of g and its fidelity to the data results in a minimization of the following criterion:

$$\frac{1}{n}\sum_{i=1}^{n}[Y_i - g(x_i)]^2 + \lambda \int_0^1 [g^{(m)}(u)]^2 \, du,$$

where λ is a *smoothing parameter* controlling the trade-off. Larger λ implies that we put more weight on the smoothness of g in estimating the regression function. Schoenberg (1964) showed that, when $n \geq m$, the solution of the following minimization problem

$$\min_{g \in W_2^m([0,1])} \frac{1}{n}\sum_{i=1}^{n}[Y_i - g(x_i)]^2 + \lambda \int_0^1 [g^{(m)}(u)]^2 du \tag{2.18}$$

was unique, and it was a $(2m-1)$-th-order polynomial spline function with knots $\{x_i, i = 1, 2, \ldots, n\}$, where $W_2^m([0,1])$ denotes the following Sobolev function space:

$$\begin{aligned} W_2^m([0,1]) &= \Big\{g : g, g', \ldots, g^{(m-1)} \text{ are continuous on [0,1], and} \\ &\qquad \int_0^1 [g^{(m)}(u)]^2 \, du < \infty\Big\}. \end{aligned}$$

The solution of problem (2.18) is denoted by $\widehat{f}_\lambda$, and it is called the $(2m-1)$-th-order *smoothing spline* estimator of f.

The smoothing spline estimator $\widehat{f}_\lambda$ can be expressed by a linear combination of a set of basis functions of the related function space. Several different sets of basis functions are suggested in the literature (cf. Anselone and Laurent 1968, Demmler and Reinsch 1975, Kimeldorf and Wahba 1970, 1971, Reinsch 1967, 1971). Here, we present an expression of $\widehat{f}_\lambda$ based on the basis functions suggested by Craven and Wahba (1979). When $n \geq m$, the smoothing spline estimator $\widehat{f}_\lambda$ can be expressed by

$$\widehat{f}_\lambda(x) = \sum_{r=0}^{m} \theta_r k_r(x) + (-1)^{(m-1)} \sum_{j=1}^{n} \alpha_j k_{2m}(x, x_j),$$

where

$$\begin{aligned} \theta &= (\theta_0, \theta_1, \ldots, \theta_m)' = (T'M^{-1}T + \Delta)^{-1}T'M^{-1}\mathbf{Y}; \\ \alpha &= (\alpha_1, \alpha_2, \ldots, \alpha_n)' = M^{-1}(\mathbf{Y} - T\theta); \\ \mathbf{Y} &= (Y_1, Y_2, \ldots, Y_n)'; \end{aligned}$$

M, T, Δ are some matrices that do not depend on $\mathbf{Y}$; $k_r(x) = B_r((x))/r!$; (x) in $B_r((x))$ denotes the fractional part of x; $\{B_r(x), r = 0, 1, \ldots\}$ are Bernoulli polynomials; and $k_r(s,t)$ is the *Bernoulli kernel* defined by

$$k_r(s,t) = -\sum_{\nu=-\infty,\nu\neq 0}^{\infty} \frac{1}{(2\pi i\nu)^r} e^{2\pi i\nu(s-t)}, \; r = 1, 2, \ldots.$$

Therefore, the smoothing spline estimator $\widehat{f}_\lambda(x)$ is also a *linear estimator*, namely, a linear combination of all observations of the response variable Y, so are all local polynomial kernel estimators discussed in the previous two subsections.

According to Schoenberg (1964), $\widehat{f}_\lambda$ is also the solution of the following problem:

$$\min_{g\in W_2^m([0,1]);J_m(g)\leq\rho} \sum_{i=1}^{n}[Y_i - g(x_i)]^2,$$

where $J_m(g) = \int_0^1 [g^{(m)}(u)]^2 du$ and $\rho = \sigma^2/(n\lambda)$. Therefore, when λ tends to infinity, ρ tends to zero and $\widehat{f}_\lambda$ converges to the $(m-1)$-th order polynomial estimator of f. In the other extreme case with $\lambda = 0$, the objective function in (2.18) reduces to the residual sum of squares, and consequently $\widehat{f}_\lambda$ is the function in $W_2^m([0,1])$ that interpolates all the data points. In the first case mentioned above, $\widehat{f}_\lambda$ might oversmooth the data, whereas in the second it might undersmooth the data.

Silverman (1984) built a connection between the smoothing spline estimator $\widehat{f}_\lambda$ and the kernel estimators discussed in Section 2.3. He explained that, in the case when n is large, λ is small and all the design points are a certain distance away from the border of the design space,

$$\widehat{f}_\lambda(x) \approx \sum_{j=1}^{n} Y_j \, K\left(\frac{x-x_j}{h(x_j)}\right) \Big/ \, [p(x_j)h(x_j)],$$

where $p(\cdot)$ is the probability density function of the design points $\{x_j, j = 1, 2, \ldots, n\}$, $h(u) = (\lambda/p(u))^{1/4}$, $K(u) = 0.5\sin(|u|/\sqrt{2} + \pi/4)\exp(-|u|/\sqrt{2})$, and "$\approx$" denotes asymptotic equality. Obviously, the right-hand side of the above expression has the form of a kernel estimator, and its bandwidth is related to the smoothing parameter λ.

For a given point x in the design space, it can be checked that $\widehat{f}_\lambda(x)$ depends on all observations available. If one observation changes, the entire smoothing spline estimator would be affected by such a change. Therefore, smoothing spline estimation is a *global* smoothing method, different from the local smoothing procedures discussed in the previous two sections.

2.5.2 Selection of the smoothing parameter

The bandwidth selection procedures discussed in Subsection 2.4.4 (cf. equations (2.14)–(2.17)) can be used for choosing the smoothing parameter λ. For example, suppose that $\widehat{f}_{\lambda,-i}(x_i)$ denotes the smoothing spline estimator of $f(x_i)$ without using the ith observation (x_i, Y_i). Then λ can be chosen by minimizing the following CV score (Wahba and Wold 1975):

$$CV(\lambda) = \frac{1}{n}\sum_{i=1}^{n}\left(Y_i - \widehat{f}_{\lambda,-i}(x_i)\right)^2. \tag{2.19}$$

Craven and Wahba (1979) showed that

$$Y_i - \widehat{f}_{\lambda,-i}(x_i) = \left[Y_i - \widehat{f}_\lambda(x_i)\right] \Big/ (1 - a_{ii}),$$

where a_{ii} is the ith diagonal element of the hat matrix $\mathbf{H}$ defined by $(\widehat{f}_\lambda(x_1), \widehat{f}_\lambda(x_2), \ldots, \widehat{f}_\lambda(x_n))' = \mathbf{HY}$, for $i = 1, 2, \ldots, n$. Based on this result and the fact that $Y_i - \widehat{f}_{\lambda,-i}(x_i)$ is numerically unstable if $(1 - a_{ii})$ is close to 0, they suggest using the following *generalized cross-validation* (GCV) function:

$$GCV(\lambda) = \frac{1}{n}\sum_{i=1}^{n}\left[Y_i - \widehat{f}_\lambda(x_i)\right]^2 \Big/ \left[\frac{1}{n}\mathrm{tr}(I_{n\times n} - \mathbf{H})\right]^2. \tag{2.20}$$

Note that $GCV(\lambda) = CV(\lambda)$ if all diagonal elements of $\mathbf{H}$ are the same. By the GCV criterion, the smoothing parameter λ can be chosen by minimizing $GCV(\lambda)$.

The selected λ by GCV, which is denoted by $\widehat{\lambda}$, has some good properties. For example, Craven and Wahba (1979) proved that, under some regularity conditions,

$$\lim_{n\to\infty} \frac{R_n(\widehat{\lambda})}{\min_{\lambda>0} R_n(\lambda)} = 1,$$

where $R_n(\lambda) = E[\sum_{i=1}^{n}(\widehat{f}_\lambda(x_i) - f(x_i))^2/n]$ measures the average distance between the true regression function f and its smoothing spline estimator $\widehat{f}_\lambda$. Other properties of $\widehat{\lambda}$ have been discussed by several authors, including Li (1985, 1986, 1987) and Speckman (1985).

2.5.3 Multivariate smoothing spline estimation

Suppose that $\{(x_{1i}, x_{2i}, \ldots, x_{pi}, Y_i), i = 1, 2, \ldots, n\}$ are n observations generated from the p-dimensional model (2.7). Then the smoothing spline estimator of f is defined by the solution of the following minimization problem:

$$\min_{g\in W_2^m([0,1]^p)} \frac{1}{n}\sum_{i=1}^{n}[Y_i - g(x_{1i}, x_{2i}, \ldots, x_{pi})]^2 + \lambda J_{m,p}(g), \tag{2.21}$$

where $W_2^m([0,1]^p)$ is the Sobolev function space in the p-dimensional real space $[0,1]^p = \{(x_1, x_2, \ldots, x_p) : x_j \in [0,1], \text{ for } j = 1, 2, \ldots, p\}$, defined similarly to $W_2^m([0,1])$ used in problem (2.18), and

$$J_{m,p}(g) =$$

$$\sum_{\alpha_1+\cdots+\alpha_p=m} \frac{m!}{\alpha_1!\cdots\alpha_p!}\int_{-\infty}^{\infty}\cdots\int_{-\infty}^{\infty}\left(\frac{\partial^m g(u_1, \ldots, u_p)}{\partial u_1^{\alpha_1}\cdots\partial u_p^{\alpha_p}}\right)^2 du_1\cdots du_p.$$

Duchon (1977), Wahba (1979), and several other authors have studied various properties of the multivariate smoothing spline estimator. They have also discussed the CV and GCV smoothing parameter selections in multivariate cases. For real-life applications, one major limitation of the multivariate smoothing spline procedure is the expensive computation involved, especially when p is relatively large. To deal with this issue, several simplified multivariate regression models and the corresponding smoothing spline estimation procedures have been suggested in the literature, including additive spline estimators (Buja *et al.* 1989, Hastie and Tibshirani 1990, Stone 1985) and smoothing spline ANOVA (Gu and Wahba 1993). A detailed discussion about this topic can be found in Gu (2002).

There are a number of variants of the objective function in problem (2.21). For example, Cox (1983) studied robust smoothing spline estimation by replacing the square loss function used in problem (2.21) with a robust loss function (see also O'Sullivan (1983) and the references cited there). We next briefly introduce another variant that is for fitting *semiparametric* regression models defined below. Suppose that the response variable Y has a linear relationship with p_1 explanatory variables $\mathbf{x_1} = (x_1, \ldots, x_{p_1})'$, and it has a nonparametric relationship with the remaining explanatory variables $\mathbf{x_2} = (x_{p_1+1}, \ldots, x_p)'$, where $1 \leq p_1 \leq p$. Then the relationship between $\mathbf{x} = (\mathbf{x_1}', \mathbf{x_2}')'$ and Y can be described by the following semiparametric regression model:

$$Y_i = \mathbf{x}'_{1i}\beta + f(\mathbf{x_{2}}_i) + \varepsilon_i, \text{ for } i = 1, 2, \ldots, n,$$

where β is a p_1-dimensional vector of the regression coefficients and f is a nonparametric function. They can be estimated by the solutions to b and g, respectively, of the following minimization problem:

$$\min_{\mathbf{b} \in R^{p_1}; g \in W_2^m([0,1]^{p-p_1})} \frac{1}{n} \sum_{i=1}^{n} [Y_i - \mathbf{x}'_{1i}\mathbf{b} - g(\mathbf{x_{2}}_i)]^2 + \lambda J_{m,p-p_1}(g). \tag{2.22}$$

The solution to g is called the *partial spline* estimator of f (see e.g., Heckman 1986, Shiau and Wahba 1988, Wahba 1984). The partial spline estimation procedure (2.22) can be used for fitting certain jump regression models and is further discussed in Subsection 3.2.4.

For a quite complete discussion about smoothing spline estimation, please read Wahba (1990).

2.5.4 Regression spline estimation

Besides smoothing spline procedures, there are some *regression spline* procedures suggested in the literature for curve and surface estimation. Suppose that $0 < t_1 < t_2 < \ldots < t_s < 1$ are s given knots in the design interval $[0, 1]$. A function $S(x)$ defined on $[0, 1]$ is called a *spline of order* k, with $k \geq 1$ an integer, if it satisfies the following two conditions.

- $S(x)$ is a polynomial of degree at most k in each interval $[t_j, t_{j+1}]$, with $j = 0, 1, \ldots, s$, $t_0 = 0$, and $t_{s+1} = 1$.

- $S(x)$ has continuous $(k-1)$-th-order derivative in the entire design space.

Obviously, the $(2m-1)$-th-order natural polynomial spline defined in Subsection 2.5.1 is a $(2m-1)$-th-order spline and, in addition, satisfies the boundary conditions that its derivatives of order $m, m+1, \ldots, 2m-1$ all vanish in the boundary regions $[0, t_1)$ and $(t_s, 1]$.

For a given set of knots $\{t_j, j = 1, 2, \ldots, s\}$, a kth-order spline $S(x)$ can be expressed by a linear combination of the following power and *truncated power* basis functions (see e.g., Phillips and Taylor 1996, Chapter 6):

$$1, x, \ldots, x^k, (x-t_1)_+^k, \ldots, (x-t_s)_+^k,$$

where

$$(x-t_j)_+^k = \begin{cases} (x-t_j)^k, & \text{if } x > t_j \\ 0, & \text{otherwise,} \end{cases}$$

for $j = 1, 2, \ldots, s$. That is,

$$S(x) = \sum_{m=0}^{k} c_m x^m + \sum_{j=1}^{s} d_j (x-t_j)_+^k, \tag{2.23}$$

where $\{c_m, m = 0, 1, \ldots, k;\ d_j, j = 1, 2, \ldots, s\}$ are the coefficients.

If $\{(x_i, Y_i),\ i = 1, 2, \ldots, n\}$ are n observations generated from the regression model (1.1), then the regression function f can be estimated by a kth-order spline with knots $\{t_j, j = 1, 2, \ldots, s\}$ as follows. Let $\{\widehat{c}_m, m = 0, 1, \ldots, k;\ \widehat{d}_j, j = 1, 2, \ldots, s\}$ be least squares estimators of $\{c_m, m = 0, 1, \ldots, k;\ d_j, j = 1, 2, \ldots, s\}$. Namely, they are the solutions to $\{c_m, m = 0, 1, \ldots, k;\ d_j, j = 1, 2, \ldots, s\}$ of the following least squares problem:

$$\min_{c_0, \ldots, c_k; d_1, \ldots, d_s} \sum_{i=1}^{n} [Y_i - S(x_i)]^2,$$

where $S(x)$ is defined in equation (2.23). Then the regression spline (RS) estimator of $f(x)$ is defined by

$$\widehat{f}_{RS}(x) = \sum_{m=0}^{k} \widehat{c}_m x^m + \sum_{j=1}^{s} \widehat{d}_j (x-t_j)_+^k.$$

In regression spline estimation, the number of knots s is usually chosen much smaller than the sample size n. It plays the role of a smoothing parameter, similar to the smoothing parameter λ used in smoothing spline estimation, in the sense that, when s is chosen smaller, the resulting regression spline estimator would be smoother, and vice versa.

The power and truncated power basis functions are easy to use. If we decide to delete a specific knot t_{j^*} from function estimation, then it can be accomplished easily by deleting the corresponding truncated power basis function $(x - t_{j^*})_+^k$ from the set of power and truncated power basis functions.

Another popular basis of all the kth-order splines with knots $\{t_j, j = 1, 2, \ldots, s\}$ is based on *B-splines*, which are defined as follows. Let us first extend the sequence of s knots $\{t_j, j = 1, 2, \ldots, s\}$ so that they are part of the following infinite sequence of knots:

$$\ldots, t_{-1}, t_0, t_1, \ldots, t_s, t_{s+1}, t_{s+2}, \ldots,$$

where $t_0 = 0$, $t_{s+1} = 1$, $\lim_{m\to\infty} t_{-m} = -\infty$, and $\lim_{m\to\infty} t_m = \infty$. Then the B-splines of order 0 are defined by

$$B_j^0(x) = \begin{cases} 1, & \text{if } t_j \leq x < t_{j+1} \\ 0, & \text{otherwise,} \end{cases}$$

where $j = 0, \pm 1, \ldots$. For any integer $k \geq 1$, the B-splines of order k are defined by

$$B_j^k(x) = \frac{x - t_j}{t_{j+k} - t_j} B_j^{k-1}(x) + \frac{t_{j+k+1} - x}{t_{j+k+1} - t_{j+1}} B_{j+1}^{k-1}(x) \tag{2.24}$$

for $j = 0, \pm 1, \ldots$. For example, when $t_j, t_{j+1}, t_{j+2}, t_{j+3}$, and t_{j+4} are 0, 0.25, 0.5, 0.75, and 1.0, the B-splines $B_j^0(x)$, $B_j^1(x)$, $B_j^2(x)$, and $B_j^3(x)$ are displayed in Figures 2.5(a)–2.5(d), respectively. It can be seen that (1) the B-spline $B_j^k(x)$ takes nonzero values only on interval (t_j, t_{j+k+1}) and (2) $B_j^k(x)$ is smoother when k is larger.

For a given set of knots $\{t_j, j = 1, 2, \ldots, s\}$ in the design interval $[0, 1]$, there is only a finite number of B-splines of order k defined in equation (2.24) which would have nonzero values on some nonempty parts of the design interval $[0, 1]$. These B-splines form a basis for all splines of order k with knots $\{t_j, j = 1, 2, \ldots, s\}$ on $[0, 1]$, and thus they can be used for estimating the regression function f as we did with the power and truncated power basis functions. In addition, because the B-splines are defined recursively by equation (2.24), they are more convenient to handle numerically than the power and truncated power basis functions. For more detailed discussion about the numerical properties of the B-splines, see books on numerical mathematics, such as de Boor (1978) and Cheney and Kincaid (1994).

By using either the power and truncated power basis functions or the B-splines for estimating the regression function f, we first need to choose the order k of the related basis functions. By definition, a first-order spline is a piecewise linear function of which the pieces are connected at its knots. Although the first-order spline function itself is continuous, its first-order derivative has jumps at the knots. Consequently, the slope of the first-order spline may change abruptly at each knot. For a second-order spline, discontinuities exist in its second-order derivative, and thus its curvature may change abruptly at each knot. In applications, the most commonly used splines are *cubic splines* of which the order is three. Although discontinuities may still exist in their third derivatives, such discontinuities cannot be detected visually. Several authors, including Cheney and Kincaid (1994), point out that using splines of order larger than three seldom yields any advantage.

The regression spline estimator of f also depends on the knots of the basis functions, which we need to choose. Intuitively, most knots should be placed at positions

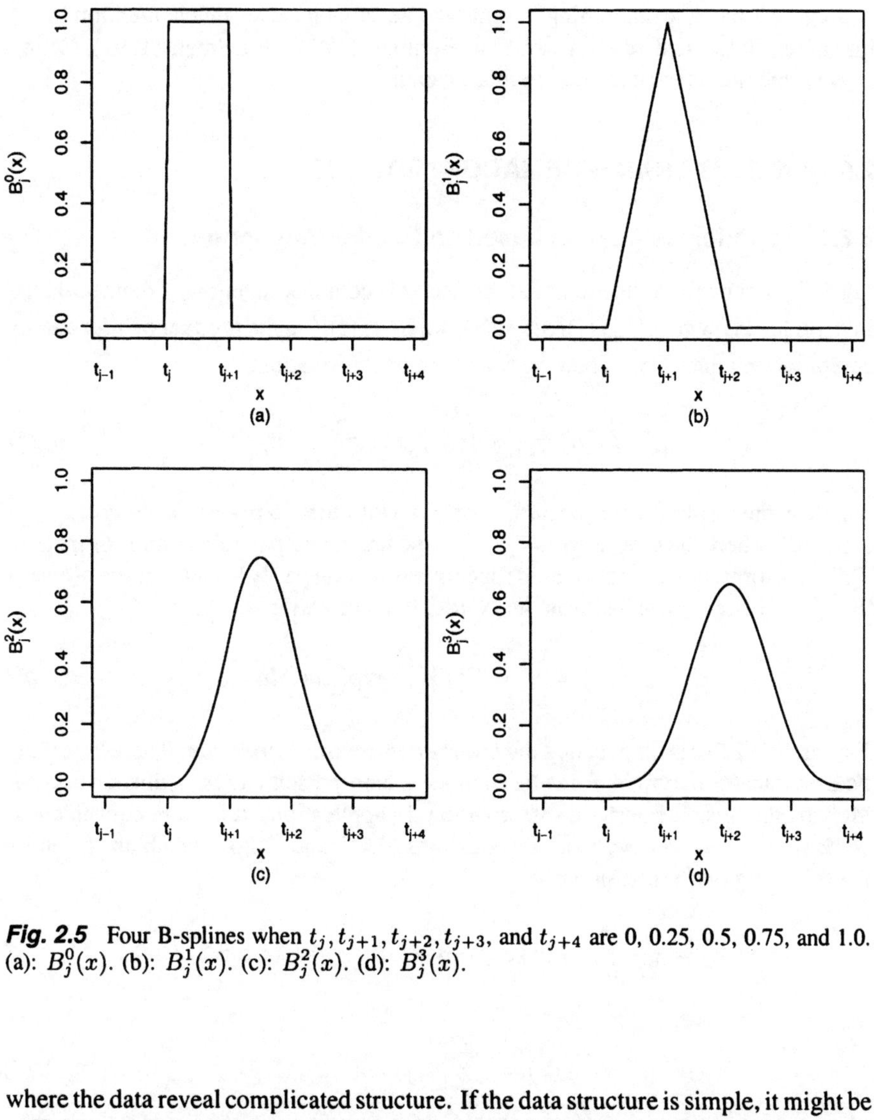

Fig. 2.5 Four B-splines when $t_j, t_{j+1}, t_{j+2}, t_{j+3}$, and t_{j+4} are 0, 0.25, 0.5, 0.75, and 1.0. (a): $B_j^0(x)$. (b): $B_j^1(x)$. (c): $B_j^2(x)$. (d): $B_j^3(x)$.

where the data reveal complicated structure. If the data structure is simple, it might be sufficient to choose the knots by hand. However, for most applications, data-driven knot selection procedures are desirable. In the literature, several forward/backward or mixed knot selection procedures have been proposed (e.g., Friedman and Silverman 1989, Kooperberg and Stone 1991). For example, backward procedures need an initial selection of the knots, and the initially selected knots are often a subset of the design points (e.g., one-fourth of the ordered design points $x_{(1)} \leq x_{(5)} \leq \ldots$). Then, based on some model assessment (e.g., the residual sum of squares) and model selection (e.g., the Akaike's information criterion) criteria, some knots are deleted or merged step by step, until the estimated regression function becomes worse, as judged by the model selection criterion.

Regression spline smoothing in multivariate cases receives much attention in the literature. Interested readers can read Breiman (1993), Friedman (1991), Zhang (1994), and the references cited in those papers.

2.6 WAVELET TRANSFORMATION METHODS

2.6.1 Function estimation based on Fourier transformation

Let $L^2[0, 2\pi]$ be the collection of all measurable complex functions f defined on the interval $[0, 2\pi]$ with $\int_0^{2\pi} |f(t)|^2 dt < \infty$, where $|f(t)|^2$ is the product of $f(t)$ and its complex conjugate $\overline{f(t)}$. Then the following transformation:

$$\mathcal{F}\{f\}(u) = \frac{1}{2\pi} \int_0^{2\pi} f(x) \exp(-ixu)\, dx \tag{2.25}$$

is called the *Fourier transformation* of f. Note that $\exp(-ixu) = \cos(xu) - i\sin(xu)$, where the term with $i = \sqrt{-1}$ is the imaginary part of a complex function. This transformation is important in function estimation partly because of the property that f can be reconstructed from its Fourier transformation

$$f(x) = \int_{-\infty}^{\infty} \mathcal{F}\{f\}(u) \exp(ixu)\, du. \tag{2.26}$$

Expression (2.26) is often called the *inverse Fourier transformation*. Therefore, if the Fourier transformation of f can be estimated, then f itself can be estimated accordingly by the inverse Fourier transformation. In applications, it is more convenient to work with the discrete versions of equations (2.25) and (2.26), which are given by the following two expressions:

$$c_j = \frac{1}{2\pi} \int_0^{2\pi} f(x) \exp(-ijx)\, dx, \; j = 0, \pm 1, \pm 2, \ldots$$

and

$$f(x) = \sum_{j=-\infty}^{\infty} c_j \exp(ijx), \tag{2.27}$$

where $\{c_j, j = 0, \pm 1, \pm 2, \ldots\}$ are called the *Fourier coefficients* of f.

It can be easily checked that the functions $\{\exp(ijx), j = 0, \pm 1, \pm 2, \ldots\}$ in equation (2.27) form an orthonormal basis for $L^2[0, 2\pi]$ in the sense that

$$\int_0^{2\pi} \exp(ij_1x) \exp(-ij_2x)\, dx = \begin{cases} 0, & \text{if } j_1 \neq j_2 \\ 1, & \text{otherwise.} \end{cases}$$

Thus, equation (2.27) is the expression of f in terms of this basis. When j is large, the sin and cos functions involved in the basis function $\exp(ijx) = \cos(jx) + i\sin(jx)$ have short periods, and, consequently, the frequency that they switch between positive and negative values in a given interval is large, and vice versa. So j is a frequency parameter of the basis functions, and f is expressed as a linear combination of the basis functions with different frequencies in equation (2.27).

The Fourier coefficients $\{c_j, j = 0, \pm 1, \pm 2, \ldots\}$ satisfy the so-called *Parseval's relation*:

$$\sum_{j=-\infty}^{\infty} |c_j|^2 = \frac{1}{2\pi} \int_0^{2\pi} |f(x)|^2 \, dx.$$

Therefore, c_j tends to zero when j goes to $\pm\infty$.

Now, suppose that we have a dataset $\{(x_i, Y_i), i = 1, 2, \ldots, n\}$ from model (1.1) and that we would like to estimate the regression function f from the dataset on the design interval $[0, 1]$. Since f is defined on the interval $[0, 1]$, the basis functions in equation (2.27) need to be transformed into functions $\{\exp(i2\pi jx), j = 0, \pm 1, \pm 2, \ldots\}$, which are also defined on $[0, 1]$. Clearly, these functions form an orthonormal basis for $L^2[0, 1]$. Therefore, f can be expressed in terms of this basis similarly to equation (2.27). Because f is a real function, only the basis functions with $j \geq 0$ need to be included in such an expression. With a fixed sample size n, it is impossible to estimate all infinite number of the Fourier coefficients. A natural strategy to overcome this difficulty is to discard all the terms with $j > d - 1$, where $d \leq n$ is a positive integer. Then the remaining Fourier coefficients $\{c_j, j = 0, 1, \ldots, d - 1\}$ can be estimated by their least squares estimators $\{\widehat{c}_j, j = 0, 1, \ldots, d - 1\}$, which are solutions to $\{c_j, j = 0, 1, \ldots, d - 1\}$ of the following minimization problem:

$$\min_{c_0, \ldots, c_{d-1} \in R} \sum_{\ell=1}^{n} \left[Y_\ell - \sum_{j=0}^{d-1} c_j \exp(i2\pi j x_\ell) \right]^2 .$$

Therefore, $f(x)$ can be estimated by

$$\widehat{f}_F(x) = \sum_{j=0}^{d-1} \widehat{c}_j \exp(i2\pi jx). \tag{2.28}$$

The role of the parameter d is similar to that of the smoothing parameter λ used in smoothing spline estimation. When d is chosen larger, the resulting estimator will become noisier, and vice versa. More detailed discussion about this function estimation procedure can be found in Eubank (1988).

2.6.2 Univariate wavelet transformations

A major disadvantage of the estimator $\widehat{f}_F$ defined by equation (2.28) is that any local change in the data would affect all the Fourier coefficients and thus the estimator in the entire design space. That is, $\widehat{f}_F$ is a global smoothing procedure, and its global smoothing nature results from the fact that the Fourier basis functions are all smooth functions with support $[0, 1]$. Consequently, a large amount of terms are needed in equation (2.28) to preserve any singularity in f such as jumps and angles.

However, it is often impossible to include many terms in (2.28) because the number of Fourier coefficients should not exceed the number of data points, in order to define $\widehat{f}_F$ uniquely. In reality, even when the sample size is large, we usually do not want to include too many terms in (2.28); otherwise, little noise would be removed by (2.28) and the estimator $\widehat{f}_F$ would be too noisy.

Data smoothing procedures based on *wavelet transformations* are designed to overcome this limitation. In this subsection, some basic terminologies and the main properties of the wavelet transformations are discussed. For practical purposes, our discussion focuses on discrete wavelet transformations only. More complete discussions about this topic can be found in Bachman *et al.* (2000), Chui (1992), Daubechies (1992), and Hernández and Weiss (1996), among many others.

A wavelet transformation is based on two basic functions: the *scaling function* (or the *father wavelet*) $\phi(x)$ and the *primary wavelet function* (or the *mother wavelet*) $\psi(x)$. A simple example is the Haar father and mother wavelets shown in Figures 2.6(a) and 2.6(b).

The *wavelet functions* are defined by

$$\psi_{j,k}(x) = 2^{j/2}\psi(2^j x - k), \text{ for } j, k = 0, \pm 1, \ldots. \tag{2.29}$$

Namely, the wavelet function $\psi_{j,k}(x)$ is obtained from the mother wavelet function by a *binary dilation* (i.e., dilation by 2^j) and a *dyadic translation*. As an example, the Haar wavelet functions $\psi_{1,0}(x)$ and $\psi_{1,1}(x)$ are displayed in Figures 2.6(c) and 2.6(d), respectively. Note that the four wavelet functions displayed in Figure 2.6 are orthonormal, and any function that is constant on each quarter-interval can be expressed as a linear combination of them.

More generally, when the range of the indices is $0 \le k \le 2^j - 1$ and $j \ge 0$, it can be checked that the Haar wavelet functions $\{\psi_{j,k}\}$ together with the Haar father wavelet function ϕ form an orthonormal basis for $L^2[0,1]$. Therefore, if it is reasonable to assume that the true regression function f in model (1.1) belongs to $L^2[0,1]$, then it can be expressed by a linear combination of the basis functions $\{\phi, \psi_{j,k}, 0 \le k \le 2^j - 1, j \ge 0\}$ as follows:

$$f(x) = w_{-1,0}\phi(x) + \sum_{j=0}^{\infty}\sum_{k=0}^{2^j-1} w_{j,k}\psi_{j,k}(x), \text{ for } x \in [0,1], \tag{2.30}$$

where the *wavelet coefficients* $\{w_{-1,0}, w_{j,k}, 0 \le k \le 2^j - 1, j \ge 0\}$ can be calculated by

$$\begin{aligned} w_{-1,0} &= \int_0^1 f(t)\phi(t)\,dt; \\ w_{j,k} &= \int_0^1 f(t)\psi_{j,k}(t)\,dt,\ 0 \le k \le 2^j - 1, j \ge 0. \end{aligned} \tag{2.31}$$

The transformation defined by equation (2.31) is called the *discrete wavelet transformation*, and the one defined by equation (2.30) is called the *inverse discrete wavelet transformation*.

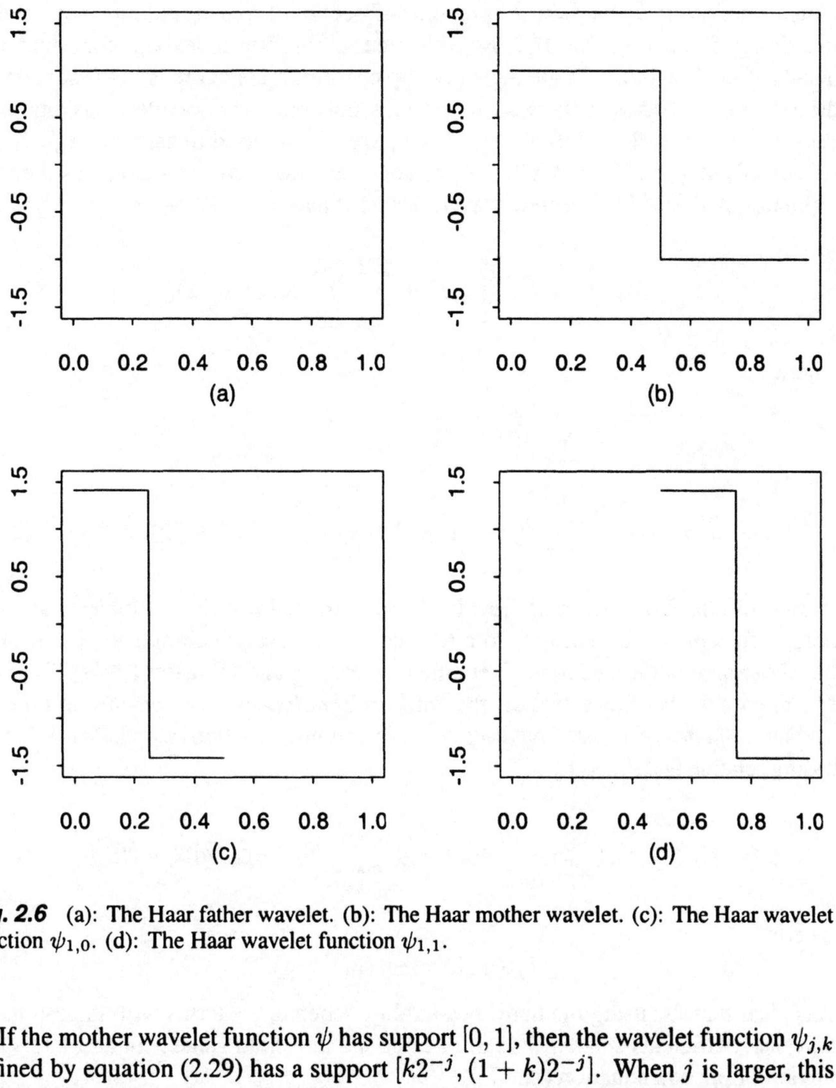

Fig. 2.6 (a): The Haar father wavelet. (b): The Haar mother wavelet. (c): The Haar wavelet function $\psi_{1,0}$. (d): The Haar wavelet function $\psi_{1,1}$.

If the mother wavelet function ψ has support $[0, 1]$, then the wavelet function $\psi_{j,k}$ defined by equation (2.29) has a support $[k2^{-j}, (1+k)2^{-j}]$. When j is larger, this interval becomes shorter; thus, the wavelet basis functions are different from the Fourier basis functions in that they are *local*. If the function f changes its value at a given point $x^* \in [0, 1]$, then this change only affects the coefficients of the wavelet functions whose supports include x^* (cf. equations (2.30) and (2.31)). This "local" property of the wavelet functions makes wavelet transformation procedures appropriate for approximating functions having local singularities, such as jumps and angles, by using only a limited number of wavelet functions.

Suppose we wish to estimate f based on a sample of size $n = 2^J$, where J is a positive integer. Then, at most n wavelet basis functions can be used for expressing f. Otherwise, the wavelet coefficients cannot be estimated uniquely by the least squares estimation procedure from the data, similar to the situation of the Fourier transformation estimator $\widehat{f}_F$ in the previous subsection. From the definition of wavelet functions (cf. equation (2.29)), those with large values of the index j are for estimating details of the function f. With a given sample size n, it is not realistic to estimate all the details of f. Based on these considerations, only the first n wavelet basis functions $\{\phi, \psi_{j,k}, 0 \leq k \leq 2^j - 1, 0 \leq j \leq J-1\}$ are usually used in estimating f. More specifically, if $\{(x_i, Y_i), i = 1, 2, \cdots, n\}$ are observations from model (1.1), then the estimator of f based on discrete wavelet transformations is defined by

$$\widehat{f}_W(x) = \widehat{w}_{-1,0}\phi(x) + \sum_{j=0}^{J-1}\sum_{k=0}^{2^j-1} \widehat{w}_{j,k}\psi_{j,k}(x), \tag{2.32}$$

where

$$\begin{aligned} \widehat{w}_{-1,0} &= \frac{1}{n}\sum_{i=1}^{n} \phi(x_i)Y_i; \\ \widehat{w}_{j,k} &= \frac{1}{n}\sum_{i=1}^{n} \psi_{j,k}(x_i)Y_i, \ 0 \leq k \leq 2^j - 1, 0 \leq j \leq J-1. \end{aligned} \tag{2.33}$$

Because the data have noise involved, the estimated wavelet coefficients are also noisy. To remove some noise from the estimated wavelet coefficients, the idea of thresholding is suggested in the literature (cf. Donoho and Johnston 1994). There are two major thresholding schemes: the *hard-thresholding* and the *soft-thresholding*.

The estimator of f based on discrete wavelet transformations and the hard-thresholding scheme is defined by

$$\widehat{f}_{WH}(x) = T_H(\widehat{w}_{-1,0}, \lambda)\phi(x) + \sum_{j=0}^{J-1}\sum_{k=0}^{2^j-1} T_H(\widehat{w}_{j,k}, \lambda)\psi_{j,k}(x), \tag{2.34}$$

where

$$T_H(w, \lambda) = wI(|w| > \lambda).$$

It is clear that, by using the hard-thresholding scheme, the terms with the estimated wavelet coefficients smaller than or equal to the threshold value λ in absolute value are all dropped in equation (2.34).

The estimator of f based on discrete wavelet transformations and the soft-thresholding scheme is defined by

$$\widehat{f}_{WS}(x) = T_S(\widehat{w}_{-1,0}, \lambda)\phi(x) + \sum_{j=0}^{J-1}\sum_{k=0}^{2^j-1} T_S(\widehat{w}_{j,k}, \lambda)\psi_{j,k}(x), \tag{2.35}$$

where

$$T_S(w, \lambda) = \text{sign}(w)(|w| - \lambda)_+$$

and $(x)_+ = x$ if $x > 0$ and 0 otherwise. By the soft-thresholding scheme, all terms in equation (2.32) with the estimated wavelet coefficients smaller than or equal to the threshold value λ in absolute value are dropped; the terms with the estimated wavelet coefficients larger than λ in absolute value are kept, but the corresponding estimated wavelet coefficients are shrunk by the amount λ to avoid jumps or other singularities caused by thresholding.

The Haar mother wavelet function ψ is discontinuous, and consequently all the Haar wavelet functions $\psi_{j,k}$ are discontinuous. This attribute of the Haar wavelet basis functions makes function estimation based on the Haar wavelet transformations appropriate for some "irregular" functions, such as jump functions. However, if the function to be estimated is "regular", then "regular" wavelet basis functions are preferred. Based on the theory of *multi-resolution analysis*, several other wavelet function systems have been constructed, including Daubechies' wavelet function system, spline wavelet function system, Franklin wavelet function system, Battle-Lemarié wavelet function system, etc. (Debnath 2002, Chapter 7).

2.6.3 Bivariate wavelet transformations

The wavelet transformation discussed in the previous subsection can be generalized to a p-dimensional case without much difficulty, where $p > 1$ is an integer. For simplicity, we briefly discuss discrete wavelet transformation in the two-dimensional case only; this serves the book well because the focus of the book is on two-dimensional jump surface estimation. More detailed discussion about two-dimensional wavelet transformation can be found in Chang *et al.* (2000), Huang and Cressie (2000), Mallat (1989), and the references cited there.

Two-dimensional wavelet transformations are based on a univariate father wavelet ϕ and the corresponding univariate mother wavelet ψ, both defined on $[0, 1]$. The functions $\{\phi, \psi_{j,k}, 0 \leq k \leq 2^j - 1, j \geq 0\}$ form an orthonormal basis for $L^2[0, 1]$, where $\psi_{j,k}$ are defined by equation (2.29). Then the four basic two-dimensional wavelet functions are defined in the *separable* form as follows:

$$
\begin{aligned}
\Phi(x, y) &= \phi(x)\phi(y) \\
\Psi^{(1)}(x, y) &= \phi(x)\psi(y) \\
\Psi^{(2)}(x, y) &= \psi(x)\phi(y) \\
\Psi^{(3)}(x, y) &= \psi(x)\psi(y).
\end{aligned} \tag{2.36}
$$

Other two-dimensional wavelet functions are defined by

$$\Psi^{(m)}_{j,k_1,k_2}(x, y) = 2^j \Psi^{(m)}(2^j x - k_1, 2^j y - k_2),$$

for $j, k_1, k_2 = 0, \pm 1, \pm 2, \ldots$, and $m = 1, 2, 3$.

Similar to the one-dimensional case, the following functions

$$\{\Phi(x, y), \Psi^{(1)}_{j,k_1,k_2}(x, y), \Psi^{(2)}_{j,k_1,k_2}(x, y), \Psi^{(3)}_{j,k_1,k_2}(x, y), 0 \leq k_1, k_2 \leq 2^j - 1, j \geq 0\}$$

form an orthonormal basis of the real function space $L^2([0, 1] \times [0, 1])$, consisting of all real functions f defined on $[0, 1] \times [0, 1]$ with $\int_0^1 \int_0^1 f^2(x, y)dxdy < \infty$. So any real function $f \in L^2([0,1] \times [0,1])$ can be expressed by this basis as follows:

$$f(x,y) = w^{(0)}_{-1,0,0}\Phi(x,y) + \sum_{j=0}^{\infty}\sum_{k_1=0}^{2^j-1}\sum_{k_2=0}^{2^j-1}\sum_{m=1}^{3} w^{(m)}_{j,k_1,k_2}\Psi^{(m)}_{j,k_1,k_2}(x,y). \quad (2.37)$$

Here, $w^{(0)}_{-1,0,0}$ and $\{w^{(m)}_{j,k_1,k_2}, 0 \leq k_1, k_2 \leq 2^j-1, j \geq 0, m = 1,2,3\}$ are the wavelet coefficients, which are defined by the inner products of f and the corresponding basis functions, similarly as in equation (2.31).

In the special case when $\phi(x)$ and $\psi(x)$ are the Haar father and mother wavelets, $\Phi(x,y)$, $\Psi^{(1)}(x,y)$, $\Psi^{(2)}(x,y)$, and $\Psi^{(3)}(x,y)$ are displayed by the four images in Figure 2.7, with white indicating -1 and black indicating 1. From the plots, the function $\Psi^{(1)}(x,y)$ corresponds to a difference operator in the y direction, $\Psi^{(2)}(x,y)$ corresponds to a difference operator in the x direction, and $\Psi^{(3)}(x,y)$ corresponds to a difference operator in the diagonal direction. Thus, at level $j \geq 0$, the wavelet coefficients can be classified into three categories: the vertical coefficients, the horizontal coefficients, and the diagonal coefficients. Expression (2.37) corresponds to a decomposition of f in these three spatial orientations.

Now suppose that we have a two-dimensional dataset $\{(x_{1i}, x_{2i}, Y_i), i = 1,2,\ldots, N\}$ generated from the following two-dimensional regression model:

$$Y_i = f(x_{1i}, x_{2i}) + \varepsilon_i,\ i = 1,2,\ldots,N,$$

where N is the sample size and is assumed to be a power of 2. Then the regression function f can also be expressed by equation (2.37) under some regularity conditions, and the wavelet coefficients can be estimated by $\{\widehat{w}^{(0)}_{-1,0,0}, \widehat{w}^{(m)}_{j,k_1,k_2}, 0 \leq k_1, k_2 \leq 2^j - 1, j \geq 0, m = 1,2,3\}$, which are defined similarly to those in one-dimensional cases (cf. equation (2.33)). When j is large, the estimated wavelet coefficients are noisy because only a few observations are actually used in the estimation. Therefore, we usually select a positive integer J, and only the wavelet coefficients up to the level J are included in estimating f. The estimated regression function is then defined by

$$\widehat{f}_W(x,y) = \widehat{w}^{(0)}_{-1,0,0}\Phi(x,y) + \sum_{j=0}^{J}\sum_{k_1=0}^{2^j-1}\sum_{k_2=0}^{2^j-1}\sum_{m=1}^{3} \widehat{w}^{(m)}_{j,k_1,k_2}\Psi^{(m)}_{j,k_1,k_2}(x,y). \quad (2.38)$$

If the design points are equally spaced in the design space with n_1 rows and n_2 columns, then there are about $(n_1/2^j) \times (n_2/2^j)$ observations used in estimating the wavelet coefficients at level j; J is typically chosen large enough such that $n_1/2^J \ll n_1, n_2/2^J \ll n_2$ and $\min(n_1/2^J, n_2/2^J) > 1$.

The hard- and soft-thresholding schemes defined in equations (2.34) and (2.35) can be applied directly to the estimated wavelet coefficients in equation (2.38). The resulting estimators are denoted by $\widehat{f}_{WH}$ and $\widehat{f}_{WS}$, respectively, as in one-dimensional cases.

The four basic wavelet functions defined in the separable form (cf. equation (2.36)) are simple to use; also, many one-dimensional properties of the wavelet transformations carry over to two-dimensional cases. However, more general definitions of

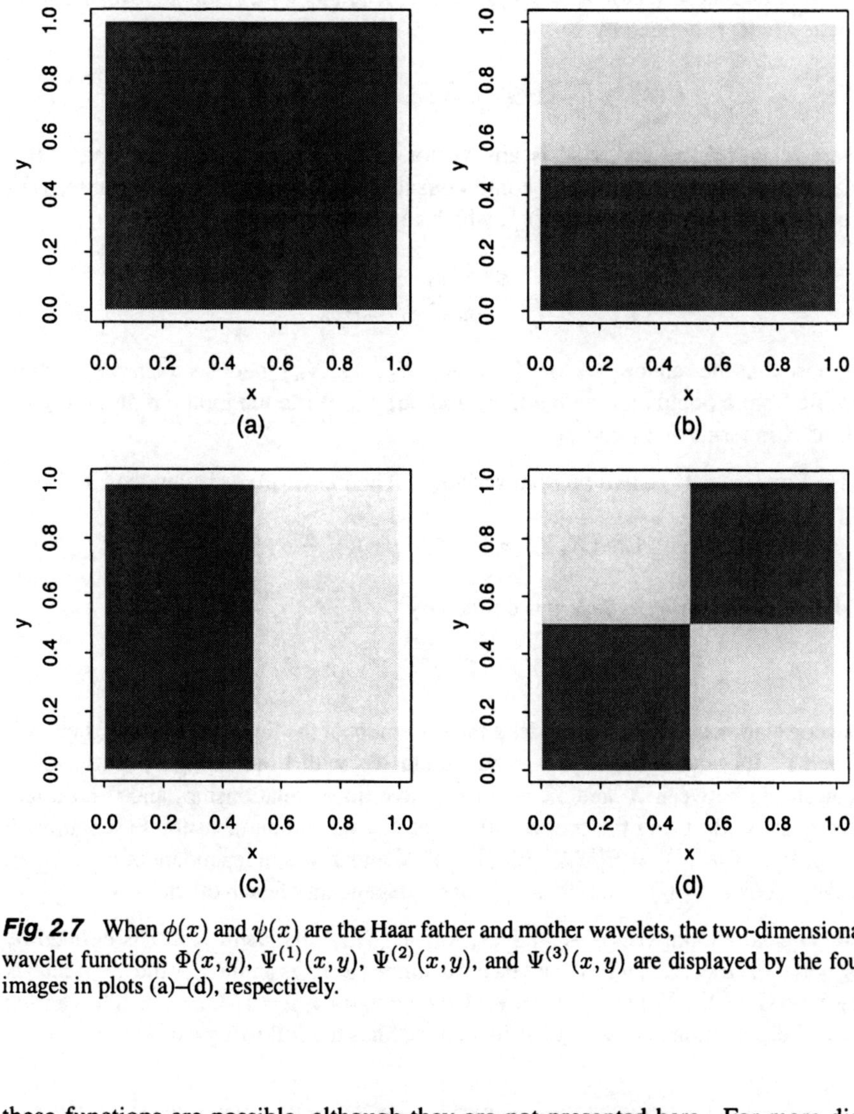

Fig. 2.7 When $\phi(x)$ and $\psi(x)$ are the Haar father and mother wavelets, the two-dimensional wavelet functions $\Phi(x, y)$, $\Psi^{(1)}(x, y)$, $\Psi^{(2)}(x, y)$, and $\Psi^{(3)}(x, y)$ are displayed by the four images in plots (a)–(d), respectively.

these functions are possible, although they are not presented here. For more discussion about this topic, please read Karlsson and Vetterli (1990) and Ruskai *et al.* (1992).

Problems

Problems 2.1–2.6 below are only for those who are not familiar with introductory level statistics, such as the materials covered by Section 2.2.

2.1 Suppose that a random variable X has a normal distribution $N(\mu, \sigma^2)$. Verify that $\mu_X = \mu$ and $\sigma_X = \sigma$.

2.2 For a random vector $\mathbf{X} = (X_1, X_2, \ldots, X_n)'$, its joint cumulative distribution function (cdf) is defined by

$$F(\mathbf{x}) = P(X_1 \leq x_1, X_2 \leq x_2, \ldots, X_n \leq x_n),$$

where $\mathbf{x} = (x_1, x_2, \ldots, x_n)'$ is any vector in the n-dimensional real space R^n. The random vector is absolutely continuous if it has a nonnegative, integrable, joint probability density function (pdf) f, which can be computed by

$$f(\mathbf{x}) = \frac{\partial^n F(x_1, x_2, \ldots, x_n)}{\partial x_1 \partial x_2 \cdots \partial x_n}.$$

Suppose that the random vector $\mathbf{X} = (X_1, X_2, \ldots, X_n)'$ denotes a simple random sample from a population with cdf F_1 and pdf f_1. Write the joint cdf and the joint pdf of $\mathbf{X}$ in terms of F_1 and f_1.

2.3 Let X and Y be two random variables. Their *covariance* is defined by

$$\text{Cov}(X, Y) = \text{E}[(X - \mu_X)(Y - \mu_Y)],$$

and their *correlation coefficient* is defined by

$$\text{Cor}(X, Y) = \text{Cov}(X, Y)/(\sigma_X \sigma_Y).$$

The correlation coefficient provides a measurement of the linear relationship between X and Y. Its value is between -1 and 1 inclusive, with 1 indicating a positive linear relationship between X and Y, -1 a negative linear relationship, and 0 no linear relationship. By using the property that, for any univariate measurable functions h and g, $\text{E}[h(X)g(Y)] = \text{E}[h(X)]\text{E}[g(Y)]$ if X and Y are independent of each other, prove that $\text{Cor}(X, Y) = 0$ if X and Y are independent of each other.

2.4 For a random vector $\mathbf{X} = (X_1, X_2, \ldots, X_n)'$, its *mean vector* is defined by $\mu_{\mathbf{X}} = (\mu_{X_1}, \mu_{X_2}, \ldots, \mu_{X_n})'$, and its *covariance matrix* is defined by the $n \times n$ matrix $\Sigma_{\mathbf{X}}$ with $\text{Cov}(X_i, X_j)$ being its (i, j)-th element, for $i, j = 1, 2, \ldots, n$. It has a joint normal distribution $N(\mu_{\mathbf{X}}, \Sigma_{\mathbf{X}})$ if its joint pdf has the following expression:

$$f(\mathbf{x}) = \frac{1}{(2\pi)^{n/2}|\Sigma_{\mathbf{X}}|^{1/2}} \exp\left[-\frac{1}{2}(\mathbf{x} - \mu_{\mathbf{X}})'\Sigma_{\mathbf{X}}^{-1}(\mathbf{x} - \mu_{\mathbf{X}})\right],$$

where $|\Sigma_{\mathbf{X}}|$ denotes the determinant of $\Sigma_{\mathbf{X}}$. Now suppose that $\mathbf{X} = (X_1, X_2, \ldots, X_n)'$ is a simple random sample from a population with the normal distribution $N(\mu, \sigma^2)$.

2.4.1. Write the mean vector and the covariance matrix of $\mathbf{X}$ in terms of μ and σ^2.

2.4.2. Write the joint pdf of $\mathbf{X}$. Is the distribution of $\mathbf{X}$ jointly normal?

2.5 Suppose that $\mathbf{a}'\mathbf{X} = \sum_{i=1}^n a_i X_i$ is a linear combination of the random variables $X_1, X_2, \ldots, X_n$, where $\mathbf{a} = (a_1, a_2, \ldots, a_n)'$ is a coefficient vector. Then the mean and variance operators have the following properties:

$$\begin{aligned} \text{E}(\mathbf{a}'\mathbf{X}) &= \mathbf{a}'\text{E}(\mathbf{X}) = \sum_{i=1}^{n} a_i \text{E}(X_i) \\ \text{Var}(\mathbf{a}'\mathbf{X}) &= \mathbf{a}'\text{Var}(\mathbf{X})\mathbf{a} = \sum_{i=1}^{n}\sum_{j=1}^{n} a_i a_j \text{Cov}(X_i, X_j). \end{aligned}$$

By using these properties, prove that the sample mean $\overline{X}$ of a simple random sample $(X_1, X_2, \ldots, X_n)$ is an unbiased point estimator of the population mean μ, and its variance equals σ^2/n, where σ^2 is the population variance.

2.6 Suppose that $X_1, X_2, \ldots, X_n$ are i.i.d. random variables with standard normal distribution. Then the random variable $Y = X_1^2 + X_2^2 + \cdots + X_n^2$ has a *chi-square distribution* with n degrees of freedom, denoted by $Y \sim \chi^2(n)$. If X and Y are two independent random variables, $X \sim N(0,1)$ and $Y \sim \chi^2(n)$, then the random variable $T = X/\sqrt{Y/n}$ has a *t-distribution* with n degrees of freedom. The density curve of a t-distribution is bell-shaped and symmetric about 0. Compared to the density curve of the standard normal distribution, it has heavier tails in the sense that $P(T > a) > P(Z > a)$, for any $a > 0$; T denotes a random variable with a t-distribution, and Z denotes a random variable with the standard normal distribution. When the degree of freedom of a t-distribution tends to infinity, the t-distribution converges to the standard normal distribution. Now suppose that $(X_1, X_2, \ldots, X_n)$ is a simple random sample from a population with the normal distribution $N(\mu, \sigma^2)$. By some standard techniques about random variable and random vector transformations (e.g., Casella and Berger 1990), it can be shown that $T = (\overline{X} - \mu)/(s/\sqrt{n})$ has a t-distribution with $n-1$ degrees of freedom. Based on this result, construct a 95% confidence interval for μ.

2.7 For the Nadaraya-Watson kernel estimator $\widehat{f}_{NW}$ defined in expression (2.3), verify expression (2.5) and give all necessary conditions on h_n, K, and f.

2.8 Compare the three kernel estimators $\widehat{f}_{NW}$, $\widehat{f}_{PC}$, and $\widehat{f}_{GM}$ introduced in Subsection 2.3.1, and discuss their differences and connections.

2.9 For the local linear kernel estimator $\widehat{f}_{LK}$ defined in expression (2.10), verify expressions (2.10)–(2.12), (2.16), and (2.17).

2.10 Show that the Nadaraya-Watson kernel estimator $\widehat{f}_{NW}(x)$ is biased in the case of Figure 2.4(b) when the design points on one side of x are significantly denser than the design points on the other side and the bias depends on $f'(x)$. As a comparison, show that the bias of the local linear kernel estimator $\widehat{f}_{LK}(x)$ in the same case is much smaller and that this bias depends little on $f'(x)$.

2.11 The sea-level pressure data shown in Figure 1.3 can be downloaded from the ftp site of this book (see the front cover for address). Please apply the three kernel smoothing procedures $\widehat{f}_{NW}$, $\widehat{f}_{PC}$, and $\widehat{f}_{GM}$, introduced in Subsection 2.3.1, and the local linear kernel smoothing procedure $\widehat{f}_{LK}$, introduced in Subsection 2.4.1, to this data. In all these procedures, use the Epanechnikov kernel function given in Subsection 2.3.1, and choose their bandwidths by the cross-validation procedure introduced in Subsection 2.4.4 (cf. equation (2.14)). Comment on the four estimators, especially around the year 1960 where the data appear to have a large discontinuity.

2.12 Many commonly used statistical software packages have procedures or functions for smoothing spline function estimation. For example, in the R package, both functions sreg() and smooth.spline() are for this purpose. Please use R or another software package to do the following *simulation* study: (1) generate 100 observations from model (1.1) with $f(x) = \sin(2\pi x) + x$, $\{x_i = i/100, i = 1, 2, \ldots, 100\}$ and $\varepsilon_1 \sim N(0,1)$; (2) estimate f by the cubic smoothing spline procedure (i.e., $m = 2$ in (2.18)) with the smoothing parameter λ equal to one of the following values $10^{-6}, 10^{-5}$, and 10^{-4}; and (3) estimate f by the cubic smoothing spline procedure with λ determined by the GCV procedure. Repeat the above simulation after changing the sample size from 100 to 1,000. Discuss differences among the results in the various cases.

2.13 For the sea-level pressure data (see Problem 2.11 for downloading information), find the cubic regression spline estimator with knots at years 1935, 1948, 1959, and 1965. Try both the power and truncated power basis functions and the B-spline basis functions. Compare the two estimators using these two different types of basis functions. If you were asked to recommend a new set of knots, what would be your recommendation?

2.14 Derive formulas for the least squares estimators $\{\widehat{c}_j, j = 0, 1, \ldots, d-1\}$ used in expression (2.28).

2.15 Verify that the Haar basis functions $\{\phi(x), \psi_{j,k}(x), 0 \leq k \leq 2^j - 1, j \geq 0\}$ (cf. Figure 2.6) form an orthonormal basis of the function space consisting of all measurable real functions f defined on the interval $[0, 1]$ with $\int_0^1 f^2(x)\, dx < \infty$.

3

Estimation of Jump Regression Curves

3.1 INTRODUCTION

This chapter introduces some recent methodologies on fitting the one-dimensional regression model (1.1), where the nonparametric regression function f may have jumps in the design interval $[0, 1]$. A one-dimensional *jump regression function* f can be described by the following equation:

$$f(x) = g(x) + \sum_{j=1}^{p} d_j I(x > s_j), \quad \text{for } x \in [0, 1], \tag{3.1}$$

where g is a continuous function in the entire design interval, p is the number of jump points, $\{s_j, j = 1, 2, \ldots, p\}$ are the jump positions, and $\{d_j, j = 1, 2, \ldots, p\}$ are the corresponding jump magnitudes. If $p = 0$, then f is continuous in the entire design interval. In this book, the function g is sometimes called the *continuity part* of f, and the summation $\sum_{j=1}^{p} d_j I(x > s_j)$ is called the *jump part* of f.

The jump part of f is determined by three factors: the number of jumps, the jump positions, and the jump magnitudes. Therefore, the jump regression function f can be classified into eight categories, depending on whether each factor is known (Qiu 1991; Qiu *et al.* 1991). However, if the jump magnitudes are unknown, they are not difficult to estimate after the number and positions of the jumps are determined. Therefore, only four categories are really of interest in our research; they correspond to the following four cases: (1) the number and positions of the jumps are both known; (2) the number of jumps is unknown, but the potential jump positions are known; (3) the number of jumps is known, but the jump positions are unknown; and (4) the number and positions of the jumps are both unknown.

In the first case, both the number and positions of jumps are known. In such a case, the regression function f can be estimated as usual in the design subintervals separated by the true jump positions. This case is not difficult to handle and will not be further discussed in this chapter.

The second case, in which the exact number of jumps is unknown but the potential jump positions are known, might be possible in applications. For example, suppose that we have blood pressure readings of a patient every minute of a day. The patient receives a medical treatment of different intensities at, say, 10 specific time points of the day, and the medical treatment above certain intensity level is believed to have a practically significant effect on the blood pressure reading. Then, the 10 time points at which the patient receives the medical treatment are the potential jump positions of the blood pressure reading. However, the exact number of jumps may be unknown for two reasons: first, the medical treatment at a low intensity level may not cause any practically significant change in the blood pressure reading; second, the researcher often does not know the threshold value of the intensity level above which the medical treatment could have a practically significant effect. Qiu (1991) and Qiu *et al.* (1991) have provided some discussions about estimation of the jump regression function f in case (2). Based on these discussions, methodologies for handling cases (3) and (4) can be adapted without much difficulty to handle case (2). For this reason, jump regression analysis in case (2) will not be further discussed in this chapter.

In the literature, a number of methodologies have been recently suggested for fitting the regression model (1.1) when the regression function f has expression (3.1). Most of these methods are appropriate to use only in cases (3) and (4). For this reason and the reasons explained above, our discussion in this chapter will focus on introducing methodologies about one-dimensional jump regression analysis in cases (3) and (4).

This chapter has five sections. In the next section, we introduce several jump detection procedures that are appropriate to use when the number of jumps is assumed known but the jump positions are unknown (i.e., the case (3) mentioned previously). Jump detection when both the number and positions of jumps are unknown (i.e., the case (4) mentioned previously) is discussed in Section 3.3. In Section 3.4, we introduce two jump-preserving curve estimation procedures by which jumps do not need to be detected explicitly before regression curves are estimated. In Section 3.5, we make several remarks about the procedures discussed in this chapter and about some open research problems on one-dimensional jump regression analysis.

3.2 JUMP DETECTION WHEN THE NUMBER OF JUMPS IS KNOWN

In this section, we discuss the case when the number of jumps p is assumed known, but the jump positions $\{s_j, j = 1, 2, \ldots, p\}$ and the jump magnitudes $\{d_j, j = 1, 2, \ldots, p\}$ are all unknown in expression (3.1). Various methods have been suggested in the literature for estimating the regression function f in such cases. Some of these methods are introduced below, in several subsections organized by the smoothing techniques on which they are based.

3.2.1 Difference kernel estimation procedures

Suppose we wish to estimate f using n observations $\{(x_i, Y_i), i = 1, 2, \ldots, n\}$ generated from model (1.1); we assume that the regression function f has expression (3.1) and that the number of jumps p is known. For simplicity, let us first discuss the case when $p = 1$. In such a case, our major objective is to estimate the jump position s_1, the corresponding jump magnitude d_1, and the continuity part g of the regression function f, based on n observations.

The difference kernel estimator is defined as follows. Let K_1 and K_2 be two kernel functions satisfying the following conditions:

- the support of K_1 is included in $[0, 1]$, and the support of K_2 is included in $[-1, 0)$;
- $K_j(x) \geq 0$ for any $x \in [-1, 1]$ and for $j = 1, 2$; and
- $\int_{-1}^{1} K_j(x)dx = 1$ for $j = 1, 2$.

Then K_1 is a density kernel function with right-sided support, and K_2 is a density kernel function with left-sided support. For a given $x \in [h_n, 1 - h_n]$, we define

$$\begin{aligned} M_{DKE}(x) &= \frac{1}{nh_n}\sum_{i=1}^{n} Y_i K_1\left(\frac{x_i - x}{h_n}\right) - \frac{1}{nh_n}\sum_{i=1}^{n} Y_i K_2\left(\frac{x_i - x}{h_n}\right) \\ &= \frac{1}{nh_n}\sum_{i=1}^{n} Y_i \left[K_1\left(\frac{x_i - x}{h_n}\right) - K_2\left(\frac{x_i - x}{h_n}\right)\right], \end{aligned} \tag{3.2}$$

where h_n is a positive bandwidth parameter. In equation (3.2), the expression $\frac{1}{nh_n}\sum_{i=1}^{n} Y_i\, K_1(\frac{x_i - x}{h_n})$ is a simplified version of the Nadaraya-Watson kernel estimator

$$\sum_{i=1}^{n} Y_i K_1\left(\frac{x_i - x}{h_n}\right) \Big/ \sum_{i=1}^{n} K_1\left(\frac{x_i - x}{h_n}\right),$$

since $\lim_{n\to\infty} \frac{1}{nh_n}\sum_{i=1}^{n} K_1(\frac{x_i - x}{h_n}) = \int_0^1 K_1(u)du = 1$ when h_n is chosen such that $\lim_{n\to\infty} h_n = 0$ and $\lim_{n\to\infty} 1/(nh_n) = 0$. Similarly, $\frac{1}{nh_n}\sum_{i=1}^{n} Y_i K_2(\frac{x_i - x}{h_n})$ is a simplified version of

$$\sum_{i=1}^{n} Y_i K_2\left(\frac{x_i - x}{h_n}\right) \Big/ \sum_{i=1}^{n} K_2\left(\frac{x_i - x}{h_n}\right).$$

Obviously, $M_{DKE}(x)$ is a difference of two *one-sided* kernel estimators. The first kernel estimator in equation (3.2) is right-sided; it is a weighted average of the observations in the right-sided neighborhood $[x, x + h_n]$. Similarly, the second kernel estimator in (3.2) is left-sided; it is a weighted average of the observations in the left-sided neighborhood $[x - h_n, x)$.

Intuitively, if the true jump position s_1 is not in the interval $[x - h_n, x + h_n]$, then the two one-sided kernel estimators in equation (3.2) should both estimate $f(x)$ well, and, consequently, $M_{DKE}(x)$ should be small. On the other hand, if x is the true jump position, then the first kernel estimator in (3.2) is a good estimator of $f_+(x)$ and the second kernel estimator is a good estimator of $f_-(x)$, where $f_+(x)$ and $f_-(x)$ denote the right and left limits of f at the point x. Thus $M_{DKE}(x)$ should be close to the jump magnitude $d_1 = f_+(x) - f_-(x)$. As a result, the function M_{DKE} includes information about the true jump position and can be used as a jump detection criterion. These results are summarized in the following theorem (Qiu 1999).

Theorem 3.1 *In model (1.1), suppose the following: the regression function f has continuous second-order derivative in the entire design interval $[0, 1]$, except the jump position s_1 at which it has the right and left second-order derivatives; the random errors $\{\varepsilon_i, i = 1, 2, \ldots, n\}$ are independent and identically distributed (i.i.d.) with mean zero and finite variance σ^2; and the design points are equally spaced on the design interval. The two one-sided kernel functions K_1 and K_2 are assumed to be Lipschitz continuous density functions on their supports. The bandwidth h_n satisfies the conditions that $h_n = o(1)$ and $1/(nh_n) = o(1)$. For simplicity, it is also assumed that $K_2(x) = K_1(-x)$ for $x \in [-1, 0)$. Then*

(i) when $x \in (0, 1)$ is the true jump position,

$$E(M_{DKE}(x)) = [f_+(x) - f_-(x)] + [f'_+(x) - f'_-(x)] \frac{K_{11}^{(1)}}{K_{01}^{(1)}} h_n + [f''_+(x) - f''_-(x)] \frac{K_{21}^{(1)}}{2K_{01}^{(1)}} h_n^2 + o(h_n^2); \tag{3.3}$$

(ii) when $x \in (0, 1)$ is not the true jump position,

$$E(M_{DKE}(x)) = o(h_n^2); \tag{3.4}$$

(iii) in both cases,

$$Var(M_{DKE}(x)) = \frac{2\sigma^2 K_{02}^{(1)}}{\left(K_{01}^{(1)}\right)^2 nh_n} + o\left(\frac{1}{nh_n}\right), \tag{3.5}$$

where $K_{j_1 j_2}^{(1)} = \int_0^1 u^{j_1} K_1^{j_2}(u)du$, for $j_1, j_2 = 0, 1, 2$.

This theorem indicates the usefulness of M_{DKE} as a jump detection criterion. From equation (3.3), it can be seen that possible jumps in the derivatives of f at the jump position s_1 also have some impact on the jump detection criterion M_{DKE}. However, this kind of impact is much smaller than the impact of the jump in f itself. From equation (3.4), it seems that the first two derivatives of f have no effect on $M_{DKE}(x)$ when x is a continuity point of f. It is therefore natural to use M_{DKE} for jump detection.

We would like to point out that expressions (3.3)–(3.5) give large-sample properties of M_{DKE} only, and its finite-sample properties are not demonstrated well by these expressions. For example, in finite-sample cases, the value of $M_{DKE}(x)$ is affected by the slope of f at x even when x is a continuity point of f, as shown in the following example.

Example 3.1 Suppose that the true regression function is $f(x) = x + (1.5 - x)I(x \geq 0.5)$ for $x \in [0, 1]$ and that the observations are noise free (i.e., $\sigma = 0$ in (1.1)). Then f is continuous on $[0, 1]$ except for one jump of size 1 at the position $x = 0.5$. This function is shown in Figure 3.1(a). In expression (3.2), suppose that $n = 100$, $h_n = 0.1$, $K_1(x) = 1.5(1 - x^2)I(x \in [0, 1])$, and $K_2(x) = 1.5(1 - x^2)I(x \in [-1, 0))$. Then the function M_{DKE} defined by expression (3.2) is displayed in Figure 3.1(b). From the plot, it can be seen that (i) $M_{DKE}(x)$ is close to the true jump magnitude when x is at the true jump position (i.e., $x = 0.5$), (ii) it decreases when x moves away from the true jump position, (iii) it is relatively small when x is certain distance away from the true jump position, and (iv) the value of $M_{DKE}(x)$ depends on the slope of f at x (i.e., $M_{DKE}(0.2)$ is larger than $M_{DKE}(0.8)$ due to the fact that the slope of f at $x = 0.2$ is larger than the slope of f at $x = 0.8$).

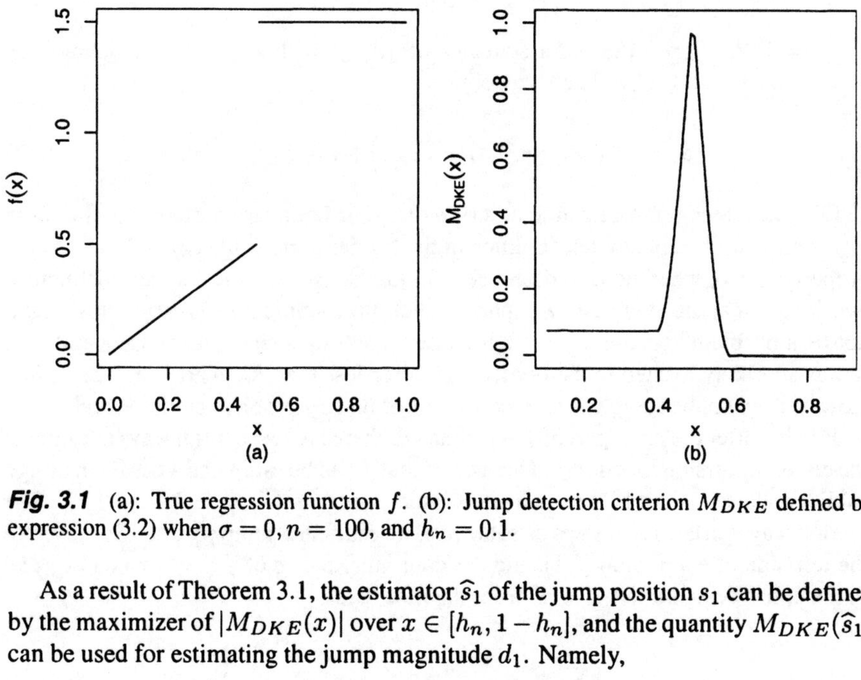

Fig. 3.1 (a): True regression function f. (b): Jump detection criterion M_{DKE} defined by expression (3.2) when $\sigma = 0, n = 100$, and $h_n = 0.1$.

As a result of Theorem 3.1, the estimator $\widehat{s}_1$ of the jump position s_1 can be defined by the maximizer of $|M_{DKE}(x)|$ over $x \in [h_n, 1 - h_n]$, and the quantity $M_{DKE}(\widehat{s}_1)$ can be used for estimating the jump magnitude d_1. Namely,

$$\begin{aligned} \widehat{s}_1 &= \arg \max_{x \in [h_n, 1-h_n]} |M_{DKE}(x)|; \\ \widehat{d}_1 &= M_{DKE}(\widehat{s}_1). \end{aligned} \tag{3.6}$$

Since this estimation procedure is based on the difference of two one-sided kernel estimators, Qiu *et al.* (1991) called such a procedure the difference kernel estimation procedure and the corresponding estimators the *difference kernel estimators* (DKEs), which is identified by the subscript of M_{DKE}.

Several authors, including Müller (1992), Qiu (1991), Qiu *et al.* (1991), and Wu and Chu (1993a), have suggested similar DKE procedures; most of them used the Gasser-Müller kernel estimator, instead of the Nadaraya-Watson kernel estimator, in defining M_{DKE}. Some properties are also derived in these papers. Müller (1992) derived the asymptotic distribution of $\widehat{s}_1$ and provided a confidence interval formula for s_1. Qiu *et al.* (1991) proved that $\widehat{s}_1$ and $\widehat{d}_1$ in (3.6) are both L_2 and almost surely consistent under some regularity conditions, and Wu and Chu (1993a) provided similar results in the case when there are $p \geq 1$ jumps. Their procedure is introduced in some detail in the next section.

When it is assumed that there are $p \geq 1$ jumps and p is known, the jump positions $\{s_j, j = 1, 2, \ldots, p\}$ and the jump magnitudes $\{d_j, j = 1, 2, \ldots, p\}$ can be estimated in a similar way. Let s_j^* be the maximizer of $|M_{DKE}(x)|$ over the range

$$x \in [h_n, 1 - h_n] \Bigg\backslash \left(\bigcup_{\ell=1}^{j-1} [s_\ell^* - h_n, s_\ell^* + h_n] \right)$$

for $j = 1, 2, \ldots, p$. The order statistics of $\{s_j^*, j = 1, 2, \ldots, p\}$ are denoted by $s_{(1)}^* \leq s_{(2)}^* \leq \ldots \leq s_{(p)}^*$. Then we define

$$\widehat{s}_j = s_{(j)}^*, \; \widehat{d}_j = M_{DKE}\left(s_{(j)}^*\right), \; j = 1, 2, \ldots, p. \tag{3.7}$$

Obviously, the jump detection criterion (3.2) has boundary problems in the sense that it may not be able to detect jumps in the border regions $[0, h_n)$ and $(1 - h_n, 1]$ properly. Also, when used to detect several jumps as above, it may not distinguish two jumps that are less than h_n apart. When the sample size is large, this would not be a problem because h_n is often chosen small in such cases; consequently, no jumps are likely located in the border regions or less than h_n apart from each other. However, when the sample size is small, this boundary problem could be real.

Finally, after the jump part of f is estimated, there are two natural ways to estimate the entire regression function f. One way is that f can be estimated as usual in design subintervals separated by the estimated jump positions $\{\widehat{s}_j, j = 1, 2, \ldots, p\}$. The second way works in two steps as follows. The estimated jump part of f is moved to the left-side of expression (3.1), and the continuity part g of f is estimated as usual in the entire design interval $[0, 1]$ from the new data

$$\left\{ Y_i - \sum_{j=1}^{p} \widehat{d}_j I(x_i > \widehat{s}_j), \; i = 1, 2, \ldots, n \right\}.$$

Then, the estimated regression function is defined by

$$\widehat{f}(x) = \widehat{g}(x) + \sum_{j=1}^{p} \widehat{d}_j I(x > \widehat{s}_j), \text{ for } x \in [0, 1],$$

where $\widehat{g}$ denotes the estimated continuity part of f.

Kang *et al.* (2000) compared these two approaches for estimating f after the jumps are detected. They concluded that the second approach is more appropriate to use because information included in the observations on neighboring intervals is shared in estimating f around the detected jump positions, especially when g has certain smoothness (e.g., g is kth-order smooth with $k \geq 1$). The first approach does not have this information-sharing property.

A modified version of M_{DKE} was suggested by Gijbels *et al.* (1999); their jump detection procedure is based on

$$D(x, h_n) = \frac{1}{nh_n^2} \sum_{i=1}^{n} Y_i K' \left(\frac{x_i - x}{h_n} \right),$$

where K is a kernel function with support $[-v, v]$, $v > 0$ is a constant, K' is the first-order derivative of K, and h_n is a bandwidth. Among some regularity conditions, the kernel function K is assumed to satisfy the following conditions: it has continuous second-order derivative on its support, $K(0) > K(x) \geq 0$ for all $x \neq 0$, $K'(0) = 0$, and $K''(0) < 0$. In the case when $p = 1$, their jump estimator is defined by the maximizer of $|D(x, h_n)|$ over an interval slightly smaller than the design interval $[0, 1]$, after the bandwidth h_n is properly selected. Obviously, $D(x, h_n)$ is the first-order derivative of the kernel estimator $\frac{1}{nh_n} \sum_{i=1}^{n} Y_i K(\frac{x_i - x}{h_n})$, so it can be regarded as a continuous version of $M_{DKE}(x)$ as defined in expression (3.2).

Gijbels *et al.* also provided some nice results for this criterion. They proved that their jump estimator is consistent in probability and that the convergence rate is $O(n^{-1})$. Based on this jump detection procedure and the bootstrap resampling techniques, Gijbels *et al.* (2004) suggested an algorithm for constructing confidence intervals of the true jump positions and confidence band of the true regression function f.

Bandwidth selection may depend on the purpose of the estimation. The cross-validation and other parameter selection procedures discussed in Subsections 2.4.4 and 2.5.2 can be used for selecting the bandwidth h_n used in the DKE procedures, if our major goal is to find an estimator $\widehat{f}$ of the regression function f such that their distance, which can be measured by a distance measure such as the mean squared error $\frac{1}{n} \sum_{i=1}^{n} [\widehat{f}(x_i) - f(x_i)]^2$, is as small as possible. Toward this end, Wu and Chu (1993b) provided some discussion about bandwidth selection by the cross-validation procedure. However, in some applications, our major goal is jump detection instead of curve estimation. In such cases, we are mainly concerned about the accuracy of the detected jumps. Our numerical studies show that the bandwidth should be chosen relatively small for this purpose, compared with the bandwidth selected for curve estimation. A bootstrap bandwidth selection procedure based on the accuracy of detected jumps was recently proposed by Gijbels and Goderniaux (2004).

3.2.2 Jump detection based on local linear kernel smoothing

As discussed in Section 2.4, the local linear kernel estimator has several good properties compared to the local constant kernel estimator. The jump detection criterion

M_{DKE} in expression (3.2) is defined by a difference of two one-sided local constant kernel estimators of the regression function f; thus, it is natural to define the following jump detection criterion based on a difference of two one-sided local linear kernel estimators (DLK) of f:

$$M_{DLK}(x) = \widehat{f}_+(x) - \widehat{f}_-(x).$$

In this definition, $x \in [h_n, 1-h_n]$, $\widehat{f}_+(x)$ and $\widehat{f}_-(x)$ are the right- and left-sided local linear kernel estimators of $f(x)$ defined by

$$\widehat{f}_+(x) = \sum_{i=1}^{n} \frac{w_2^{(1)} - w_1^{(1)}(x_i - x)}{w_0^{(1)}w_2^{(1)} - \left(w_1^{(1)}\right)^2} Y_i K_1\left(\frac{x_i - x}{h_n}\right)$$

and

$$\widehat{f}_-(x) = \sum_{i=1}^{n} \frac{w_2^{(2)} - w_1^{(2)}(x_i - x)}{w_0^{(2)}w_2^{(2)} - \left(w_1^{(2)}\right)^2} Y_i K_2\left(\frac{x_i - x}{h_n}\right),$$

$w_j^{(r)} = \sum_{i=1}^n (x_i - x)^j K_r(\frac{x_i - x}{h_n})$ for $r = 1, 2$ and $j = 0, 1,$ and 2, and K_1 and K_2 are two one-sided kernel functions as defined in the previous subsection (cf. equation (2.10) for the formula of a conventional local linear kernel estimator). Suppose we also assume the following: the two kernel functions are chosen such that $K_2(x) = K_1(-x)$ for all $x \in [-1, 0)$; the design points are equally spaced in the design interval $[0, 1]$; and x is a design point in $[h_n, 1-h_n]$. Then $M_{DLK}(x)$ can be written as

$$M_{DLK}(x) = \frac{1}{w_0^{(1)}w_2^{(1)} - \left(w_1^{(1)}\right)^2} \sum_{i=1}^{n} \left[\left(w_2^{(1)} - w_1^{(1)}(x_i - x)\right) K_1\left(\frac{x_i - x}{h_n}\right) - \left(w_2^{(1)} + w_1^{(1)}(x_i - x)\right) K_2\left(\frac{x_i - x}{h_n}\right)\right] Y_i. \tag{3.8}$$

The jump position s_1 is again estimated by the maximizer of $|M_{DLK}(x)|$, and the jump magnitude d_1 is estimated by $M_{DLK}(\widehat{s}_1)$, in the case when $p = 1$. Similar to Theorem 3.1, we have the following results about M_{DLK} (Qiu 1999).

Theorem 3.2 *Under the conditions stated in Theorem 3.1,*

(i) if $x \in (0, 1)$ is the true jump position, then

$$E(M_{DLK}(x)) = [f_+(x) - f_-(x)] + [f_+''(x) - f_-''(x)] \frac{\left(K_{21}^{(1)}\right)^2 - K_{11}^{(1)}K_{31}^{(1)}}{2\left(K_{01}^{(1)}K_{21}^{(1)} - \left(K_{11}^{(1)}\right)^2\right)} h_n^2 + o(h_n^2); \tag{3.9}$$

(ii) if $x \in (0, 1)$ is not the true jump position, then

$$E(M_{DLK}(x)) = o(h_n^2); \tag{3.10}$$

(iii) in both cases,

$$Var(M_{DLK}(x)) = \frac{2\sigma^2\left[\left(K_{21}^{(1)}\right)^2 K_{02}^{(1)} - 2K_{11}^{(1)}K_{21}^{(1)}K_{12}^{(1)} + \left(K_{11}^{(1)}\right)^2 K_{22}^{(1)}\right]}{\left[K_{01}^{(1)}K_{21}^{(1)} - \left(K_{11}^{(1)}\right)^2\right] nh_n} \tag{3.11}$$

where $K_{j_1j_2}^{(1)} = \int_0^1 u^{j_1} K_1^{j_2}(u)\,du$, *for* $j_1 = 0, 1, 2, 3$ *and* $j_2 = 1, 2$.

By comparing expressions (3.3) and (3.9), it can be seen that the criterion M_{DKE} is affected by a possible jump in the first-order derivative of f at the jump position s_1, but the criterion M_{DLK} is not affected by such a jump in large-sample cases. In finite sample situations, Example 3.1 has shown that the first-order derivative of f has some effect on $M_{DKE}(x)$ even when x is a continuity point of f, although this effect is invisible in expression (3.4), since (3.4) gives a large-sample property of M_{DKE}. In fact, the next example shows that, when the sample size is finite and the jump size is relatively small, the effect of the first-order derivative of f on M_{DKE} could be a major obstacle for jump detection, whereas the effect of the derivative on M_{DLK} is negligible.

Example 3.2 Suppose that the true regression function is $f(x) = 7x - I(x \geq 0.5)$ for $x \in [0, 1]$ and that the observations have no random noise involved as in Example 3.1. Then f has one jump at $x = 0.5$ with size -1. Compared to the regression function in Example 3.1, this function has larger slope across the design interval. If $n = 100$ and h_n, K_1, and K_2 are chosen to be the same as those in Example 3.1, then the two jump detection criteria M_{DKE} and M_{DLK} are as shown in Figures 3.2(a) and 3.2(b), respectively. It can be seen from the plots that M_{DKE} is dominated by the slope effect of f and consequently loses its capability of jump detection (e.g., the value of $|M_{DKE}(0.2)|$ is larger than the value of $|M_{DKE}(0.5)|$). As a comparison, the slope effect on M_{DLK} is minimal.

However, if we compare variances of these two criteria, $M_{DKE}(x)$ has better properties than $M_{DLK}(x)$ if they use the same bandwidth. For example, if we choose $K_1(x) = 1.5(1 - x^2)I(x \in [0, 1])$ and $K_2(x) = K_1(-x)$ in both expressions (3.5) and (3.11), then it can be checked that

$$\begin{aligned} \text{Var}(M_{DKE}(x)) &= \frac{2.4\sigma^2}{nh_n}, \\ \text{Var}(M_{DLK}(x)) &= \frac{4.49\sigma^2}{nh_n}, \text{ for any } x \in (0, 1). \end{aligned}$$

So the jump detection criterion M_{DLK} is noisier than the criterion M_{DKE} if they use the same bandwidth.

In (3.2), the two one-sided kernel estimators used for defining M_{DKE} should behave similarly to the conventional local constant kernel estimators at the two endpoints of the design interval $[0, 1]$, since the conventional local constant kernel estimators at the two endpoints of $[0, 1]$ are constructed from some observations located on a single side of the endpoints. As discussed in Section 2.3, the bias of a conventional local

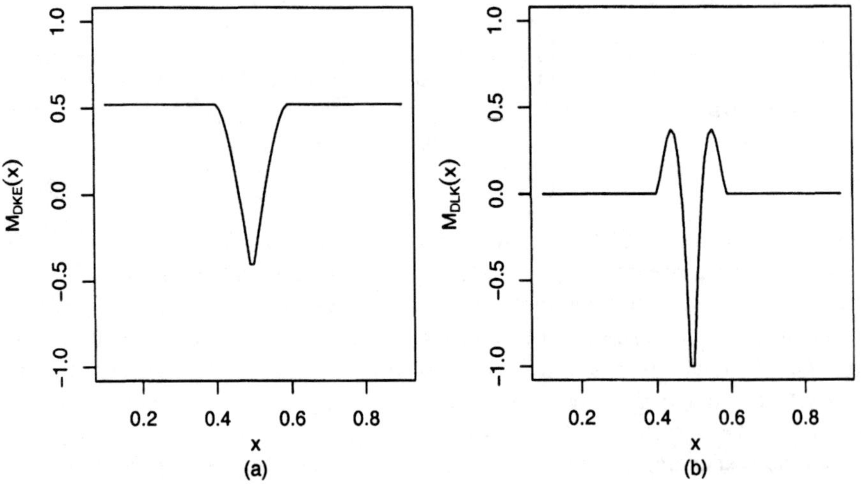

Fig. 3.2 (a): Jump detection criterion M_{DKE}. (b): Jump detection criterion M_{DLK}. The true regression function is $f(x) = 7x - I(x \geq 0.5)$ for $x \in [0, 1]$ and the observations have no random noise involved.

constant kernel estimator is much larger in the border regions of the design interval than in the interior region, and its bias in the border regions is mainly contributed by the slope of f in these regions. This explains why M_{DKE} is also much affected by the slope of f. As a comparison, the conventional local linear kernel estimator does not suffer the "boundary problem", as explained in Subsection 2.4.2, making M_{DLK} immune to the slope effect. It can be checked that M_{DLK} inherits some other good properties, including the design adaptation property, from the conventional local linear kernel estimator. Loader (1996) provided more discussion about this method, including the asymptotic distributions of the estimators of the jump position and magnitude when it is assumed that there is only one jump in f.

When the regression function f is assumed to have p jumps in the design interval and p is known, estimators of the p jump positions and the corresponding jump magnitudes by M_{DLK} can be defined similarly to those in expression (3.7). Their statistical properties, such as the strong consistency and asymptotic normality, can also be derived similarly to those of the estimators by M_{DKE}.

Sometimes, we are also interested in detecting jumps in derivatives of f. More specifically, suppose that f itself is kth-order smooth; however, its $(k + 1)$-th-order derivative may have jumps in the design interval, and we are interested in detecting these jumps. For simplicity, let us assume that $k = 0$; that is, we are actually interested in detecting jumps in the first-order derivative of f. Jump detection in higher order derivatives of f can be similarly discussed.

By the same idea of the DKE jump detection procedures, jumps in the first-order derivative of f can be detected by a criterion that is a difference of two estimators of f'_+ and f'_-, respectively. For example, if the one-sided kernel estimators are used for estimating the right and left first-order derivatives of f, then the jump detection criterion is defined by

$$M_{DKE}^{(1)}(x) = \frac{1}{nh_n^2}\sum_{i=1}^{n} Y_i K_1'\left(\frac{x_i - x}{h_n}\right) - \frac{1}{nh_n^2}\sum_{i=1}^{n} Y_i K_2'\left(\frac{x_i - x}{h_n}\right),$$

where K_1' and K_2' are the first-order derivatives of K_1 and K_2, respectively. Of course, the two kernel functions K_1 and K_2 need to satisfy some regularity conditions, such as those mentioned in the third last paragraph of Subsection 3.2.1 about the procedure by Gijbels *et al.* (1999).

If the local linear kernel estimators are used for estimating the right and left first-order derivatives of f, then these estimators can be obtained at the same time as when we compute the one-sided estimators of f for defining M_{DLK} (cf. expression (2.10) for formulas of the conventional local linear kernel estimators of f and f'). The resulting jump detection criterion is

$$\begin{aligned} M_{DLK}^{(1)}(x) &= \frac{1}{w_0^{(1)} w_2^{(1)} - \left(w_1^{(1)}\right)^2} \sum_{i=1}^{n} \left[\left(-w_1^{(1)} + w_0^{(1)}(x_i - x)\right) K_1\left(\frac{x_i - x}{h_n}\right) \right. \\ &\quad \left. - \left(w_1^{(1)} + w_0^{(1)}(x_i - x)\right) K_2\left(\frac{x_i - x}{h_n}\right)\right] Y_i \end{aligned}$$

in the case when $K_2(x) = K_1(-x)$.

3.2.3 Estimation of jump regression functions based on semiparametric modeling

Suppose that the regression function f in model (1.1) is continuous on $[0,1]$ and its first-order derivative has a jump at an unknown position $\tau \in (0,1)$. Then, such a function can be expressed by

$$f(x) = \beta\phi_\tau(x) + g(x), \text{ for } x \in [0,1], \tag{3.12}$$

where $\phi_\tau(x) = (x - \tau)_+$ (cf. Subsection 2.5.4 for its definition), β is an unknown coefficient representing the jump magnitude in the first-order derivative of f at τ, and g is an unknown function with continuous first-order derivative on $[0,1]$. The expression (3.12) is in a semiparametric form, in which the first term $\beta\phi_\tau(x)$ is the parametric part with parameters β and τ and the second term $g(x)$ is the nonparametric part. Eubank and Speckman (1994) considered the above model and suggested estimating both the parametric and nonparametric parts of f by using their one-dimensional semiparametric function estimation approach, originally discussed in Eubank and Speckman (1991).

Let K be a kernel function with support $[-1,1]$. It is assumed to satisfy the following conditions:

$$\int_{-1}^{1} K(u)\,du = 1; \int_{-1}^{1} uK(u)\,du = 0; \int_{-1}^{1} u^2K(u)\,du = C_0 > 0.$$

Then a simplified version of the conventional Nadaraya-Watson kernel estimator of $f(x)$ for $x \in [0,1]$ (see some explanation given in Subsection 3.2.1) is

$$\widehat{f}(x) = \frac{1}{nh_n}\sum_{i=1}^{n} Y_i K\left(\frac{x_i - x}{h_n}\right),$$

where h_n is a bandwidth. Let $\widehat{\mathbf{f}} = (\widehat{f}(x_1), \widehat{f}(x_2), \ldots, \widehat{f}(x_n))'$ and let S be the $n \times n$ matrix of kernel weights with its (i,j)-th element defined by

$$\frac{1}{nh_n} K\left(\frac{x_j - x_i}{h_n}\right),$$

for $i, j = 1, 2, \ldots, n$. Then $\widehat{\mathbf{f}}$ has the following matrix expression:

$$\widehat{\mathbf{f}} = S\mathbf{Y},$$

where $\mathbf{Y} = (Y_1, Y_2, \ldots, Y_n)'$.

This matrix form is now used to derive estimators of β and τ. By the above notation with $\phi_\tau = (\phi_\tau(x_1), \phi_\tau(x_2), \ldots, \phi_\tau(x_n))'$, $S(\mathbf{Y} - \beta\phi_\tau)$ should be a good estimator of $\mathbf{g} = (g(x_1), g(x_2), \ldots, g(x_n))'$ if both β and τ are known. If the jump position τ is known and the coefficient β is unknown, then β can be estimated efficiently by the solution to β of the following minimization problem:

$$\min_{\beta \in R} \|(I - S)(\mathbf{Y} - \beta\phi_\tau)\|^2,$$

where $\|\cdot\|$ is the Euclidean norm; the objective function is actually a residual sum of squares of the estimator of $\mathbf{g}$. The resulting estimator of β has the expression

$$\widehat{\beta}(\tau) = \frac{\phi_\tau{}'(I - S)^2\mathbf{Y}}{\phi_\tau{}'(I - S)^2\phi_\tau},$$

which depends on τ. After β is estimated by $\widehat{\beta}(\tau)$, $\mathbf{g}$ can be estimated by $S(\mathbf{Y} - \widehat{\beta}(\tau)\phi_\tau)$, and an estimator of $\mathbf{f} = (f(x_1), f(x_2), \ldots, f(x_n))'$ can be defined by

$$\begin{aligned}\widehat{\mathbf{f}}_\tau &= S\left(\mathbf{Y} - \widehat{\beta}(\tau)\phi_\tau\right) + \widehat{\beta}(\tau)\phi_\tau \\ &= S\mathbf{Y} + P_\tau(I - S)\mathbf{Y},\end{aligned}$$

where

$$P_\tau = \frac{(I - S)\phi_\tau\phi_\tau'(I - S)}{\phi_\tau'(I - S)^2\phi_\tau}.$$

If the jump position τ is unknown, then it is natural to estimate it by the minimizer of the following residual sum of squares with respect to τ:

$$\begin{aligned}\|\mathbf{Y} - \widehat{\mathbf{f}}_\tau\|^2 &= \|(I - P_\tau)(I - S)\mathbf{Y}\|^2 \\ &= \|(I - S)\mathbf{Y}\|^2 - \|P_\tau(I - S)\mathbf{Y}\|^2.\end{aligned}$$

The resulting estimator is denoted by $\widehat{\tau}$. Because the term $\|(I-S)\mathbf{Y}\|^2$ in the above expression does not depend on τ, the estimator $\widehat{\tau}$ is also the maximizer of

$$\|P_\tau(I-S)\mathbf{Y}\|^2 = \widehat{\beta}^2(\tau)\phi_\tau'(I-S)^2\phi_\tau$$

with respect to τ.

Eubank and Speckman (1994) showed that the term $\phi_\tau'(I-S)^2\phi_\tau$ is essentially independent of τ, and thus $\widehat{\tau}$ is asymptotically equivalent to the maximizer of $\widehat{\beta}^2(\tau)$ with respect to τ. They also derived the asymptotic distributions of the estimators $\widehat{\tau}, \widehat{\beta}(\tau)$ and $\widehat{\beta}(\widehat{\tau})$, by which confidence intervals for τ and β can be constructed and by which hypothesis tests about these parameters can be performed.

3.2.4 Estimation of jump regression functions by spline smoothing

There are some existing procedures for estimating the jump regression function f by spline smoothing. In this part, we introduce two such procedures: the partial spline procedure by Shiau (1985, 1987) and the regression spline procedure by Koo (1997).

Shiau (1985, 1987) considered the case when jumps exist in both f and its derivatives. She first assumed that the d_k-th-order derivative of f has a jump at the location α_k for $k = 1, 2, \ldots, K$ and that both the jump locations $\{\alpha_k, k = 1, 2, \ldots, K\}$ and the associated orders of the derivatives $\{d_k, k = 1, 2, \ldots, K\}$ are known. She then suggested estimating the regression function f in the framework of smoothing spline estimation (cf. Section 2.5) as follows. Suppose that

$$0 \leq \max_{1 \leq k \leq K} d_k \leq m - 1,$$

where m is a known integer. Then a function space $\mathbf{H}$ is defined by

$$\mathbf{H} = \mathbf{H}_0 \oplus \mathbf{H}_1,$$

where

$$\begin{aligned} \mathbf{H}_1 \;\; = \;\; & \Big\{ g : g^{(\nu)}(0) = 0, \; g^{(\nu)} \text{ is absolutely continuous on } [0,1], \\ & \text{for } \nu = 0, 1, \ldots, m-1; \int_0^1 [g^{(m)}(u)]^2 du < \infty \Big\}. \end{aligned}$$

$\mathbf{H}_0$ is a function space defined below, and $\mathbf{H}_0 \oplus \mathbf{H}_1$ is a summation of $\mathbf{H}_0$ and $\mathbf{H}_1$, meaning that any function in $\mathbf{H}$ is a summation of a function in $\mathbf{H}_0$ and another function in $\mathbf{H}_1$. The function space $\mathbf{H}_0$ is defined by

$$\mathbf{H}_0 = \mathbf{H}_p \oplus \mathbf{H}_d,$$

where $\mathbf{H}_p$ denotes the set of all polynomials of degree less than or equal to $m-1$ and $\mathbf{H}_d$ is a function space spanned by the following truncated power basis functions (cf. Subsection 2.5.4 for their definitions):

$$\gamma_k(x) = \frac{(x-\alpha_k)_+^{d_k}}{d_k!}, \text{ for } k = 1, 2, \ldots, K.$$

Clearly, the function space $\mathbf{H}_1$ consists of functions g in the Sobolev space $W_2^m([0,1])$ (cf. Subsection 2.5.1 for its definition) satisfying the boundary conditions $g^{(\nu)}(0) = 0$ for $\nu = 0, 1, \ldots, m-1$. It can be checked that $W_2^m([0,1]) = \mathbf{H}_p \oplus \mathbf{H}_1$. Therefore,

$$\mathbf{H} = W_2^m([0,1]) \oplus \mathbf{H}_d,$$

which implies that any function f in $\mathbf{H}$ is a summation of a function in $W_2^m([0,1])$, which is mainly for estimating the continuity part of f, and another function in $\mathbf{H}_d$, which is mainly for estimating the jump part of f.

In this context, the jump regression function f can be estimated by the solution to g of the following minimization problem:

$$\min_{g=g_1+g_2;g_1\in W_2^m([0,1]);g_2\in\mathbf{H}_d} \frac{1}{n}\sum_{i=1}^{n}[Y_i - g(x_i)]^2 + \lambda\int_0^1 [g_1^{(m)}(u)]^2\, du, \quad (3.13)$$

where $\lambda > 0$ is a smoothing parameter, which can be chosen by the CV or GCV procedure introduced in Subsection 2.5.2 (cf. equations (2.19) and (2.20)). The resulting estimator is called a *partial spline estimator* by Shiau. By comparing expressions (3.13) with (2.22) in Subsection 2.5.3, it can be seen that the minimization problem (3.13) is a special case of the minimization problem (2.22). So the partial spline estimator defined here is also a special case of the partial spline estimator discussed in Subsection 2.5.3. Shiau showed that the solution of (3.13) exists and is unique under some regularity conditions. She also derived an expression for this solution.

Shiau next considered the case when there is only one jump in f but the jump position $\alpha \in (0,1)$ is unknown. She suggested five approaches for estimating α in such a case. The first approach works in several steps, as follows. For fixed λ and α, suppose that the solution of (3.13) is $\widehat{f}_{\lambda,\alpha}$. Then, for a given λ, the value of α can be estimated by the minimizer with respect to α of the objective function in expression (3.13) after g is replaced by $\widehat{f}_{\lambda,\alpha}$. The corresponding estimators of α and f after this step are denoted by $\widehat{\alpha}(\lambda)$ and $\widehat{f}_{\lambda,\widehat{\alpha}(\lambda)}$, respectively. Then, the optimal value of λ is determined by GCV to be $\widehat{\lambda}$. Finally, the jump position α and the regression function f are estimated by $\widehat{\alpha}(\widehat{\lambda})$ and $\widehat{f}_{\widehat{\lambda},\widehat{\alpha}(\widehat{\lambda})}$.

In the case when the regression function f has $p \geq 1$ jumps, with p known, in the design interval, the jump positions $\{s_j, j = 1, 2, \ldots, p\}$ are known, and the jump magnitudes $\{d_j, j = 1, 2, \ldots, p\}$ are unknown, Qiu (1991) suggested a *trimmed spline estimator* of f. The trimmed spline estimation procedure has three steps. First, obtain L_2 consistent estimators of the jump magnitudes; this is not difficult because the jump positions are assumed to be known, and the values of M_{DKE} (cf. equation (3.2)) at the jump positions, for example, can be used for estimating the jump magnitudes. In the second step, the smoothing spline estimator of f is obtained by expression (2.18) from the new data $\{Y_i - \sum_{j=1}^{p} \widehat{d}_j I(x_i > s_j)\}$, where $\widehat{d}_j$ denotes the L_2 consistent estimator of d_j, for $j = 1, 2, \ldots, p$. This spline estimator can be expressed as a linear combination of the basis functions $\{\exp(2\pi i\nu(x_j - x)), j = 1, 2, \ldots, n; \nu = \pm 1, \pm 2, \ldots\}$. Then, in the third step, a certain amount of terms in both tails (with respect to ν) of the linear combination is dropped, resulting in the trimmed spline estimator of f. Qiu proved that this estimator is uniformly L_2 consistent under some regularity conditions and that this property still holds when the method is used for handling the case when the number of jumps is known but the jump positions are not completely known.

Koo (1997) considered the case that the regression function f itself has jumps at some unknown locations. He suggested a multiple ordered regression spline (MORS) procedure for estimating f in such a case. For a given set of knots $\{t_1, t_2, \ldots, t_K\}$, the following linear spline is used for approximating f:

$$s(x) = \beta_0 + \beta_{-1}x + \sum_{j=1}^{K} \sum_{\nu_j=1}^{2} \beta_{j,\nu_j}(x - t_j)_+^{\nu_j - 1},$$

where $\{(x - t_j)_+^{\nu_j - 1}, \nu_j = 1, 2, j = 1, 2, \ldots, K\}$ are the truncated power functions defined in Subsection 2.5.4 and $\{\beta_0, \beta_{-1}, \beta_{j,\nu_j}, \nu_j = 1, 2, j = 1, 2, \ldots, K\}$ are the unknown coefficients. For a given j, if $\nu_j = 1$, then $(x - t_j)_+^{\nu_j - 1}$ is a step function with a jump of size 1 at t_j; so the linear spline function $s(x)$ may have jumps at its knots. Once the knots are given, the unknown coefficients in $s(x)$ can be estimated by the least squares estimation procedure, as discussed in Subsection 2.5.4.

Koo also suggested a knot selection procedure that is a hybrid of the stepwise addition and stepwise deletion procedures in multiple regression, along with knot-merging. This procedure has several steps, the first of which involves forward knot addition. Initially, a linear regression is fitted in the entire design interval. Then, knots are added one by one from the design points whose indices are multiples of M, where M is an integer closest to the number $-\log_2[-(1/n)\log(1-\alpha)]/5$ with $0.05 \leq \alpha \leq 0.1$ a given constant. Each time, a candidate knot is selected and added to achieve the maximum reduction in the residual sum of squares of the regression function estimator. This knot addition process ends when one of the following three conditions is satisfied.

(1) The number of basis functions reaches the specified upper-bound $p_{max} = \min\{3n^{1/3}, n/4, 50\}$.

(2) The AIC criterion

$$AIC = n\log(RSS) + 2q$$

begins to increase, where RSS is the residual sum of squares of the current model, and q is the number of basis functions in the current model.

(3) No potential new knots can be found.

The second step involves knot deletion for minimizing the Bayes information criterion (BIC):

$$BIC = n\log(RSS) + \log(n)q.$$

Each time, a knot achieving the minimum increase in RSS is deleted. This process ends when further deletion would increase BIC. The third step involves knot-merging. Two adjacent knots t_j and t_{j+1} are considered for merging if both basis functions

$(x - t_j)_+$ and $(x - t_{j+1})_+$ are in the current model. They are replaced by a design point between them, which is chosen to attain the smallest RSS of the model after merging. If more than one such adjacent pairs of knots exist, then the pair resulting in the smallest RSS of the model after merging is selected and merged. This process stops when either $-n \log(RSS^{**}/RSS^*) < \tau(q)$ or no such adjacent pair of knots can be found, where RSS^{**} is the RSS value of the current model, RSS^* is the RSS value of the model obtained immediately before the merging process, $\tau(q)$ is the qth quantile of the distribution of the random variable $\chi_4^2 - \chi_2^2$, and χ_4^2 and χ_2^2 are two independent chi-squared random variables with four and two degrees of freedom, respectively. Finally, at the fourth step, the truncated power basis functions are successively deleted until further deletion would increase the BIC of the current model.

3.2.5 Jump and cusp detection by wavelet transformations

In this part, we introduce the wavelet transformation procedure suggested by Wang (1995) for detecting jumps and sharp cusps in a regression function. More recent developments in this direction can be found in Antoniadis and Gijbels (2002) and the references cited there.

A function g is said to have an α-cusp at a point x if, as h tends to zero,

$$|g(x+h) - g(x)| \geq c|h|^\alpha,$$

where $0 \leq \alpha < 1$ and $c > 0$ are two constants. Obviously, if the function g has an α-cusp at x and $\alpha = 0$, then g actually has a jump at x.

Wang (1995) considered the following white noise model:

$$Y(dx) = g(x)dx + \tau W(dx), \; x \in [0,1], \tag{3.14}$$

where W is a standard Wiener process, τ is a noise level parameter, and g is a function defined on $[0, 1]$ which may have jumps or sharp cusps. For readers who are not familiar with differential notation for stochastic processes, model (3.14) has the following connection with the regression model (1.1) when the error terms in model (1.1) are i.i.d. and normally distributed. Suppose that $\{(x_i, Y_i), i = 1, 2, \ldots, n\}$ are n observations obtained from model (1.1). Then a regression process $\{Y_n(x) : x \in [0, 1]\}$ can be defined by: $Y_n(0) = 0$ and $Y_n(x_i) = \sum_{j=1}^{i} Y_j$, for $i = 1, 2, \ldots, n$, with interpolation by a Wiener process W for $x_i \leq x < x_{i+1}$, where $i = 0, 1, \ldots, n$, $x_0 = 0$, and $x_{n+1} = 1$. Then this regression process is a white noise process defined by model (3.14), with $g(x) = f(x_i)$ when $x_i \leq x < x_{i+1}$ for $i = 0, 1, \ldots, n$; $\tau = \sigma n^{-1/2}$, where σ is the standard deviation of the error terms in model (1.1).

Suppose that ψ is a Daubechies mother wavelet (cf. the last paragraph in Subsection 2.6.2) and that $\psi_s(x) = s^{-1/2}\psi(x/s)$, where s is a scale parameter. Then the wavelet transformation of g is defined by

$$Tg(s, x) = \int \psi_s(x - u) g(u)\, du.$$

The wavelet transformation $Tg(s,x)$ is a continuous version of the discrete wavelet transformations discussed in Subsection 2.6.2. As a matter of fact, the discrete wavelet transformations defined in expression (2.31) can be obtained by setting $s = 2^{-j}$ and $x = k/2^j$ in the above expression. By the theoretical properties of the wavelet transformation, Wang (1995) pointed out that (1) $Tg(s,x)$ is of order $s^{3/2}$ as s tends to zero, if g is differentiable at x, and (2) the maximum of $|Tg(s,x)|$ over a neighborhood of x of size proportional to s converges to zero, with a rate no faster than $s^{\alpha+1/2}$ if g has an α-cusp at x.

Similarly, the wavelet transformation of Y is defined by

$$TY(s,x) = \int \psi_s(x-u)Y(du) = Tg(s,x) + \tau TW(s,x),$$

where $TW(s,x) = \int \psi_s(x-u)W(du)$ is the wavelet transformation of the white noise $W(dx)$. Clearly, $TY(s,x)$ is also a continuous version of the estimated wavelet coefficients $\widehat{w}_{j,k}$ defined in expression (2.33) if Y equals the regression process Y_n mentioned above. In fact, it can be checked that $n^{1/2}\widehat{w}_{j,k}$ approximately equals $TY_n(2^{-j}, k2^{-j})$.

Wang's major strategy for detecting sharp cusps is to select scales s_τ of order $(\tau^2|\log\tau|^\eta)^{1/(2\alpha+1)}$ for some constant $\eta > 1$ (as $\tau \longrightarrow 0$), such that $Tg(s_\tau, x)$ and $\tau TW(s_\tau, x)$ are balanced at places where g is differentiable. With such scales, if g has a sharp cusp around a point x, then $TY(s_\tau, x)$ will be dominated by $Tg(s_\tau, x)$, and consequently its value at such a place would be significantly larger than its values at places where g is differentiable. So, if it is assumed that g has only one α-cusp at θ and is differentiable elsewhere, then the estimator of θ can be defined by

$$\widehat{\theta} = \arg\max_{0\le x\le 1}\{|TY(s_\tau, x)|\}.$$

For this estimator, Wang proved that

$$\lim_{n\to\infty} P\left(s_\tau^{-1}(\widehat{\theta}-\theta) \in supp(\psi)\right) = 1,$$

where $supp(\psi)$ denotes the support of the mother wavelet ψ. Since ψ has a compact support, the above expression implies the strong consistency of $\widehat{\theta}$ with convergence rate s_τ. By the connection between model (3.14) and regression model (1.1) discussed above, if the regression function f in model (1.1) is assumed to have a single α-cusp at θ and be differentiable elsewhere, then $\widehat{\theta}$ can also be used for estimating the position of the α-cusp of f. Wang pointed out that the convergence rate of $\widehat{\theta}$ in the regression setup could reach $n^{-1/3}(\log n)^{\eta/3}$.

With the assumption that g is differentiable in the entire design interval, Wang proved that

$$\lim_{\tau\to 0} P\left(\max_{0\le x\le 1}|TY(s_\tau, x)| \ge C_{\tau,\gamma}\right) = \gamma,$$

where γ is a given significance level and

$$C_{\tau,\gamma} = \tau(2|\log s_\tau|)^{-1/2}\left\{2|\log s_\tau| + \log\left(\left(\int [\psi'(u)]^2 du\right)^{1/2} \Big/ (2\pi)\right) - \log\left(-\frac{\log(1-\gamma)}{2}\right)\right\}.$$

This result can be used for testing whether g has any sharp cusps in the design interval.

If it is assumed that (1) g has p cusps in the design interval with an α_j-cusp at θ_j for $j = 1, 2, \ldots, p$; (2) $\alpha_1 \leq \alpha_2 \leq \ldots \leq \alpha_p$; and (3) g is differentiable elsewhere, then Wang suggested estimating p and $\{\theta_j, j = 1, 2, \ldots, p\}$ by the following algorithm.

- For a given significance level $\gamma > 0$, we first define $\widehat{\theta}_1$ to be the maximizer of $|TY(s_\tau, x)|$ over $[0, 1]$ if

$$\max_{0 \leq x \leq 1} |TY(s_\tau, x)| \geq C_{\tau,\gamma}.$$

- For $j > 1$, define $\widehat{\theta}_j$ to be the maximizer of $|TY(s_\tau, x)|$ over

$$[0, 1] \Big\backslash \bigcup_{\ell=1}^{j-1} \left\{\widehat{\theta}_\ell + s_\tau x : x \in supp(\psi)\right\}$$

  if the corresponding maximum value is larger than or equal to $C_{\tau,\gamma}$. This process ends if the maximum value is smaller than $C_{\tau,\gamma}$.

Suppose that a total of $\widehat{p}$ maximums are found in the above process and the corresponding locations are $\widehat{\theta}_1, \widehat{\theta}_2, \ldots, \widehat{\theta}_{\widehat{p}}$. Then $\widehat{p}$ and $\{\widehat{\theta}_1, \widehat{\theta}_2, \ldots, \widehat{\theta}_{\widehat{p}}\}$ are defined as estimators of p and $\{\theta_1, \theta_2, \ldots, \theta_p\}$, respectively. For these estimators, Wang proved that

$$\lim_{\tau \to 0; \gamma \to 0} P(\widehat{p} = p) = 1$$

and that there exists a constant $A > 0$ such that

$$\lim_{\tau \to 0; \gamma \to 0} P\left[\frac{1}{s_\tau^2}\sum_{j=1}^{\widehat{p}} (\widehat{\theta}_j - \theta_j)^2 \leq A\right] = 1.$$

3.3 JUMP ESTIMATION WHEN THE NUMBER OF JUMPS IS UNKNOWN

Procedures introduced in Section 3.2 provide estimators of the jump position if the regression function f is assumed to have a single jump in the design interval. Some of them, such as Wang's (1995) procedure, also provide estimators of the number of jumps if this number is unknown. In the literature, there are several other jump detection procedures that are appropriate to use for jump detection when the number of jumps p is unknown. In this section, we introduce four such procedures.

3.3.1 Jump detection by comparing three local estimators

The first procedure that we introduce is one suggested by Hall and Titterington (1992). For simplicity of presentation, we introduce this procedure only in the case when the design points are equally spaced on the design interval $[0, 1]$. Namely, the design points are assumed to be $x_i = i/n$ for $i = 1, 2, \ldots, n$. In their paper, Hall and Titterington also discussed jump detection in the case when the design points were unequally spaced on the design interval, although we will not discuss that here.

At a given design point i/n for $2m + 1 \leq i \leq n - 2m$, Hall and Titterington considered three local linear estimators of f:

$$\widehat{f}_c(i/n) = \sum_{j=-m}^{m} c_j Y_{i+j};$$

$$\widehat{f}_r(i/n) = \sum_{j=0}^{2m} r_j Y_{i+j};$$

$$\widehat{f}_l(i/n) = \sum_{j=-2m}^{0} l_j Y_{i+j},$$

in three neighborhoods of i/n, with $2m + 1$ design points in each neighborhood. When the design points are equally spaced in the design interval as they are here, we sometimes call the number of design points in the neighborhood the bandwidth or *window size*. So $\widehat{f}_c(i/n)$ is a local linear estimator of f constructed from the observations in a *central* neighborhood of size $2m + 1$, $\widehat{f}_r(i/n)$ is a local linear estimator of f constructed from the observations in a *right* neighborhood of the same size, and $\widehat{f}_l(i/n)$ is a local linear estimator constructed from the observations in a *left* neighborhood of the same size.

It can be easily checked that the conventional local smoothing estimators of f discussed in Chapter 2 are all local linear estimators, so any of them can be used for defining the three local linear estimators mentioned above. For example, if the Nadaraya-Watson kernel estimator (cf. equation (2.3) for its definition) is used for defining $\widehat{f}_c(i/n)$, then

$$c_j = \frac{K(j/(2m))}{\sum_{\ell=-m}^{m} K(\ell/(2m))}, \quad \text{for } j = 0, \pm 1, \ldots, \pm m. \tag{3.15}$$

Since the design points are assumed to be equally spaced, it is natural to take $l_j = r_{-j}$ for all $j = -2m, \ldots, 0$. Hall and Titterington (1992) suggested choosing the coefficients $\{c_j, j = 0, \pm 1, \ldots, \pm m\}$ and $\{r_j, j = 0, 1, \ldots, 2m\}$ such that $E[\widehat{f}_c(i/n)]$ and $E[\widehat{f}_r(i/n)]$ have the same q leading terms in their Taylor expansions, where $q \geq 1$ is a given integer number. Hall and Titterington suggested choosing q small; in their numerical examples, q was chosen to be 1. To meet the above requirement, they suggested determining the coefficients $\{r_j, j = 0, 1, \ldots, 2m\}$ in the following way once the coefficients $\{c_j, j = 0, \pm1, \ldots, \pm m\}$ are given (e.g., by equation (3.15)). Let

$$s_k = \sum_{j=0}^{2m} j^k, \; k = 0, 1, \ldots, 2q,$$

$$S = \begin{bmatrix} s_0 & s_1 & \cdots & s_q \\ s_1 & s_2 & \cdots & s_{q+1} \\ \vdots & \vdots & \ddots & \vdots \\ s_q & s_{q+1} & \cdots & s_{2q} \end{bmatrix},$$

$$v_k = \sum_{j=-m}^{m} j^k c_j, \; k = 0, 1, \ldots, q,$$

and

$$\mathbf{u} = S^{-1}\mathbf{v},$$

where $\mathbf{u} = (u_0, u_1, \ldots, u_q)'$ and $\mathbf{v} = (v_0, v_1, \ldots, v_q)'$. Then it can be checked that $\mathrm{E}[\widehat{f}_c(i/n)]$ and $\mathrm{E}[\widehat{f}_r(i/n)]$ have the same q leading terms in their Taylor expansions if $\{r_j, j = 0, 1, \ldots, 2m\}$ are defined by

$$r_j = \sum_{\ell=0}^{q} u_\ell j^\ell, \; j = 0, 1, \ldots, 2m.$$

These three estimators can now be compared for detecting jump points. Intuitively, if there are no jumps in the interval $[(i - 2m)/n, (i + 2m)/n]$, then the difference among the three local linear estimators $\widehat{f}_c(i/n)$, $\widehat{f}_r(i/n)$, and $\widehat{f}_l(i/n)$ should be small because all of them are good estimators of $f(i/n)$. If the design point $x_i = i/n$ is a jump point, on the other hand, then the difference among the three estimators should be relatively large. For example, when the true regression function f is the one displayed by the solid line in Figure 3.3(a) and no random noise is involved in the data, then the three local linear estimators are displayed in the same plot by the dotted, short-dashed, and wide-dashed curves, respectively. The corresponding quantity of $|\widehat{f}_r - \widehat{f}_l|$ is displayed in Figure 3.3(b). From the plots, it can be seen that $\widehat{f}_c$ lies between $\widehat{f}_r$ and $\widehat{f}_l$ at the jump point i/n, and $|\widehat{f}_r - \widehat{f}_l|$ reaches the local maxima at the jump point. Between m and $2m$ units to the left of the jump point, $|\widehat{f}_l - \widehat{f}_c|$ equals 0, but $\widehat{f}_r$ is different from both $\widehat{f}_c$ and $\widehat{f}_l$. A similar result is true on the right side of the jump point.

On the basis of the relationship among $\widehat{f}_c$, $\widehat{f}_r$, and $\widehat{f}_l$ discussed above, Hall and Titterington (1992) suggested the following two-step procedure for jump detection. In the first step, all possible jump points are flagged. A design point i/n is diagnosed as a possible jump point and then flagged if it satisfies the following three conditions:

(1) $\left(\widehat{f}_r(i/n) - \widehat{f}_c(i/n)\right)\left(\widehat{f}_c(i/n) - \widehat{f}_l(i/n)\right) > 0$;

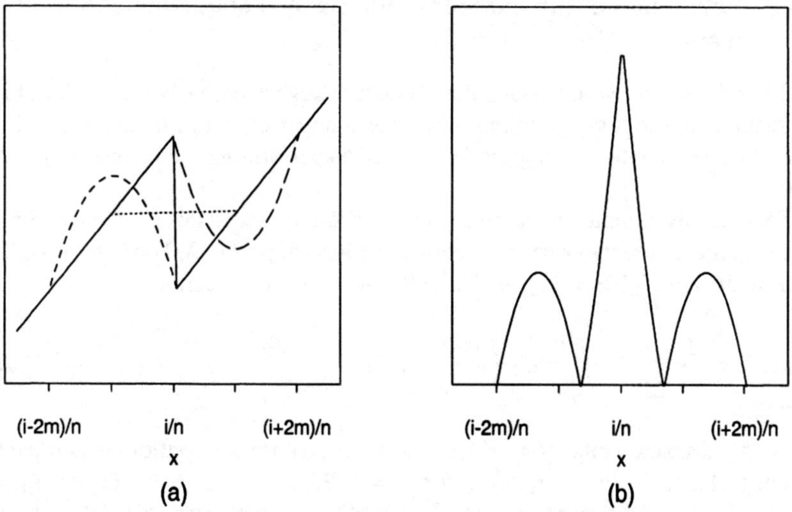

Fig. 3.3 (a): Solid line denotes the true regression function, dotted line denotes $\widehat{f}_c$, short-dashed curve denotes $\widehat{f}_r$, and wide-dashed curve denotes $\widehat{f}_l$. (b): The curve denotes $|\widehat{f}_r - \widehat{f}_l|$.

(2) $\left|\widehat{f}_r(i/n) - \widehat{f}_l(i/n)\right| > \left|\widehat{f}_r((i+j)/n) - \widehat{f}_l((i+j)/n)\right|$ for all $j = \pm m, \ldots, \pm 2m$;

(3) $U(i/n) = \left|\left(\widehat{f}_l(i/n) - \widehat{f}_r(i/n)\right)\left(\widehat{f}_c(i/n) - \widehat{f}_r(i/n)\right)\left(\widehat{f}_c(i/n) - \widehat{f}_l(i/n)\right)\right| > u$, where

$$\begin{aligned} u &= 6(\tau z)^3, \\ \tau^2 &= \sigma^2 \left\{ \sum_{j=0}^{2m} r_j^2 - \frac{2}{2m+1} \sum_{j=0}^{m} r_j + \frac{1}{2m+1} \right\}, \\ z &= \Phi^{-1}\left[0.5\left(1 + \sqrt{1-p}\right)\right], \end{aligned}$$

$0 < p < 1$ is a given significance level and Φ is the cumulative distribution function of the standard normal distribution.

Hall and Titterington (1992) noticed that if the point i/n is flagged as a possible jump point, then some of its neighboring design points might also be flagged by the above procedure. So in the second step, they suggest scanning all the flagged design points from left to right. If the point i/n is flagged, then we look ahead locally in the window of i/n and identify the position at which U is maximum. Then that position is regarded as a jump point, and any other flagged points in the window are all disregarded.

A similar diagnostic procedure is suggested by Hall and Titterington for detecting jumps in the first-order derivative of f. In that procedure, $|\widehat{f}_r - \widehat{f}_l|$ and $|\widehat{f}_r + \widehat{f}_l - 2\widehat{f}_c|$ are used as diagnostic functions.

3.3.2 Estimation of the number of jumps by a sequence of hypothesis tests

In this subsection, we introduce the procedure suggested by Wu and Chu (1993a) for estimating the jump positions when the number of jumps is unknown. In this procedure, the number of jumps is determined by performing a sequence of hypothesis tests.

We begin by defining the jump detection criterion used in this procedure. Suppose that K_1 and K_2 are two kernel functions, K_1 has support $[-\lambda, 1]$ where $\lambda \in [0, 1]$ is a constant, and $K_2(x) = K_1(-x)$ for all $x \in R$. Then we define

$$J(x) = \frac{1}{h_n}\sum_{i=1}^{n} Y_i \int_{t_{i-1}}^{t_i} K_1\left(\frac{u-x}{h_n}\right) du - \frac{1}{h_n}\sum_{i=1}^{n} Y_i \int_{t_{i-1}}^{t_i} K_2\left(\frac{u-x}{h_n}\right) du, \tag{3.16}$$

where h_n is a bandwidth, $\{(x_i, Y_i), i = 1, 2, \ldots, n\}$ are observations generated from model (1.1), $t_i = (x_i + x_{i+1})/2$ for $i = 1, 2, \ldots, n-1$, $t_0 = 0$, and $t_n = 1$. Obviously, J is a difference of two Gasser-Müller kernel estimators (cf. Subsection 2.3.1 for its definition) with kernel functions K_1 and K_2 respectively. If the number of jumps p is known, then the jump positions $\{s_j, j = 1, 2, \ldots, p\}$ can be estimated by the jump detection criterion J in the interval $[a, b]$ in a way similar to the use of M_{DKE} in equation (3.7), where $0 < a < b < 1$ are two constants and the interval $[a, b]$ is large enough to include all jumps. The estimated jump positions are denoted by $\{\widehat{s}_j, j = 1, 2, \ldots, p\}$. Please notice that, unlike the two one-sided kernel functions used in defining M_{DKE} in Subsection 3.2.1, K_1 and K_2 used in equation (3.16) are not one-sided. Wu and Chu (1993a) suggested choosing these two kernel functions by minimizing the asymptotic mean squared errors of the detected jump points. By this consideration, the selected K_1 is defined by:

$$\begin{aligned} K_1(x) &= \left(0.4857 + 3.8560x + 2.8262x^2 - 19.1631x^3 + 11.9952x^4\right) \\ &\quad I\left(x \in [-0.2012, 1]\right). \end{aligned}$$

Estimating the jump magnitudes requires a new criterion, since K_1 and K_2 are not one-sided. That is, $|J(\widehat{s}_j)|$ are generally smaller than the true jump magnitudes $|d_j|$, for $j = 1, 2, \ldots, p$, because observations on two different sides of each jump point are averaged in each of the two kernel estimators involved in J. To estimate $\{d_j, j = 1, 2, \ldots, p\}$ properly, Wu and Chu define

$$S(x) = \frac{1}{g_n}\sum_{i=1}^{n} Y_i \int_{t_{i-1}}^{t_i} K_3\left(\frac{u-x}{g_n}\right) du - \frac{1}{g_n}\sum_{i=1}^{n} Y_i \int_{t_{i-1}}^{t_i} K_4\left(\frac{u-x}{g_n}\right) du,$$

where K_3 and K_4 are two kernel functions different from K_1 and K_2 and g_n is a bandwidth chosen larger than the bandwidth h_n used in (3.16). The kernel function K_3 is assumed to have support $[-w, 1]$ with $w \in [0, 1]$, and $K_4(x) = K_3(-x)$ for all $x \in R$; Wu and Chu suggest choosing $K_3(x) = I(x \in [0, 1])$. Then the jump magnitudes d_j are estimated by $\widehat{d}_j = S(\widehat{s}_j)$ for $j = 1, 2, \ldots, p$.

The above estimation procedure is based on the assumption that the number of jumps p is known. When p is unknown, Wu and Chu suggested determining p by performing a sequence of hypothesis tests as follows. First, they prove that if the regression function f is continuous on the entire design interval and some other regularity conditions are satisfied, then

$$\lim_{n\to\infty} P\left(\widehat{d}_1 < a_n + b_n x\right) = \exp\left[-2\exp(-x)\right], \tag{3.17}$$

where

$$\begin{aligned} a_n &= \left[\frac{1}{ng_n}\sigma^2 c_w^2 \int_{-\infty}^{\infty} \left[K_3(u) - K_4(u)\right]^2 du\right]^{1/2} \left[g_{ab} + g_{ab}^{-1}\log\left(3/\sqrt{4\pi}\right)\right], \\ b_n &= \left[\frac{1}{ng_n}\sigma^2 c_w^2 \int_{-\infty}^{\infty} \left[K_3(u) - K_4(u)\right]^2 du\right]^{1/2} g_{ab}^{-1}, \\ c_w &= \left[\int_0^1 K_3(u)du - \int_0^w K_4(u)du\right]^{-1}, \end{aligned}$$

and

$$g_{ab} = \left[2\log\left(\frac{b-a}{g_n}\right)\right]^{1/2}.$$

The asymptotic distribution defined by equation (3.17) can be used for testing $H_0 : p = 0$ versus $H_1 : p > 0$. For a given positive integer j, let

$$A_j = [a, b] - \bigcup_{k=1}^{j-1} [\widehat{s}_k - 2h_n, \widehat{s}_k + 2h_n] = \bigcup_{k=1}^{j} A_{j,k},$$

where $A_{j,k}$ are disjoint subintervals of A_j for $k = 1, 2, \ldots, j$. If the true number of jumps p equals j, then Wu and Chu claim that

$$\lim_{n\to\infty} P\left(\widehat{d}_{j+1} < x\right) \approx \Pi_{k=1}^{j+1} \exp\left[-2\exp\left(-\frac{x - a_{n,j,k}}{b_{n,j,k}}\right)\right],$$

where $a_{n,j,k}$ and $b_{n,j,k}$ are defined the same as a_n and b_n after the endpoints a and b of the interval $[a, b]$ are replaced by the endpoints of the subinterval $A_{j+1,k}$ in the definitions of a_n and b_n, and "$\approx$" denotes asymptotic equality. This asymptotic result can be used for testing $H_0 : p = j$ versus $H_1 : p > j$. Therefore, for a given dataset, a sequence of hypothesis tests can be performed, and this testing process stops after the first "fail to reject H_0" is obtained. If we get the first "fail to reject H_0" at the $(\widehat{p}+1)$-th test, then the number of jumps is estimated by $\widehat{p}$. After the number of jumps is determined, the jump positions and jump magnitudes can be estimated accordingly by the estimation procedure described above, under the assumption that the number of jumps is known.

3.3.3 Jump detection by DAKE

We begin by defining a jump detection criterion M_{DAKE} in the same way as the criterion M_{DKE} in equation (3.2); only the kernel functions are defined differently. Let K_1 and K_2 be two one-sided density kernel functions with supports $[1, 2]$ and $[-2, -1]$, respectively. These two one-sided kernel functions are different from the two one-sided kernel functions used in equation (3.2) in two ways: (1) the supports of the former kernel functions are one unit away from the origin, and (2) closures of the supports of the latter kernel functions include the origin. Then, for any $x \in [2h_n, 1-2h_n]$, $M_{DAKE}(x)$ is a difference of two one-sided kernel estimators of $f(x)$ constructed from the observations in the two one-sided intervals $[x + h_n, x + 2h_n]$ and $[x - 2h_n, x - h_n]$, respectively. These two one-sided intervals are removed from the point x by a distance of h_n. Similarly to M_{DKE} discussed in Subsection 3.2.1, M_{DAKE} should be a good criterion for jump detection; the jump detection procedure based on M_{DAKE} is suggested by Qiu (1994). It is called the *difference apart kernel estimation (DAKE)* procedure and is identified by the subscript of M_{DAKE}.

A major feature of the DAKE procedure is that the two kernel functions K_1 and K_2 used in this procedure are required to vanish on $[0, 1)$ and $(-1, 0]$ respectively. Thus, if x is a jump point, then all values of $M_{DAKE}(s)$ for $s \in (x - h_n, x + h_n)$ are good estimators of the jump magnitude at x. As a comparison, the value of $M_{DKE}(x)$ should also be close to the jump magnitude at x in such a case, but the absolute value of $M_{DKE}(s)$ would decrease when s moves away from x. Therefore, the criterion M_{DAKE} is more *robust* to the jump positions than the criterion M_{DKE}.

To detect jumps by the criterion M_{DAKE}, we first define

$$a_i = \frac{i}{m}, \quad i = 1, 2, \cdots, m,$$

where

$$m = \left[\frac{1}{2h_n}\right] + 1,$$

and $[x]$ denotes the integer part of x. Then $\{a_i, i = 1, 2, \cdots, m\}$ is a sequence of equally spaced points on the design interval $[0, 1]$, and the distance between any two neighboring points in $\{a_i, i = 1, 2, \cdots, m\}$ is $1/m$, which is slightly smaller than $2h_n$. So for a given point x, there must be at least one and no more than two points in $\{a_i, i = 1, 2, \cdots, m\}$ which are located in the neighborhood $(x - h_n, x + h_n)$. The values of M_{DAKE} at such points are close to the jump magnitude of f at x if x is a jump point. Because of this property, Qiu (1994) suggested detecting jumps at the points $\{a_i, i = 1, 2, \cdots, m\}$ only, instead of at all design points as some other jump detection procedures do. By doing so, much computation is saved without missing any jumps.

Next, we need to identify which points are jump points. Let us first assume that a lower bound $B_0 > 0$ of all absolute jump magnitudes is known, that is, $\min_{1 \le j \le p} |d_j| \ge B_0$. Suppose that $a_{i_1} < a_{i_2} < \cdots < a_{i_r}$ are the points in

$\{a_i, i = 1, 2, \cdots, m\}$ satisfying

$$\left|M_{DAKE}\left(a_{i_j}\right)\right| \geq \frac{B_0}{2}, \text{ for } j = 1, 2, \cdots, r.$$

Then the points $\{a_{i_j}, j = 1, 2, \cdots, r\}$ can be flagged as jump candidates. But if a_{i_j} is judged as a jump candidate, some of its neighboring points in $\{a_i, i = 1, 2, \cdots, m\}$ would probably be judged as jump candidates as well, even if they are actually continuity points of f. Thus we need to cancel some of the jump candidates in $\{a_{i_j}, j = 1, 2, \ldots, r\}$, which can be accomplished by the following modification procedure. If there are two integers $r_1 < r_2$ such that

$$\begin{cases} a_{i_{j+1}} - a_{i_j} = \frac{1}{m}, j = r_1, r_1 + 1, \ldots, r_2 - 1 \\ a_{i_{r_1}} - a_{i_{r_1 - 1}} > \frac{1}{m} \\ a_{i_{r_2+1}} - a_{i_{r_2}} > \frac{1}{m}, \end{cases} \tag{3.18}$$

then we say that all candidates $\{a_{i_j}, j = r_1, r_1 + 1, \ldots, r_2\}$ form a *tie*. For the jump candidates in a tie, we replace all of them by a new candidate which is defined by the middle point $(a_{i_{r_2}} + a_{i_{r_1}})/2$ of the tie. After this modification, the current candidates consist of two types of points: those in $\{a_{i_j}, j = 1, 2, \cdots, r\}$ which do not belong to any ties and the middle points of all ties. The current jump candidates are denoted by $b_1 < b_2 < \cdots < b_q$. Obviously, they have the property that

$$b_{j+1} - b_j > \frac{1}{m}, \text{ for } j = 1, 2, \cdots, (q - 1).$$

Then q, $\{b_j, j = 1, 2, \cdots, q\}$, and $\{M_{DAKE}(b_j), j = 1, 2, \cdots, q\}$ are defined as the estimators of the number of jumps p, the jump positions $\{s_j, j = 1, 2, \cdots, p\}$, and the jump magnitudes $\{d_j, j = 1, 2, \cdots, p\}$, respectively. Such estimators are called the *difference apart kernel estimators* (DAKEs).

Some asymptotic properties for these estimators can be derived. Under some regularity conditions, it can be shown that

(i)

$$\lim_{n \to \infty} \|M_{DAKE}\|_{\bigcup_{i=0}^{p} [s_i + 2h_n, s_{i+1} - 2h_n]} = 0, \ a.s.$$

where $s_0 = 0$ and $s_{p+1} = 1$; and

(ii) for any $i = 1, 2, \ldots, p$,

$$\lim_{n \to \infty} \|M_{DAKE} - d_i\|_{(s_i - h_n, s_i + h_n)} = 0, \ a.s.$$

By combining these two properties, we reach the following conclusions. When n is large enough, (1) all points in $\{a_{i_j}, j = 1, 2, \cdots, r\}$ should be located in $\bigcup_{i=1}^{p}(s_i - 2h_n, s_i + 2h_n)$, (2) each interval $(s_i - h_n, s_i + h_n)$ includes at least one point in $\{a_{i_j}, j = 1, 2, \cdots, r\}$, and (3) each interval $(s_i - 2h_n, s_i + 2h_n)$ includes at most three points in $\{a_{i_j}, j = 1, 2, \cdots, r\}$. After the modification procedure, each interval

$(s_i - h_n, s_i + h_n)$ includes one and only one jump candidate. Therefore, q equals p in such cases, and consequently q is an almost surely consistent estimator of p. Additionally, for the estimators of jump positions and jump magnitudes, Qiu (1994) proved that they are all almost surely consistent under some regularity conditions.

In most applications, the lower bound B_0 of the absolute jump magnitudes is unknown. Theoretically speaking, this lower bound can be replaced by a positive constant B_n, which converges to zero in the rate of $n^{-\nu} \log n$, where $\nu > 0$ is a constant specified by Qiu. It has been shown by Qiu that the resulting estimators are still almost surely consistent under some regularity conditions. In applications, B_n can be chosen to be

$$B_n = \frac{\widehat{\sigma} Z_{1-\alpha_n/2} \sqrt{\int_{-2}^{2} (K_1(u) - K_2(u))^2 du}}{\sqrt{n h_n}},$$

where $Z_{1-\alpha_n/2}$ is the $1 - \alpha_n/2$ quantile of the standard normal distribution and $\widehat{\sigma}$ is a consistent estimator of σ. For a given point x, if it is not a jump point, then it can be shown that

$$P(|M_{DAKE}(x)| > B_n) \approx \alpha_n.$$

As long as α_n is chosen such that $Z_{1-\alpha_n/2} = O(\log n)$ and $1/Z_{1-\alpha_n/2} = o(1)$, the resulting estimators are still almost surely consistent.

3.3.4 Jump detection by local polynomial regression

Qiu and Yandell (1998) suggested a jump detection procedure based on local polynomial estimation. This procedure, which is quite convenient to implement, is introduced in some details below.

Suppose that $\{(x_i, Y_i), i = 1, 2, \ldots, n\}$ are n observations generated from the regression model (1.1) in which the design points are equally spaced on the design interval $[0, 1]$. The first $(m-1)$ derivatives of the regression function f are assumed to be continuous on the design interval, and the mth order derivative of f is assumed to have the expression

$$f^{(m)}(x) = g(x) + \sum_{j=1}^{p} d_j I(x > s_j), \text{ for } x \in [0, 1], \tag{3.19}$$

where g is a continuous function in the entire design interval, p is the number of jump points in $f^{(m)}$, $\{s_j, j = 1, 2, \cdots, p\}$ are the jump positions, and $\{d_j, j = 1, 2, \cdots, p\}$ are the jump magnitudes. We are interested in estimating p, $\{s_j, j = 1, 2, \cdots, p\}$, and $\{d_j, j = 1, 2, \cdots, p\}$.

For this purpose, at a given design point x_i, for $\ell + 1 \leq i \leq n - \ell$ where $\ell \ll n$ is a given positive integer, let us consider its neighborhood:

$$N(x_i) = \{x_{i-\ell}, x_{i-\ell+1}, \ldots, x_i, \ldots, x_{i+\ell-1}, x_{i+\ell}\}$$

with window size $k = 2\ell + 1$, an odd positive integer. In this neighborhood, a local polynomial function of order $(m+1)$ is fitted by the least squares (LS) estimation procedure. The fitted polynomial function is denoted by

$$\widehat{Y}^{(i)}(x) = \widehat{\beta}_0^{(i)} + \widehat{\beta}_1^{(i)} x + \cdots + \widehat{\beta}_{m+1}^{(i)} x^{m+1}, \text{ for } x \in N(x_i).$$

The major idea of Qiu and Yandell's (1998) procedure is that the estimated regression coefficients $\{\widehat{\beta}_{m+1}^{(i)}, i = \ell+1, \ldots, n-\ell\}$ are sensitive to jumps in $f^{(m)}$ and thus can be used for detecting jumps in $f^{(m)}$. More specifically, if $f^{(m)}$ has no jump in the neighborhood $N(x_j)$ of a given design point x_j, then $\widehat{\beta}_{m+1}^{(j)}$ should be close to $f^{(m+1)}(x_j)/(m+1)!$ for large enough n and k, when $f^{(m+1)}$ exists in $N(x_j)$. On the other hand, if x_j is a jump point of $f^{(m)}(x)$, then $\{\widehat{\beta}_{m+1}^{(i)}, i = \ell+1, \ldots, n-\ell\}$ should have an abrupt change at $\widehat{\beta}_{m+1}^{(j)}$. Therefore, the estimated regression coefficients $\{\widehat{\beta}_{m+1}^{(i)}, i = \ell+1, \ldots, n-\ell\}$ carry information about both the continuity and jump parts of $f^{(m)}$. After excluding the information about the continuity part of $f^{(m)}$ and preserving the information about its jump part, Qiu and Yandell suggest a jump detection algorithm based on $\{\widehat{\beta}_{m+1}^{(i)}, i = \ell+1, \ldots, n-\ell\}$. An important feature of this algorithm is that it is easy to implement because it is based on the estimated local LS coefficients, which can be computed easily by any existing statistical software package.

In most applications, we are interested in detecting jumps in the regression function f itself or in its first-order derivative f'. We now introduce Qiu and Yandell's procedure in these two special cases.

When we are interested in detecting jumps in f (i.e., $m = 0$ in equation (3.19)), Qiu and Yandell suggest fitting the LS line

$$\widehat{Y}^{(i)}(x) = \widehat{\beta}_0^{(i)} + \widehat{\beta}_1^{(i)} x, \text{ for } x \in N(x_i)$$

in the neighborhood $N(x_i)$ of each design point x_i, for $\ell+1 \leq i \leq n-\ell$. Like most local smoothing jump detection procedures, Qiu and Yandell's procedure requires the following assumption (A 3.1) in finite sample cases.

(A 3.1) For any given design point x_i, $i = \ell+1, \ldots, n-\ell$, if it is a jump point, then no other jumps exist in $N(x_{i-k}) \cup N(x_i) \cup N(x_{i+k})$, where $k = 2\ell+1$.

The assumption (A 3.1) implies that jump locations are not very close to each other. Alternatively, there are enough data (n large, k/n small) to distinguish nearby jumps.

Under the assumption (A 3.1) and some other regularity conditions, Qiu and Yandell show that

(i) if there is no jump in $N(x_i)$, then

$$\widehat{\beta}_1^{(i)} \sim B_1(x_i) := g'(x_i);$$

(ii) if a jump exists in $N(x_i)$ and the jump location is at $x_{i-\ell+r}, 0 \leq r \leq 2\ell$, then

$$\widehat{\beta}_1^{(i)} \sim B_1(x_i) := g'_+(x_i) + h_1(r)C_0 - \gamma(r)C_1,$$

where "$\sim$" means that quantities on its two sides are equal if noise in the data and some high order terms are ignored; ":=" means "defined by"; C_0 and C_1 are the jump magnitudes of f and f', respectively, at the jump location; $\gamma(r)$ is a positive function taking values between 0 and 1; $g'_+(x_i) = g'(x_i)$ if $r \neq \ell$; and

$$h_1(r) = \frac{6nr}{k(k+1)}\left(1 - \frac{r}{k-1}\right).$$

Example 3.3 Suppose that $f(x) = 5x^2 + I(x \in [0.5, 1])$. Then f has a jump at $x = 0.5$ with size 1. Let us consider a sample of size 100 and let k be 7. Then

$$\widehat{\beta}_1^{(i)} \sim B_1(x_i) = 10x_i + \frac{6(100)(53-i)}{7 \times 8}\left(1 - \frac{53-i}{6}\right) I(47 \leq i \leq 53),$$

for $4 \leq i \leq 97$. The quantities $\{B_1(x_i), 4 \leq i \leq 97\}$ are shown in Figure 3.4(a).

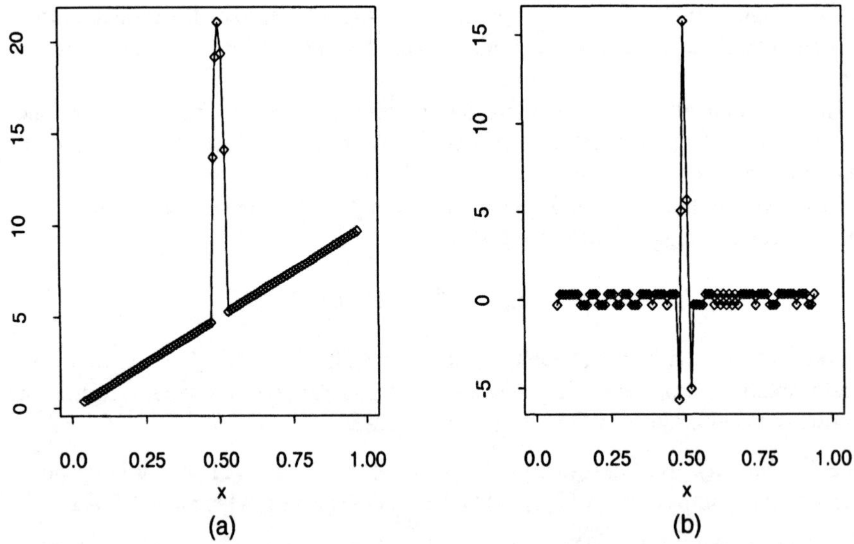

Fig. 3.4 If $f(x) = 5x^2 + I(x \in [0.5, 1])$, $n = 100$, and $k = 7$, then $\widehat{\beta}_1^{(i)} \sim B_1(x_i)$, $4 \leq i \leq 97$. The quantities $\{B_1(x_i), 4 \leq i \leq 97\}$ include information about both the continuity and the jump parts of f [represented by diamonds in plot (a)]. With the use of the difference operator defined in equation (3.20), the resulting quantities $\{J_1(x_i)\}$ include information mainly about the jump part of f [shown in plot (b)].

From Figure 3.4(a), it can be seen that $\{\widehat{\beta}_1^{(i)}, i = \ell+1, \ldots, n-\ell\}$ carry useful information about the jump in f. This information is mainly included in the "jump factor" $h_1(r)C_0$ of $B_1(x_i)$. It can be checked that $h_1(r)$ has a maximum value at $r = \ell$ of $1.5n(k-1)/[k(k+1)]$. When the window width k/n is chosen small, as is often the case, this maximum value is large.

If we have prior information about the bound of the "continuity factor" $g'(x_i)$, then points could be flagged as jump points if the LS slopes obtained in their neighborhoods are larger than the prior bound. However, that prior bound may not be available in many applications. Qiu and Yandell's (1998) strategy to overcome this difficulty is to find an operator to remove the continuity factor from $B_1(x_i)$ and to simultaneously preserve its jump factor. Since the continuity factors $g'(x_i)$ and $g'(x_j)$ are close to each other when x_i and x_j are close, a difference-type operator can remove the continuity factor. When x_i and x_j are a certain distance away from each other such that only one of $B_1(x_i)$ and $B_1(x_j)$ can have a jump factor (remember that $B_1(x_i)$ does not include a jump factor if no jump exists in $N(x_i)$), the difference between $B_1(x_i)$ and $B_1(x_j)$ includes the jump factor. Based on these considerations, Qiu and Yandell defined the following jump detection criterion:

$$\Delta_1^{(i)} = \begin{cases} \widehat{\beta}_1^{(i)} - \widehat{\beta}_1^{(i-\ell)}, & \text{if } |\widehat{\beta}_1^{(i)} - \widehat{\beta}_1^{(i-\ell)}| \leq |\widehat{\beta}_1^{(i)} - \widehat{\beta}_1^{(i+\ell)}| \\ \widehat{\beta}_1^{(i)} - \widehat{\beta}_1^{(i+\ell)}, & \text{if } |\widehat{\beta}_1^{(i)} - \widehat{\beta}_1^{(i-\ell)}| > |\widehat{\beta}_1^{(i)} - \widehat{\beta}_1^{(i+\ell)}|, \end{cases} \tag{3.20}$$

where $k \leq i \leq n-k+1$. It can be checked that, if the noise in data and some high-order terms are ignored, then

$$\Delta_1^{(i)} \sim J_1(x_i) := \begin{cases} 0, & \text{if there is no jump in } N(x_i) \\ h_2(r)C_0, & \text{if there is a jump in } N(x_i) \text{ at } x_{i-\ell+r}, \end{cases} \tag{3.21}$$

where $0 \leq r \leq k-1$, $k \leq i \leq n-k+1$, $h_2(r)$ is the one of $h_1(r) - h_1(r-\ell)$ and $h_1(r) - h_1(r+\ell)$ with smaller absolute value, and $h_1(j) = 0$ when $j < 0$ or $j > k-1$.

The quantity $h_2(r)$ has the same maximum value as $h_1(r)$. In the case of Example 3.3, $\{J_1(x_i), 7 \leq i \leq 94\}$ are shown in Figure 3.4(b). From equation (3.21) and Figure 3.4(b), we can see that $\{\Delta_1^{(i)}, k \leq i \leq n-k+1\}$ do keep the jump information and delete the continuity factors at the same time. Hence, they could be used for jump detection.

A possible alternative operator to equation (3.20) can be defined as follows. For $k \leq i \leq n-k+1$, let

$$\Delta_*^{(i)} = \frac{|\widehat{\beta}_1^{(i)} - \widehat{\beta}_1^{(i+\ell)}|(\widehat{\beta}_1^{(i)} - \widehat{\beta}_1^{(i-\ell)}) + |\widehat{\beta}_1^{(i)} - \widehat{\beta}_1^{(i-\ell)}|(\widehat{\beta}_1^{(i)} - \widehat{\beta}_1^{(i+\ell)})}{|\widehat{\beta}_1^{(i)} - \widehat{\beta}_1^{(i-\ell)}| + |\widehat{\beta}_1^{(i)} - \widehat{\beta}_1^{(i+\ell)}|}.$$

Then $\Delta_*^{(i)}$ is a weighted average of $(\widehat{\beta}_1^{(i)} - \widehat{\beta}_1^{(i-\ell)})$ and $(\widehat{\beta}_1^{(i)} - \widehat{\beta}_1^{(i+\ell)})$. From Figure 3.4(a), we can check that $\Delta_*^{(i)}$ is small when there is no jump in $N(x_i)$. When there is a jump at x_i, $\Delta_*^{(i)} \sim h_1(\ell)C_0$.

Besides the property that it avoids the effect of the continuity part of f, the original difference operator defined in equation (3.20) also narrows the regions that contain jump information. A jump affects $\widehat{\beta}_1^{(i)}$ if x_i is within $2k$ units of that jump point, while it affects $\Delta_1^{(i)}$ only when x_i is less than k units away. However, we also pay the price for that difference operator in the sense that the variance of $\Delta_1^{(i)}$ is usually larger than the variance of $\widehat{\beta}_1^{(i)}$. In applications, Qiu and Yandell (1998) suggested plotting both $\{\widehat{\beta}_1^{(i)}\}$ and $\{\Delta_1^{(i)}\}$ since the plot of $\{\widehat{\beta}_1^{(i)}\}$ is often helpful in demonstrating the jumps.

A large value of $|\Delta_1^{(i)}|$ indicates a possible jump near x_i. But how large is "large"? To answer this question, a threshold value for $|\Delta_1^{(i)}|$ needs to be defined, and the value of $|\Delta_1^{(i)}|$ is regarded as large if it exceeds this threshold value. To this end, we notice that, if there is no jump in $N(x_i)$, then $\widehat{\beta}_1^{(i)} - \widehat{\beta}_1^{(i-\ell)}$ is approximately normally distributed with mean zero, since it is a linear combination of the observations. From equation (3.20),

$$P\left(\left|\Delta_1^{(i)}\right| > u_{1i}\right) \leq P\left(\left|\widehat{\beta}_1^{(i)} - \widehat{\beta}_1^{(i-\ell)}\right| > u_{1i}\right)$$

for any $u_{1i} > 0$. Therefore, it is reasonable to consider the threshold value $u_{1i} = Z_{1-\alpha_n/2}\sigma^{(i)}$, where $\sigma^{(i)}$ is the standard deviation of $\widehat{\beta}_1^{(i)} - \widehat{\beta}_1^{(i-\ell)}$ and $Z_{1-\alpha_n/2}$ is the $1-\alpha_n/2$ quantile of the standard normal distribution. With the use of this threshold value, the chance that $|\Delta_1^{(i)}| > u_{1i}$ is smaller than or equal to α_n, when there is no jump in $N(x_i)$ and n is large enough. After some routine calculations, we have $\sigma^{(i)} = \frac{n}{k}\sqrt{\frac{6(5k-3)}{k^2-1}}\sigma$. Thus, a natural choice of the threshold value is

$$u_1 = \frac{n\widehat{\sigma} Z_{1-\alpha_n/2}}{k}\sqrt{\frac{6(5k-3)}{k^2-1}}, \tag{3.22}$$

where $\widehat{\sigma}$ is a consistent estimate of σ.

The design points $\{x_{i_j} : |\Delta_1^{(i_j)}| > u_1, j = 1, 2, \ldots, n_1\}$ are then flagged as jump candidates. Similar to the case discussed in the previous subsection about jump detection by DAKE, if the point x_{i_j} is flagged, then some of its neighboring design points will probably be flagged as well. Therefore, it is necessary to delete some of the jump candidates in $\{x_{i_j}, j = 1, 2, \ldots, n_1\}$ by the modification procedure (3.18) discussed in the previous subsection. That is, we replace all the jump candidates in a tie by the middle point of the tie.

The entire jump detection algorithm is summarized as follows.

Local Polynomial Jump Detection Algorithm for Detecting Jumps In f

1. For any given design point x_i, $\ell + 1 \leq i \leq n - \ell$, fit a least squares line in $N(x_i)$.

2. Use formula (3.20) to calculate $\Delta_1^{(i)}, k \leq i \leq n - k + 1$.

3. Use formula (3.22) to calculate the threshold value u_1.

4. Flag the design points $\{x_{i_j}, j = 1, 2, \ldots, n_1\}$ as jump candidates, where x_{i_j} satisfies $|\Delta_1^{(i_j)}| > u_1$ for $j = 1, 2, \ldots, n_1$.

5. Use the modification procedure (3.18) to determine the final jump candidates $\{b_j, j = 1, 2, \ldots, q_1\}$. We then conclude that jumps exist at $b_1 < b_2 < \ldots < b_{q_1}$.

The least squares lines fitted in step 1 of the above algorithm can be updated easily from one design point to the next one, since only two points change. Thus the whole algorithm requires $O(n)$ computation.

The estimated jumps by the above algorithm have the following large sample property (Qiu and Yandell 1998).

Theorem 3.3 *In models (1.1) and (3.1), suppose that the design points are equally spaced on the design interval* $[0, 1]$ *and that the regression function* f *has continuous first-order derivative over* $[0, 1]$, *except for the jump positions at which it has the first-order right and left derivatives. The window size* k *satisfies the conditions that* $\lim_{n\to\infty} k = \infty$ *and* $\lim_{n\to\infty} k/n = 0$. *The confidence level* α_n *used in formula (3.22) satisfies the conditions that (i)* $\lim_{n\to\infty} \alpha_n = 0$ *(ii)* $\lim_{n\to\infty} Z_{1-\alpha_n/2}/\sqrt{\log(\log(k))} = \infty$, *and (iii)* $\lim_{n\to\infty} Z_{1-\alpha_n/2}/\sqrt{k} = 0$. *Then*

(1) $\lim_{n\to\infty} q_1 = p$, *a.s.;*

(2) $\lim_{n\to\infty} b_j = s_j$, *for* $j = 1, 2, \ldots, p$, *a.s.*

The rates of these convergences can all reach $o(n^{-1}\log(n))$.

Example 3.4 In Section 1.2, we discussed the example about the sea-level pressure dataset observed by a Bombay weather station in India during 1921–1992 (cf. Figure 1.3). For this dataset, if we apply the jump detection algorithm discussed above with $k = 15$, then the slope estimators $\{\widehat{\beta}_1^{(i)}\}$ and values of the jump detection criterion $\{\Delta_1^{(i)}\}$ are presented in Figures 3.5(a) and 3.5(b), respectively. It can be seen that the plot of $\{\widehat{\beta}_1^{(i)}\}$ is helpful in demonstrating the jump location. A fitted polynomial regression function of order 4 has standard error $\widehat{\sigma} = 0.977$, leading to a jump threshold value $u_1 = 16.8$ for significance level 0.01. From Figure 3.5(b), only $|\Delta_1^{(40)}| = |-18.077|$, which corresponds to year 1960, exceeds u_1. Hence, a jump appears to exist at year 1960 with significance level of 0.01. To accommodate this detected jump point, the regression function f is then estimated in two design subintervals separated by the detected jump, by the conventional local linear kernel smoothing procedure with bandwidth $h_n = 15/72$ and the Epanechnikov kernel function $K(u) = \frac{12}{11}(1-u^2)I(-1/2 \leq u \leq 1/2)$. The estimated regression function is shown in Figure 3.6 by the solid curves. As a comparison, if the possible jump in the data is ignored, then the estimated regression function by the conventional local linear kernel smoothing procedure with the same bandwidth and kernel function is presented in the plot by the dotted curve. Apparently, a possible jump around year 1960 is blurred by the conventional procedure.

Jump detection in derivatives of f can be developed in an analogous manner. For convenience we only consider jump detection in the first-order derivative here (i.e., $m = 1$ in expression (3.19)). As pointed out at the beginning of this subsection, the following quadratic functions need to be fitted by the LS estimation procedure for this purpose

$$\widehat{Y}^{(i)}(x) = \widehat{\beta}_0^{(i)} + \widehat{\beta}_1^{(i)}x + \widehat{\beta}_2^{(i)}x^2, \text{ for } x \in N(x_i),\ i = \ell+1, \cdots, n-\ell.$$

Under the assumption (A 3.1) and other regularity conditions, Qiu and Yandell show that

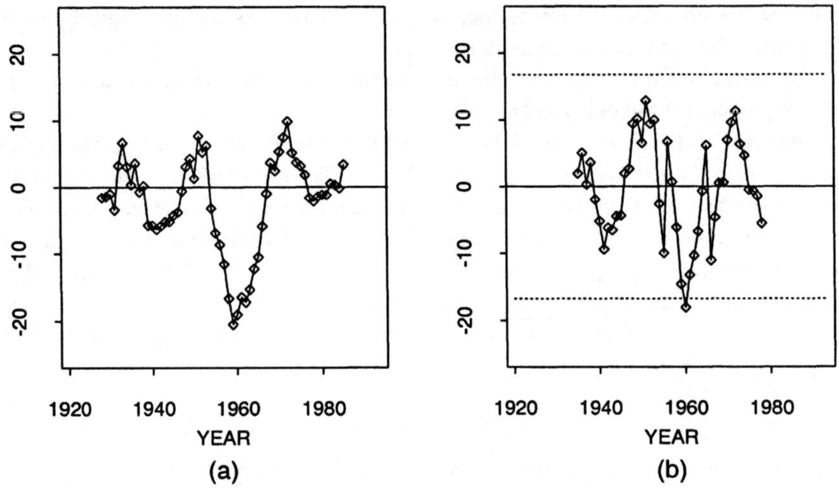

Fig. 3.5 (a): Slope estimators $\{\hat{\beta}_1^{(i)}\}$ from the Bombay sea-level pressure data. (b) Values of the jump detection criterion $\{\Delta_1^{(i)}\}$. The two dotted lines in plot (b) indicate (+) or (−) threshold value u_1 corresponding to $\alpha_n = 0.01$.

(i) if there is no jump in $N(x_i)$, then

$$\widehat{\beta}_2^{(i)} \sim B_2(x_i) := g'(x_i);$$

(ii) if a jump exists in $N(x_i)$ and the jump location is at $x_{i-\ell+r}$, for $0 \leq r \leq 2\ell$, then

$$\widehat{\beta}_2^{(i)} \sim B_2(x_i) := g'_+(x_i) + h_3(r)C_1 - \gamma(r)C_2,$$

where g is the continuity part of f'; C_1 and C_2 are the jump magnitudes of f' and f'', respectively, at the jump position; $\gamma(r)$ is a positive function taking values in [0,1];

$$h_3(r) = \frac{r(k-1-r)k\left[(k-1)(k-2) - 3r(k-1-r)\right]}{12(ks_4 - s_2^2)n^3}$$

with $s_p = \sum_{j=i-\ell}^{i+\ell}(x_j - x_i)^p$, for $p = 2, 4$; and $g'_+(x_i) = g'(x_i)$ if $r \neq \ell$.

Example 3.5 Suppose that $f'(x) = 5x^2 + I(x \in [0.5, 1])$. Then f' has a jump at $x = 0.5$ with size 1. Consider a sample of size 100 and let k be 11. Then

$$\widehat{\beta}_2^{(i)} \sim B_2(x_i) = 10x_i + h_3\left[100(0.55 - x_i)\right] I(0.45 \leq x_i \leq 0.55),$$

for $6 \leq i \leq 95$. The quantities $\{B_2(x_i), 6 \leq i \leq 95\}$ are shown in Figure 3.7(a).

Similar to formulas (3.20) and (3.21), we can construct $\{\Delta_2^{(i)}\}$ from $\{\widehat{\beta}_2^{(i)}\}$ and can define $\{J_2(x_i)\}$ similarly to $\{J_1(x_i)\}$. The quantities $\{J_2(x_i)\}$ for Example 3.5 are shown in Figure 3.7(b). It can be seen that information about the continuity part

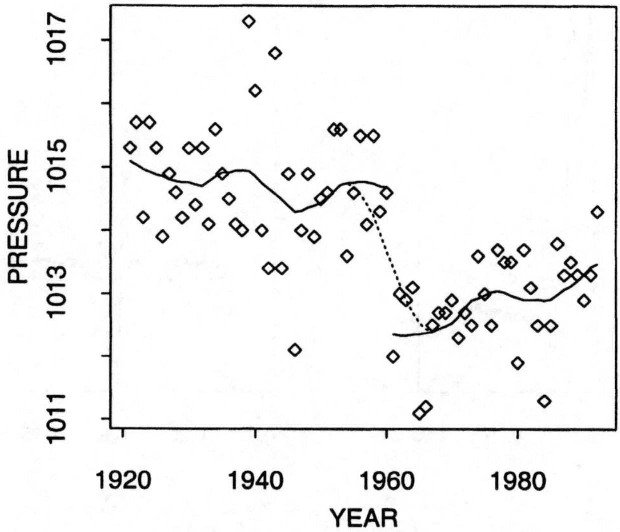

Fig. 3.6 The small diamonds denote sea-level pressures observed by a Bombay weather station in India during 1921–1992. The solid curves denote estimated regression function with a detected jump accommodated. The dotted curve denotes conventional local linear kernel estimator of the regression function.

of f' has been mostly deleted from $\{\Delta_2^{(i)}\}$, whereas information about the jump part of f' has been preserved.

Qiu and Yandell (1998) defined the following threshold value for $\{\Delta_2^{(i)}\}$, which is derived in a similar way to u_1:

$$u_2 = \widehat{\sigma} Z_{1-\alpha_n/2} \frac{\sqrt{k^2 s_4 - (k+1)s_2^2}}{k s_4 - s_2^2}.$$

The jump detection algorithm for detecting jumps in f given earlier can be used for detecting jumps in f', with $\{\Delta_2^{(i)}\}$ and u_2 substituted for $\{\Delta_1^{(i)}\}$ and u_1.

3.4 JUMP-PRESERVING CURVE ESTIMATION

As discussed in Subsection 3.2.1, the jump regression function f can be estimated by the following two steps. The first step is to detect the jump positions by one of various jump detection procedures. The second step has two options. One option is to estimate f as usual by conventional local or global smoothing procedures, in design subintervals separated by the detected jump points. The other option is to estimate the continuity part g of f from the new data $\{Y_i - \sum_{j=1}^{p} \widehat{d}_j I(x_i > \widehat{s}_j), i = 1, 2, \ldots, n\}$ and then to define the estimated regression function by $\widehat{f}(x) = \widehat{g}(x) + \sum_{j=1}^{p} \widehat{d}_j I(x >$

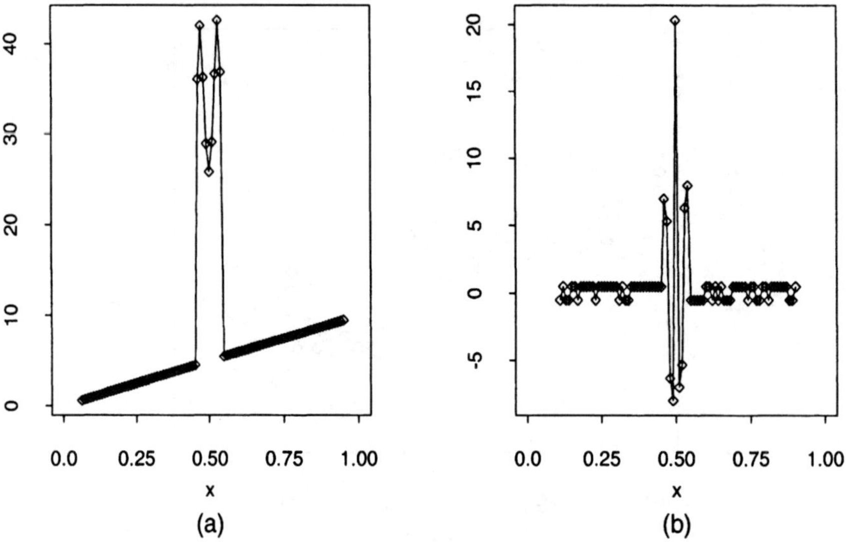

Fig. 3.7 If $f'(x) = 5x^2 + I(x \in [0.5, 1])$, $n = 100$, and k is chosen to be 11, then $\widehat{\beta}_2^{(i)} \sim B_2(x_i)$ for $6 \leq i \leq 95$. The quantities $\{B_2(x_i), 6 \leq i \leq 95\}$ are shown in plot (a) (represented by diamonds). After using the difference operator, which is similar to that in equation (3.20), we get $\{J_2(x_i)\}$, which are shown in plot (b).

$\widehat{s}_j)$, for $x \in [0, 1]$; $\{\widehat{s}_j, j = 1, 2, \ldots, p\}$ and $\{\widehat{d}_j, j = 1, 2, \ldots, p\}$ are the estimated jump positions and jump magnitudes obtained in the first step, and $\widehat{g}$ is the estimator of g.

In applications, jumps often represent structural changes of the related process; therefore, it is important to estimate their positions. However, sometimes the ultimate goal is to estimate the regression function f from noisy data. Because of the importance of the jumps, they should be preserved in the process of estimating f.

Can this goal be achieved without detecting the jumps explicitly? In the literature, several jump-preserving curve estimation procedures have been suggested, which do not detect the jumps explicitly in estimating f. In this section, we introduce three such procedures, suggested by Chu *et al.* (1998), McDonald and Owen (1986), and Qiu (2003a).

3.4.1 Jump curve estimation by split linear smoothing

McDonald and Owen (1986) suggested a smoothing algorithm with several steps for jump-preserving curve estimation. They call their smoothing procedure the *split linear smoothing algorithm.*

The first step is to consider various neighborhoods at a given design point x_i. These neighborhoods have three possible orientations: central, left, and right orientations. A central neighborhood of size k consists of x_i, with its $k/2$ closest design points on the left side (assuming that $k/2$ is an integer) and its $k/2$ closest design points on the right side. A left-sided neighborhood of size k consists of x_i, with k closest design points on the left. Similarly, a right-sided neighborhood of size k consists of x_i, with k closest design points on the right.

As the next algorithm step, in each neighborhood of x_i with orientation o and size k, a least square regression line is estimated. If the point x_i is a jump point or there is a jump point nearby, then some of the fitted regression lines provide good estimators of $f(x_i)$, while the others do not. For example, suppose x_i is a jump point and $f(x_i) = f_-(x_i)$, where $f_-(x_i)$ denotes the left limit of f at x_i; then, the estimated regression lines obtained in all left-sided neighborhoods provide good estimators of $f(x_i)$, but the estimated regression lines obtained in all central or right-sided neighborhoods do not provide good estimators.

As the last algorithm step, McDonald and Owen suggest using the following weighted average:

$$\widehat{f}(x_i) = \sum_{k,o} w(i,k,o)\widehat{f}(i,k,o) \tag{3.23}$$

as a final estimator of $f(x_i)$, where $w(i,k,o)$ is the weight, $\widehat{f}(i,k,o)$ is the estimator of $f(x_i)$ provided by the estimated regression line in the neighborhood with orientation o and size k, and $\sum_{k,o}$ denotes summation with respect to all possible orientations and window sizes.

One motivation to use neighborhoods of different sizes is to adapt locally to the presence of variable curvature of the regression function f. A small window is favored at a place where f has large curvature, whereas a relatively large window is favored at a place where f is quite flat.

It is natural that the weight $w(i,k,o)$ should be proportional to the reciprocal of the mean squared residual of the estimated regression line in the neighborhood with orientation o and size k. To reduce the tendency of a linear fit to look good simply because it is close to the data point, McDonald and Owen (1986) suggested computing the mean squared residual over all points in the neighborhood except x_i. The resulting mean squared residual is called the pseudo-mean squared error (PMSE). It is formally defined by

$$\text{PMSE}(i,k,o) = \sum_{x_j \in W(i,k,o)\setminus\{x_i\}} \frac{(Y_j - \widehat{\alpha} - \widehat{\beta}x_j)^2}{|W(i,k,o)| - 3},$$

where $W(i,k,o)$ denotes the neighborhood of x_i with orientation o and size k, $\widehat{\alpha}+\widehat{\beta}x$ is the estimated regression line in the neighborhood, and $|W(i,k,o)|$ denotes the total number of observations in the neighborhood.

McDonald and Owen pointed out that the weighting scheme should have two desirable features. One desirable feature is to put exactly zero weight on the worst fits. For example, if x_i is a jump point and $f(x_i) = f_-(x_i)$, then the estimated regression lines in all left-sided neighborhoods provide good fits, while the estimated regression lines in all right-sided neighborhoods provide bad fits which should be completely ignored. To detect bad fits, McDonald and Owen introduced a threshold value; a fit is regarded as a bad fit if the corresponding PMSE value exceeds the threshold value. The second desirable feature of the weight function is that it should smoothly decrease to zero when the PMSE value increases to the threshold value. Otherwise, spurious bumps might result from the weighting scheme. By the above considerations, the suggested weights are defined by

$$w(i,k,o) = C[\text{PMSE}(i,k,o) - u_i]^2 I(\text{PMSE}(i,k,o) < u_i),$$

where u_i is the threshold value that is selected to be the simple average of the PMSE values of all local fits at x_i, and C is a normalization constant.

McDonald and Owen pointed out that the estimated regression function by the above split linear smoothing procedure (cf. equation (3.23)) has a tendency to be noisy. To further remove noise in the estimator, they suggested applying the same smoothing procedure to the output $\{(x_i, \widehat{f}(x_i)), i = 1, 2, \ldots, n\}$. They find that this step does not blur the discontinuities much, and the resulting curve estimator looks much smoother.

3.4.2 Jump-preserving curve fitting based on local piecewise-linear kernel estimation

Qiu (2003a) pointed out that a major reason why the conventional smoothing procedures (cf. the procedures introduced in Chapter 2) cannot preserve the jumps of the true regression function f is that a *continuous* function is used by these procedures either locally (if the conventional procedures are local smoothing procedures) or globally (if the conventional procedures are global smoothing procedures) for approximating the *jump* regression function f. For instance, the local linear kernel estimator of $f(x)$ is defined by the solution to a of the minimization problem (2.9) with $m = 1$. In this case, a linear function is used locally for approximating f at x. It can be checked that the local linear kernel estimator of $f(x)$ (cf. equation (2.10) for its expression) is not statistically consistent if x is a jump point.

A natural idea to overcome the above-mentioned limitation of the local linear kernel estimator is to fit a local piecewise-linear function at x instead. Namely, we consider the following minimization problem:

$$\min_{a_{l,0};a_{l,1};a_{r,0};a_{r,1}} \sum_{i=1}^{n} \{Y_i - [a_{l,0} + a_{l,1}(x_i - x)] - [(a_{r,0} - a_{l,0})I(x_i \geq x) + (a_{r,1} - a_{l,1})(x_i - x)I(x_i \geq x)]\}^2 K\left(\frac{x_i - x}{h_n}\right), \tag{3.24}$$

where K is a kernel function with support $[-1/2, 1/2]$, and h_n is a bandwidth parameter. The minimization problem (3.24) fits the following piecewise-linear function

$$a_{l,0} + a_{l,1}(u - x) + (a_{r,0} - a_{l,0})I(u \geq x) + (a_{r,1} - a_{l,1})(u - x)I(u \geq x)$$

on the interval $u \in [x - h_n/2, x + h_n/2]$. This function has a possible jump at x.

Let $\{\widehat{a}_{l,j}(x), \widehat{a}_{r,j}(x), j = 0, 1\}$ denote the solution of (3.24). Then $\widehat{a}_{l,0}(x)$ and $\widehat{a}_{r,0}(x)$ are actually constructed from the observations in the two one-sided neighborhoods $[x - h_n/2, x)$ and $[x, x + h_n/2]$, respectively; the subscripts "l" and "r"

represent "left-sided" and "right-sided", respectively. Thus they are good estimators of $f_-(x)$ and $f_+(x)$, the left and right limits of f at x, in the case when x is a jump point. When there is no jump in $[x-h_n/2, x+h_n/2]$, both of them should estimate $f(x)$ well. In the case when x itself is not a jump point but a jump point exists in the neighborhood $[x-h_n/2, x+h_n/2]$, only one of $\widehat{a}_{l,0}(x)$ and $\widehat{a}_{r,0}(x)$ provides a good estimator of $f(x)$. Therefore, in all three cases, at least one of $\widehat{a}_{l,0}(x)$ and $\widehat{a}_{r,0}(x)$ estimates $f(x)$ well.

To choose between $\widehat{a}_{l,0}(x)$ and $\widehat{a}_{r,0}(x)$ for estimating $f(x)$, it is natural to make the decision based on the corresponding weighted residual sums of squares (RSS) of the two pieces of the fitted regression function. Let $\text{RSS}_l(x)$ and $\text{RSS}_r(x)$ be the RSS values of the left-sided and right-sided pieces of the fitted regression function, respectively. Namely,

$$\text{RSS}_l(x) = \sum_{x_i<x} \{Y_i - \widehat{a}_{l,0}(x) - \widehat{a}_{l,1}(x)(x_i - x)\}^2 K\left(\frac{x_i - x}{h_n}\right);$$

$$\text{RSS}_r(x) = \sum_{x_i\geq x} \{Y_i - \widehat{a}_{r,0}(x) - \widehat{a}_{r,1}(x)(x_i - x)\}^2 K\left(\frac{x_i - x}{h_n}\right).$$

Then the resulting estimator of $f(x)$, for $x \in [h_n/2, 1-h_n/2]$, is defined by

$$\widehat{f}(x) = \widehat{a}_{l,0}(x) I^*[\text{RSS}_r(x) - \text{RSS}_l(x)] + \widehat{a}_{r,0}(x) I^*[\text{RSS}_l(x) - \text{RSS}_r(x)], \quad (3.25)$$

where $I^*(a)$ is defined by $I^*(a) = 1$ if $a > 0$, $1/2$ if $a = 0$, and 0 if $a < 0$. Therefore, $\widehat{f}(x)$ is defined by the one between $\widehat{a}_{l,0}(x)$ and $\widehat{a}_{r,0}(x)$, which has smaller RSS value.

It can be easily checked that the minimization problem (3.24) is equivalent to fitting two different lines $a_{l,0} + a_{l,1}(u-x)$ and $a_{r,0} + a_{r,1}(u-x)$ in the two one-sided neighborhoods $[x-h_n/2, x)$ and $[x, x+h_n/2]$, respectively. Therefore, $\{\widehat{a}_{l,j}(x), \widehat{a}_{r,j}(x), j = 0, 1\}$ can also be obtained from

$$\min_{a_{l,0}, a_{l,1}} \sum_{i=1}^{n} \{Y_i - a_{l,0} - a_{l,1}(x_i - x)\}^2 K_l\left(\frac{x_i - x}{h_n}\right) \quad (3.26)$$

and

$$\min_{a_{r,0}, a_{r,1}} \sum_{i=1}^{n} \{Y_i - a_{r,0} - a_{r,1}(x_i - x)\}^2 K_r\left(\frac{x_i - x}{h_n}\right), \quad (3.27)$$

where K_l is defined by $K_l(x) = K(x)$ if $x \in [-1/2, 0)$ and 0 otherwise, and K_r is defined by $K_r(x) = K(x)$ if $x \in [0, 1/2]$ and 0 otherwise. Clearly, the procedure in (3.26) is equivalent to constructing the local linear kernel estimator of $f_-(x)$ by the observations in $[x-h_n/2, x)$, and the procedure in (3.27) is equivalent to constructing the local linear kernel estimator of $f_+(x)$ by the observations in $[x, x+h_n/2]$.

Similar to expression (2.10), it can be checked that the solutions of (3.26) and (3.27) have the following expressions:

$$\widehat{a}_{l,0}(x) = \sum_{i=1}^{n} \frac{w_{l,2} - w_{l,1}(x_i - x)}{w_{l,0}w_{l,2} - w_{l,1}^2} Y_i K_l\left(\frac{x_i - x}{h_n}\right),$$

$$\widehat{a}_{l,1}(x) = \sum_{i=1}^{n} \frac{w_{l,0}(x_i - x) - w_{l,1}}{w_{l,0}w_{l,2} - w_{l,1}^2} Y_i K_l\left(\frac{x_i - x}{h_n}\right),$$

$$\widehat{a}_{r,0}(x) = \sum_{i=1}^{n} \frac{w_{r,2} - w_{r,1}(x_i - x)}{w_{r,0}w_{r,2} - w_{r,1}^2} Y_i K_r\left(\frac{x_i - x}{h_n}\right),$$

$$\widehat{a}_{r,1}(x) = \sum_{i=1}^{n} \frac{w_{r,0}(x_i - x) - w_{r,1}}{w_{r,0}w_{r,2} - w_{r,1}^2} Y_i K_r\left(\frac{x_i - x}{h_n}\right),$$

where $w_{l,j} = \sum_{i=1}^{n}(x_i - x)^j K_l(\frac{x_i - x}{h_n})$ and $w_{r,j} = \sum_{i=1}^{n}(x_i - x)^j K_r(\frac{x_i - x}{h_n})$ for $j = 0, 1, 2$.

Example 3.6 Suppose that the true regression function is $f(x) = 2 - 3x$ when $x \in [0, 0.3)$; $f(x) = 3 - 3x - \sin[5\pi(x - 0.3)]$ when $x \in [0.3, 0.7)$; and $f(x) = 1.55 + 0.5x$ when $x \in [0.7, 1]$. This function is shown in Figure 3.8(a) by the solid curve. If no noise is involved in the data (i.e., $\sigma = 0$ in model (1.1)), the kernel function is $K(x) = \frac{12}{11}(1 - x^2)I(x \in [-0.5, 0.5])$, and the bandwidth is $h_n = 0.145$, then $\widehat{a}_{l,0}, \widehat{a}_{r,0}$, and $\widehat{f}$ are as shown in Figure 3.8(b) by the dotted, dashed, and solid curves, respectively. It can be seen from the plot that jump blurring occurs in $[x_0, x_0 + h_n/2]$ if the point x_0 is a jump point and $\widehat{a}_{l,0}$ is used for estimating f. Similarly, jump blurring occurs in $[x_0 - h_n/2, x_0)$ if x_0 is a jump point and $\widehat{a}_{r,0}$ is used for estimating f. The estimator $\widehat{f}$, however, can preserve the jumps well because it is defined as $\widehat{a}_{l,0}$ in $[x_0 - h_n/2, x_0)$ and as $\widehat{a}_{r,0}$ in $[x_0, x_0 + h_n/2]$, when x_0 is a jump point.

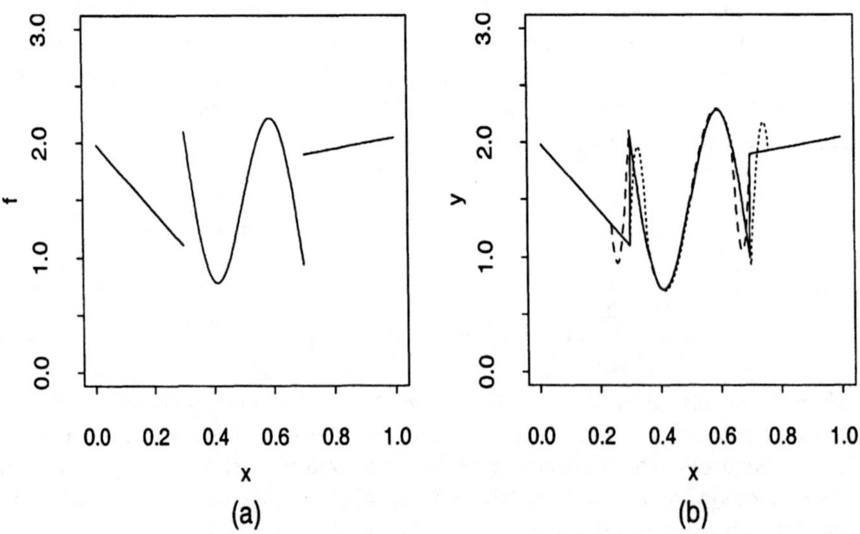

Fig. 3.8 (a): The true regression function. (b) The dotted, dashed, and solid curves denote $\widehat{a}_{l,0}$, $\widehat{a}_{r,0}$, and $\widehat{f}$ in the case when no noise is involved in the data.

When x is in the boundary regions $[0, h_n/2)$ and $(1 - h_n/2, 1]$, the estimator $\widehat{f}(x)$ is not well defined by equation (3.25). In such cases, there are several possible approaches for estimating $f(x)$ if no jumps exist in the boundary regions. For example, $\widehat{f}(x)$ could be defined by the conventional local linear kernel estimator constructed from the observations in $[0, x + h_n/2]$ or $[x - h_n/2, 1]$, depending on whether $x \in [0, h_n/2)$ or $x \in (1 - h_n/2, 1]$. It could also be defined by

$$\widehat{f}(x) = \widehat{a}_{r,0}(x), \text{ when } x \in [0, h_n/2)$$

and

$$\widehat{f}(x) = \widehat{a}_{l,0}(x), \text{ when } x \in (1 - h_n/2, 1].$$

However, if jumps exist in the boundary regions $[0, h_n/2)$ and $(1 - h_n/2, 1]$, estimation of f in these regions is still an open problem.

3.4.3 Jump-preserving smoothers based on robust estimation

Chu *et al.* (1998) discussed two local smoothing techniques for jump-preserving curve or surface estimation. The two smoothing techniques, the sigma filter and the M smoother, are both based on robust estimation. They are introduced below in one-dimensional cases, where $\{(x_i, Y_i), i = 1, 2, \ldots, n\}$ are n observations from model (1.1) in which the design points are assumed to be equally spaced on the design interval $[0, 1]$.

The sigma filter is a modification of the Nadaraya-Watson kernel estimator defined in equation (2.3). Its estimator at a given design point x_i has the following expression:

$$\widehat{f}_S(x_i) = \frac{\sum_{j=1}^{n} Y_j K((x_j - x_i)/h_n) L((Y_j - Y_i)/g_n)}{\sum_{j=1}^{n} K((x_j - x_i)/h_n) L((Y_j - Y_i)/g_n)}, \tag{3.28}$$

where K and L are two density kernel functions, and h_n and g_n are two bandwidths.

Clearly, $\widehat{f}_S(x_i)$ is a weighted average of all observations; the weights are determined not only by the distance from individual design points x_j to the given design point x_i but also by the difference between individual observations Y_j to the observation Y_i at x_i. If there is a jump around x_i and x_j and x_i are on two different sides of the jump, then the absolute difference between Y_j and Y_i is large. Consequently, Y_j would receive small weight in defining $\widehat{f}_S(x_i)$. In that way, the possible jumps can be preserved by the sigma filter (3.28). It is easy to see that the estimator $\widehat{f}_S(x_i)$ is close to the conventional Nadaraya-Watson kernel estimator when the kernel function L is a constant and g_n is large.

The M smoother also has a connection to the Nadaraya-Watson kernel estimator. As mentioned in Subsection 2.4.1 (cf. expression (2.8)), the Nadaraya-Watson kernel estimator $\widehat{f}_{NW}(x_i)$ is defined by the minimizer of

$$\min_{a \in R} \sum_{j=1}^{n} \rho(Y_j - a) K\left(\frac{x_j - x_i}{h_n}\right), \tag{3.29}$$

where $\rho(y) = y^2$ is a quadratic loss function. By using the quadratic loss function, $\widehat{f}_{NW}(x_i)$ is affected much by some extremely large or small observations, which are often called *outliers*, in the data. To make the estimator more *robust* to outliers, the loss function $\rho(y)$ needs to be chosen such that it grows slowly when y tends to $\pm\infty$. Some examples of such functions include $\rho(y) = |y|$, $\rho(y) = \max(y^2, k)$, and $\rho(y) = -\exp(-y^2/(2g^2))/(\sqrt{2\pi}g)$, where $k, g > 0$ are two constants. The estimator $\widehat{f}_M(x_i)$ of $f(x_i)$ by the M smoother is defined by the local minimizer of the objective function, denoted as $S(a)$, of the minimization problem (3.29), This estimator comes from direct descent starting at $S(Y_i)$, meaning that it is the local minimum of $S(a)$ closest to Y_i, to the right when $S'(Y_i) < 0$ and to the left when $S'(Y_i) > 0$.

Chu *et al.* (1998) suggested a numerical algorithm for computing the M estimator $\widehat{f}_M(x_i)$. They also demonstrated that both the sigma filter and the M smoother can be generalized for estimating regression surfaces with possible jumps preserved. For related discussions, please read Burt and Coakley (2000) and Rue *et al.* (2002).

3.5 SOME DISCUSSIONS

In this chapter, we introduced some recent methodologies for estimating one-dimensional jump regression functions. Most of these methods detect possible jumps first, and then estimate the regression function as usual, in design subintervals separated by the detected jumps. There are several existing procedures which avoid detecting jumps explicitly, while estimating the regression function with jumps preserved.

In a specific application, which type of methods should be used? In my viewpoint, this decision should be made based on the major goal of the application problem. In most situations, it is important to know the jump positions because jumps often represent structural changes of the related process. If this is the case, then various jump detection procedures should be considered. If the ultimate goal of the application problem is to recover the underlying regression curve from noisy data (with possible jumps preserved), then the second type of methods which avoid detecting jumps explicitly might be easier to use; these usually work like a black box, as in many conventional curve estimation procedures, with the noisy data as input and the estimated regression function as output. Compared to the first type of methods, usually less parameters need to be determined in the second type of methods.

As explained in Section 3.1, theoretically speaking, there are eight types of jump regression models, depending on whether each of the three jump factors (number, position, and magnitude of the jumps) is known. Most existing research on jump regression analysis has focused on fitting two specific types of jump regression models: those with known number of jumps, unknown jump positions, and unknown jump magnitudes and those with all three jump factors unknown. By adapting some of these procedures, the other type of jump regression models can generally be handled without much difficulty.

Most local smoothing jump detection criteria are based on a difference between a weighted average of the observations in a right-sided neighborhood of a given point and a weighted average of the observations in a left-sided neighborhood of the given point. This difference is small when there is no jump around the given point and is large otherwise. The spline smoothing and discrete wavelet transformation procedures detect jumps slightly differently, by including some "irregular" basis functions, such as the truncated power basis functions (see Subsection 2.5.4 for their definitions), in their function estimation systems. These "irregular" basis functions themselves often have the major features to detect. For example, if we are interested in detecting jumps in the regression function itself, then basis functions with step jumps are used in the partial spline smoothing procedure suggested by Shiau (1985) and the regression spline procedure suggested by Koo (1997). In such a case, wavelet functions with step jumps or steep slopes (e.g., the Haar wavelet functions) are used in the discrete wavelet transformation procedures (see e.g., Wang 1995).

Although much research effort has been made in recent years on jump detection and jump-preserving curve estimation, there are still many research problems that need to be addressed in the future. Some of these research problems are discussed below.

Bandwidth selection is always a problem for local smoothing procedures. In jump regression analysis, there are two major goals: jump detection and jump-preserving curve estimation. Our preliminary numerical studies show that bandwidths should be chosen differently for these two different goals. In a specific application, if our major goal is to recover the underlying regression curve from noisy data, then the CV and other parameter selection procedures discussed in Subsections 2.4.4 and 2.5.2 can be used for bandwidth selection. On the other hand, if our major goal is to detect jumps accurately, then, based on our numerical experience, the bandwidths should be chosen smaller than the bandwidths for curve estimation. Ideally, relatively small bandwidths should be used for jump detection and for curve estimation around the detected jumps, and relatively large bandwidths should be used for curve estimation in regions without any detected jumps. So far, we have not seen any data-driven procedures which select the bandwidths such that these goals can all be accomplished.

The local smoothing jump detection procedures have the "boundary problem" in the sense that they could not detect jumps in the border regions of the design interval. When the sample size is large, this may not be a big problem, since in such cases the border regions are often small, and it is unlikely that any jumps exist in these regions. However, when the sample size is small, the border regions are relatively large, and the boundary problem could be real. It should be pointed out that the boundary problem discussed here is different from the boundary problem discussed in Sections 2.3 and 2.4 about the conventional kernel smoothing procedures. The boundary problem discussed in Chapter 2 says that biases of the conventional kernel estimators of f are larger in the border regions than their biases in the interior region. For instance, as discussed in Subsection 2.4.2, the local linear kernel smoothing procedure has the property of automatic boundary carpentry and consequently does not suffer this type of boundary problem. But the jump detection procedure based on local linear kernel smoothing (discussed in Subsection 3.2.2) still has the boundary problem for jump detection.

In Section 3.3, we introduced several jump detection procedures in the case when the number of jumps was unknown. A major strategy of these procedures to detect jumps in such cases is to introduce a threshold parameter, which is determined by the asymptotic distribution of some test statistic. In small sample cases, however, these procedures may not perform well because the threshold parameter may not have the properties we would expect. Therefore, much research is needed on jump detection in finite sample cases, when the number of jumps is unknown.

Problems

3.1 In some research papers, the jump regression function f is described by the following equation:

$$f(x) = g(x) + \sum_{j=1}^{p} d_j^* I(s_j < x \leq s_{j+1}), \text{ for } x \in [0, 1],$$

where g is a continuous function in the entire design interval, p is the number of jump points, $\{s_j, j = 1, 2, \ldots, p\}$ are the jump positions, $s_{p+1} = 1$, and $\{d_j^*, j = 1, 2, \ldots, p\}$ are quantities related to the jump magnitudes $\{d_j, j = 1, 2, \ldots, p\}$ defined in expression (3.1). Identify the relationship between $\{d_j^*, j = 1, 2, \ldots, p\}$ and $\{d_j, j = 1, 2, \ldots, p\}$.

3.2 By using the ideas of some jump detection procedures introduced in this chapter, please suggest a procedure to estimate the number of jumps, the jump positions, and the jump magnitudes, in the case when the number of jumps is unknown but the potential jump positions are known (i.e., case (2) discussed in Section 3.1).

3.3 The expressions (3.3)–(3.5) can be obtained by the definition of $M_{DKE}(x)$ (cf. equation (3.2)) and the Taylor expansion of $f(x_i)$ at the given point x, for $i = 1, 2, \ldots, n$.

3.3.1. Verify that these expressions hold under the conditions stated in Theorem 3.1.

3.3.2. Derive the expressions (3.9)–(3.11) from Theorem 3.2 in a similar way.

3.4 Based on the formula of the conventional local linear kernel estimator $\widehat{f}_{LK}$ (cf. expression (2.10) in Subsection 2.4.1) and the definition of $M_{DLK}(x)$ given in Subsection 3.2.2, verify that formula (3.8) is true.

3.5 The sea-level pressure data shown in Figure 3.6 can be downloaded from the ftp site of this book (see front cover for address). Suppose that we know a single jump exists in the underlying regression curve. Detect this jump by the DKE and DLK procedures discussed in Subsections 3.2.1 and 3.2.2, respectively, with their bandwidths selected by the CV procedure (cf. Subsection 2.4.4 for an introduction).

3.6 The semiparametric procedure discussed in Subsection 3.2.3 is for jump detection in the first-order derivative f' of f, when it is assumed that f' has only one jump point in the design interval.

3.6.1. Can this procedure be generalized for detecting p jumps in f', where $p > 1$ is a known integer?

3.6.2. Can this procedure be modified for detecting a single jump in f itself or in its m-th order derivative $f^{(m)}$, where $m > 1$ is a known integer?

3.6.3. Can this procedure be modified for detecting p jumps in f or $f^{(m)}$, where $p > 1$ and $m > 1$ are two known integers?

3.6.4. Do you have any thoughts about the questions stated in problems 3.6.1 and 3.6.3 when p is unknown?

3.7 In Subsection 2.6.2, the wavelet functions are defined by

$$\psi_{j,k}(x) = 2^{j/2}\psi(2^j x - k), \text{ for } j, k = 0, \pm 1, \ldots,$$

where ψ is a mother wavelet.

3.7.1. If the above definition is changed to

$$\psi^*_{j,k}(x) = 2^{j/2}\psi(k - 2^j x), \text{ for } j, k = 0, \pm 1, \ldots,$$

discuss the relationship between $\{\psi_{j,k}, j, k = 0, \pm 1, \ldots\}$ and $\{\psi^*_{j,k}, j, k = 0, \pm 1, \ldots\}$.

3.7.2. In Subsection 3.2.5, the wavelet transformation of a function f is defined by

$$Tf(s, x) = \int \psi_s(x - u) f(u)\, du,$$

where $\psi_s(x - u) = s^{-1/2}\psi[(x - u)/s]$. Verify that $Tf(2^{-j}, k2^{-j}) = w^*_{j,k}$ for $j, k = 0, \pm 1, \ldots$, where

$$w^*_{j,k} = \int \psi^*_{j,k}(u) f(u)\, du.$$

3.8 In Subsection 3.3.1, it is mentioned that, for given coefficients $\{c_j, j = 0, \pm 1, \ldots, \pm m\}$, suppose the coefficients $\{r_j, j = 0, 1, \ldots, 2m\}$ are chosen by

$$r_j = \sum_{\ell=0}^{q} u_\ell j^\ell, \text{ for } j = 0, 1, \ldots, 2m,$$

where $\mathbf{u} = (u_0, u_1, \ldots, u_q)'$ as defined in Subsection 3.3.1. In this case, the two quantities $\mathrm{E}[\widehat{f}_c(i/n)]$ and $\mathrm{E}[\widehat{f}_r(i/n)]$ have the same q leading terms in their Taylor expansions if no jumps exist in the interval $[(i-2m)/n, (i+2m)/n]$, where $2m+1 \leq i \leq n - 2m$ and $q \geq 1$ is a given integer. Verify this statement.

3.9 The file sleep.dat included in the ftp site of the book (see front cover for address) provides a dataset about rat sleep. Psychiatrists believe that animal sleep is affected by lighting condition and other environmental conditions. To confirm this, an experiment exposed eight Brown Norway rats to a test lighting condition, which was a continuous three-hour lights on and three-hour lights off schedule, for three days. Sleep or wake status of each rat was determined for each 30-second interval throughout the experiment. Researchers then averaged every 10 consecutive readings for each rat, yielding percentages of time asleep for each day's 288 five-minute intervals. These percentages were then averaged over three days of experimentation and all eight rats. For a more complete description of the data, read Qiu *et al.* (1999). In the file sleep.dat, there are two columns: the first column includes middle points of 288 five-minute intervals of a day, and the second column includes the averaged percentages of sleep time of the eight rats in the 288 time intervals. Rats are nocturnal animals, who tend to go to sleep when exposed to light and who become active in dark. Thus, abrupt changes would be expected in the data at times when the light was switched. Use the jump detection procedure introduced in Subsection 3.3.1 to check the validity of this conclusion.

3.10 For the jump detection criterion J defined in expression (3.16), suppose that $x \in (0,1)$ is a jump point and that K_1 and K_2 are two continuous density kernel functions with supports $[-\lambda, 1]$ and $[-1, \lambda]$, respectively, where $\lambda \in [0,1]$ is a constant. Verify that $|\mathrm{E}(J(x))|$ is generally smaller than the absolute jump magnitude at x.

3.11 For the procedure discussed in Subsection 3.3.4, verify the following results.

3.11.1. If a given design point x_i is the only jump point in its neighborhood $N(x_i)$, then

$$\widehat{\beta}_1^{(i)} = g'_+(x_i) + h_1(r)C_0 - \gamma(r)C_1 + \xi_n,$$

where $g'_+(x_i)$ is the right first-order derivative of the continuity part g of the regression function f at x_i, C_0 and C_1 are the jump magnitudes of f and its first-order derivative f' at x_i, $\gamma(r)$ is a positive function taking values between 0 and 1, $h_1(r) = 6nr(k-1-r)/[k(k+1)(k-1)]$, and ξ_n is a random variable satisfying $\lim_{n\to\infty} \xi_n = 0$, a.s.

3.11.2. The expression (3.21) is true.

3.11.3. The function $h_2(r)$ in (3.21) has a maximum value at $r = \ell$ of $1.5n(k-1)/[k(k+1)]$.

3.11.4. If the threshold value u_1 is determined by equation (3.22) and there are no jumps in the interval $[x_{i-k}, x_{i+k}]$ for $k < i < n-k$, then

$$P\left(|\Delta_1^{(i)}| > u_1\right) = O(\alpha_n),$$

where α_n is the significance level used in (3.22).

3.12 Reproduce Figures 3.5 and 3.6.

3.13 Suppose that we are interested in detecting jumps in the mth-order derivative $f^{(m)}$ (cf. equation (3.19)) of the regression function f by the local polynomial regression procedure discussed in Subsection 3.3.4, where $m > 1$ is a given integer. Define an appropriate jump detection criterion and derive a formula for the corresponding threshold value.

3.14 The jump-preserving curve estimation procedure (3.25) discussed in Subsection 3.4.2 may have limited ability for removing noise because its estimated regression function $\widehat{f}$ equals one of the two one-sided estimators $\widehat{a}_{l,0}$ and $\widehat{a}_{r,0}$, each of which uses only half of the observations in the neighborhood of a given point. One way to improve the noise-removing ability of this procedure is to apply it once again to the output $\{(x_i, \widehat{f}(x_i)), i = 1, 2, \ldots, n\}$. Perform a simulation study to confirm that (1) this extra step does improve the noise-removing ability of the procedure (3.25), and (2) the jumps are still preserved well after this extra step.

4

Estimation of Jump Location Curves of Regression Surfaces

4.1 INTRODUCTION

This chapter discusses jump detection in two-dimensional regression analysis, using the regression model described below. If the design points are arranged by n_1 rows and n_2 columns in the design space $[0,1] \times [0,1]$, a two-dimensional regression model can be described by equation (1.3) in Chapter 1. However, in some applications, the design points are irregularly distributed in the design space. In such more general cases, a two-dimensional regression model can be described by

$$Z_i = f(x_i, y_i) + \varepsilon_i,\ i = 1, 2, \ldots, n, \tag{4.1}$$

where n is the sample size, $\{(x_i, y_i), i = 1, 2, \ldots, n\}$ are the design points in the design space, f is the two-dimensional regression function, $\{Z_i, i = 1, 2, \ldots, n\}$ are n observations of the response variable Z, and $\{\varepsilon_i, i = 1, 2, \ldots, n\}$ are independent and identically distributed (i.i.d.) random errors with mean zero and unknown variance σ^2. For simplicity of presentation, we assume that the design space is the unit square $[0,1] \times [0,1]$ in this chapter. Most procedures introduced in this chapter, however, can be used in the case when the design space is any connected region in the two-dimensional real plane R^2.

In this chapter, the regression function f is assumed to be nonparametric and continuous in the entire design space except on some curves. More specifically, we assume that there exists a partition $\{\Lambda_i, i = 1, 2, \ldots, s\}$ of the design space $[0,1] \times [0,1]$ such that:

(i) each Λ_i is a connected region in the design space;

(ii) $\Lambda_i \bigcap \Lambda_j = \emptyset$ if $i \neq j$ and $\bigcup_{i=1}^{s} \Lambda_i = [0,1] \times [0,1]$;

(iii) f is continuous in $\Lambda_i \backslash \partial\Lambda_i$, for $i = 1, 2, \ldots, s$, where $\partial\Lambda_i$ denotes the boundary point set of Λ_i; and

(iv) there exist at most finite points $\{(x_j^*, y_j^*), j = 1, 2, \ldots, s^*\}$ on $[\bigcup_{i=1}^{s} \partial\Lambda_i] \bigcap [(0,1) \times (0,1)]$ such that for each point (x_j^*, y_j^*), $j = 1, 2, \ldots, s^*$, there are $\Lambda_{j1}^*, \Lambda_{j2}^* \in \{\Lambda_i, i = 1, 2, \ldots, s\}$ satisfying (1) $(x_j^*, y_j^*) \in \partial\Lambda_{j1}^* \bigcap \partial\Lambda_{j2}^*$ and (2) $\lim_{(x,y)\to(x_j^*,y_j^*);(x,y)\in\Lambda_{j1}^*} f(x,y) = \lim_{(x,y)\to(x_j^*,y_j^*);(x,y)\in\Lambda_{j2}^*} f(x,y)$.

The above condition (iii) says that f is continuous in the interior of each region Λ_i. By condition (iv), on the boundary curves of $\{\Lambda_i, i = 1, 2, \ldots, s\}$, there exist at most finite points at which the jump magnitudes of f are zero. That is, under these conditions, the regression function f is continuous in the entire design space except on the curves defined by $D = [\bigcup_{i=1}^{s} \partial\Lambda_i] \bigcap [(0,1) \times (0,1)]$. These curves are called the *jump location curves* (JLCs) of f. The open regions $\{\Lambda_i \backslash \partial\Lambda_i, i = 1, 2, \ldots, s\}$, in which f is continuous, are sometimes called the *continuity regions* of f. The major goal of the procedures introduced in this chapter is to detect these JLCs from noisy data.

If we compare the one-dimensional (1-D) jump regression model (cf. equations (1.1) and (3.1)) with the two-dimensional (2-D) jump regression model (cf. equations (1.3) and (4.1)), there are several major differences between them, some of which are discussed below. First, in 1-D cases, the jump locations are at most a series of separated points on the design interval. In 2-D cases, however, the jump locations are some curves, and these curves do not have any mathematical expressions in most situations. Second, in 1-D cases, two different sides of a given point are well defined, and we can use the difference between two one-sided averages of the observations from the two one-sided neighborhoods for detecting a possible jump at that point, as described in the previous chapter. When this idea is generalized to 2-D cases, we can still construct two one-sided averages of the observations on two different sides of a given point, along a specific direction. But the difference between the two one-sided averages changes with that direction. Thus, it is important to choose this direction properly. Otherwise, the difference between the two one-sided averages could be small even if the point in question is on a JLC. Third, the design points in 1-D cases have a natural order among them, i.e., a design point x_i is on the left side of a design point x_j if $x_i < x_j$. In the 2-D model (4.1), this kind of ordering among the design points is hard to define unless the design points are arranged by rows and columns, as described by the equation (1.3). Fourth, in many 2-D applications, there is no pre-specified coordinate system available. Therefore, it is reasonable to require the 2-D procedures for handling such applications to be coordinate free. As a comparison, there is no such problem in 1-D cases.

Theoretically speaking, the jump detection problem exists in any p-dimensional case, where $p \geq 1$ is an integer. Because most applications in real-life involve regression curves and surfaces only, the current research in the statistical literature focuses mainly on 1-D and 2-D cases. Specifically, one driving force of the research on 2-D jump regression analysis is the connection between a 2-D regression model and an image model, as discussed in Sections 1.2 and 1.3.

This chapter has five sections. In Section 4.2, we introduce various jump detection procedures under the assumption that there is only one JLC in the design space, or that there are several JLCs but the number of JLCs is known. In Section 4.3, the idea to treat the JLCs as a point set in the design space is introduced. By this idea, curves with global nature are treated as a collection of points, which can be handled locally, in the design space. This idea makes it possible to detect arbitrary JLCs by a procedure with relatively simple computation. As a demonstration, one such procedure based on local linear kernel smoothing is discussed. The detected jumps by this procedure are some set of points in the design space, among which there are two types of deceptive jump points. One type consists of points that are located around the true JLCs, and the other type consists of points that are scattered in the entire design space. To delete these deceptive jump points, two jump modification procedures are also discussed in this section. In Section 4.4, we introduce a jump detection procedure which can detect arbitrary JLCs by searching two or more given directions at any given point. This procedure can be regarded as a generalization of the Sobel edge detector, which is popular in the image processing literature. Performance measurement of jump detection procedures is also discussed in this section, which is important for comparing different jump detection procedures properly. Finally, in Section 4.5, we make some remarks about the jump detection procedures introduced in this chapter and note some open research problems on jump detection in regression surfaces.

4.2 JUMP DETECTION WHEN THE NUMBER OF JUMP LOCATION CURVES IS KNOWN

In the statistical literature, various procedures have been suggested for detecting jumps in regression surfaces under the assumptions that there is only one JLC in the design space and that the JLC satisfies some smoothness conditions. Some of these procedures can be generalized, without much difficulty, to handle the case when there are multiple JLCs in the design space but the number of JLCs is known. We introduce several of these procedures in this section.

4.2.1 Jump detection by RDKE

In 1-D cases, each of the DKE and DLK procedures introduced in Subsections 3.2.1 and 3.2.2 uses a difference between two one-sided averages, calculated from a right-sided neighborhood and a left-sided neighborhood of a given point, as a criterion for jump detection. To generalize this idea to 2-D cases, at a given point we consider its two one-sided neighborhoods along a specific direction. The absolute difference between the two one-sided averages of the observations located in these two one-sided neighborhoods can then be calculated. Obviously, this absolute difference depends on the direction along which the two one-sided neighborhoods are defined. Intuitively, the maximized value of the absolute difference with respect to all possible directions can be used as a criterion for detecting a possible jump at the given point, because the value of this criterion should be small if no jumps exist in the related neighborhoods and relatively large if the given point is on a JLC. Qiu (1997) suggested such a jump detection procedure, which is discussed below.

Let us first discuss a simple case, where n observations $\{(x_i, y_i, Z_i), i = 1, 2, \ldots, n\}$ are generated from the 2-D regression model (4.1), and the regression function f has the following expression

$$f(x, y) = g(x, y) + C(x)I_{y>\phi(x)}, \text{ for } (x, y) \in [0, 1] \times [0, 1];$$

here, g is a continuous 2-D function in the design space $[0, 1] \times [0, 1]$, ϕ and C are two continuous 1-D functions defined on the interval $[0, 1]$, $0 < \phi(x) < 1$, and $C(x) \neq 0$ for any $x \in [0, 1]$. Obviously, the regression function f above has a single JLC in the design space which can be expressed by $\{(x, \phi(x)), x \in [0, 1]\}$. At a given $x \in [0, 1]$, the corresponding jump magnitude of f is $C(x)$.

We begin by defining K_1^* and K_2^* to be two 2-D kernel functions satisfying

(i) $K_1^*(x, y) = 0$ when $(x, y) \notin [-1/2, 1/2] \times [0, 1]$; $K_2^*(x, y) = 0$ when $(x, y) \notin [-1/2, 1/2] \times [-1, 0)$;

(ii) $K_j^*(x, y) \geq 0$, for $(x, y) \in [-1/2, 1/2] \times [-1, 1]$ and $j = 1, 2$; and

(iii) $\int_{-1/2}^{1/2} \int_{-1}^{1} K_j^*(u, v)\, dudv = 1$ for $j = 1, 2$.

The above conditions (ii) and (iii) require K_1^* and K_2^* to be two density kernel functions. Condition (i) requires the two kernel functions to have one-sided supports: K_1^* has an upper-sided support and K_2^* has a lower-sided support.

Based on K_1^* and K_2^*, we next define two *rotational kernel functions* as follows:

$$\begin{aligned} K_j(\theta, x, y) \quad = \quad & K_j^* \Big[\delta(x)\sqrt{x^2 + y^2} \cos(\arctan(y/x) - \theta), \\ & \delta(x)\sqrt{x^2 + y^2} \sin(\arctan(y/x) - \theta)\Big], \end{aligned}$$

where $j = 1, 2$, $\theta \in [-\pi/2, \pi/2]$ is a rotation parameter, $(x, y) \in R^2$, and $\delta(x) = 1$ or -1 when $x \geq 0$ or $x < 0$. The rotational kernel functions $K_j(\theta, \cdot, \cdot)$ for $j = 1, 2$ are obtained by rotating the supports of the two one-sided kernel functions K_j^*, by an angle θ counterclockwise about the origin. Namely,

$$K_j(\theta, x, y) = K_j^*(x', y'), \text{ for } j = 1, 2,$$

where the point (x, y) is obtained by rotating the point $(x', y') \in R^2$ an angle θ counterclockwise about the origin. The supports of K_j^* and $K_j(\theta, \cdot, \cdot)$ for $j = 1, 2$ are displayed in Figures 4.1(a) and 4.1(b), respectively.

Let $K_j(\theta, x, y; h_n, p_n) = K_j^*(x'/h_n, y'/p_n)$; then, we define the criterion

$$\begin{aligned} M_{RDKE}(\theta, x, y) \quad = \quad & \frac{1}{nh_np_n} \sum_{i=1}^{n} Z_i \left[K_2(\theta, x_i - x, y_i - y; h_n, p_n) - \right. \\ & \left. K_1(\theta, x_i - x, y_i - y; h_n, p_n)\right], \end{aligned} \quad (4.2)$$

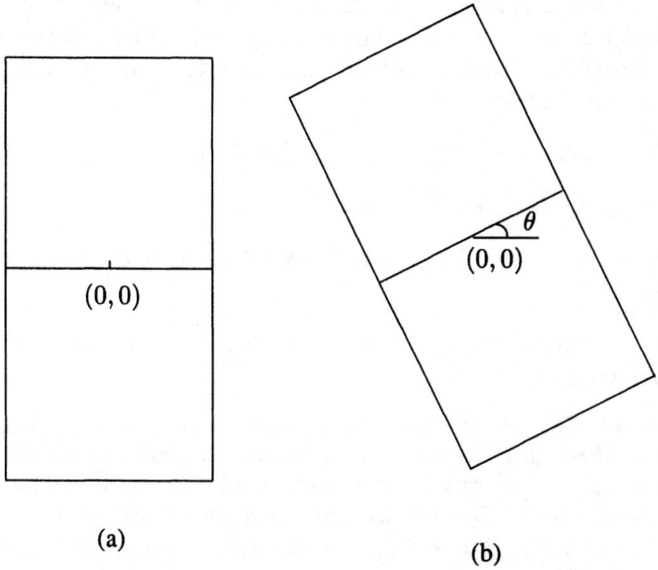

Fig. 4.1 (a): Upper- and lower-sided supports of the two kernel functions K_1^* and K_2^*. (b): Two one-sided supports of the rotational kernel functions $K_1(\theta, \cdot, \cdot)$ and $K_2(\theta, \cdot, \cdot)$, obtained by rotating the supports displayed in plot (a) by an angle θ counterclockwise about the origin.

where $(x, y) \in [b_n, 1 - b_n] \times [b_n, 1 - b_n]$, $b_n = \sqrt{h_n^2/4 + p_n^2}$, and h_n and p_n are two bandwidths. Obviously, $M_{RDKE}(\theta, x, y)$ is a difference of two one-sided kernel estimators of $f(x, y)$ constructed from the observations in the two one-sided neighborhoods of (x, y), along the direction $(\cos\theta, \sin\theta)$. If no jumps exist in the related one-sided neighborhoods of (x, y), then the value of $|M_{RDKE}(\theta, x, y)|$ should be small no matter what the value of θ is, because all one-sided kernel estimators estimate $f(x, y)$ well in this case. On the other hand, if the point (x, y) is on the JLC, namely, the point (x, y) satisfies the equation $y = \phi(x)$, then the value of $|M_{RDKE}(\theta, x, y)|$ should be close to the jump magnitude $|C(x)|$ when the JLC has a unique tangent line at (x, y) with direction $(\cos\theta, \sin\theta)$.

Based on these observations, Qiu defined estimators of $\phi(x)$ and $C(x)$ for $x \in [b_n, 1 - b_n]$ as follows. Let

$$\theta^*(x, y) = \arg \max_{-\pi/2 \leq \theta \leq \pi/2} |M_{RDKE}(\theta, x, y)|,$$

$$\widehat{\phi}(x) = \arg \max_{b_n \leq y \leq 1 - b_n} |M_{RDKE}(\theta^*(x, y), x, y)|.$$

Then $\widehat{\phi}(x)$ and $M_{RDKE}(\theta^*(x, \widehat{\phi}(x)), x, \widehat{\phi}(x))$ are defined as the estimators of $\phi(x)$ and $C(x)$, respectively. Qiu called these estimators the *rotational difference kernel estimators* (RDKEs), and he called the corresponding jump detection procedure the rotational difference kernel estimation procedure, which is identified by the subscript of M_{RDKE}.

To study the statistical properties of the RDKEs, the following assumption (A 4.1) about the design points is needed, so that the design points have some homogeneity.
(A 4.1) Suppose there exists a partition $\{\Delta_i, i = 1, 2, \ldots, n\}$ of the design space $[0,1] \times [0,1]$ such that:

(1) $\Delta_i \bigcap \Delta_j = \emptyset$, if $i \neq j$; $\bigcup_{i=1}^{n} \Delta_i = [0,1] \times [0,1]$;

(2) $(x_i, y_i) \in \Delta_i$, for $i = 1, 2, \cdots, n$;

(3) $\max_{1 \leq i \leq n} d_i = O(n^{-1/2})$, where d_i is the diameter of Δ_i for $i = 1, 2, \cdots, n$; and

(4) $\max_{1 \leq i \leq n} |S(\Delta_i) - 1/n| = O(n^{-1-\lambda})$, where $\lambda > 0$ is a constant and $S(\Delta_i)$ denotes the area of Δ_i.

Theorem 4.1 (Qiu 1997) *Let ν be a positive number and $\{\beta_n\}$ be a series of positive numbers satisfying $\lim_{n\to\infty} \beta_n = \infty$. Suppose that the design points $\{(x_i, y_i), i = 1, 2, \ldots, n\}$ satisfy the assumption (A 4.1); the kernel functions K_j^* satisfy the conditions (i)-(iii) mentioned above, and (iv) K_j^* are Lipschitz (1) continuous for $j = 1$ and 2; $E|\varepsilon_1|^p < \infty$ for $p \geq 2$; the continuity part g of f is Lipschitz (1) continuous; the JLC function ϕ has continuous second-order derivative on $[0,1]$, and also satisfies the condition that $0 < \phi(x) < 1$ for any $x \in [0,1]$; the jump magnitude function C has a positive lower-bound; and the positive bandwidths h_n and p_n satisfy the conditions that*

(1) $\frac{n^{\nu}}{\beta_n \log n}\left[h_n + p_n + \frac{1}{n^{\lambda} h_n p_n} + \frac{\sqrt{h_n^2 + p_n^2}}{\sqrt{n h_n^2 p_n^2}}\right] = o(1)$;

(2) $\frac{n^{2\nu}}{n h_n p_n \beta_n} = O(1)$; *and*

(3) $\frac{n^{\nu + (1/p) - 1}}{h_n p_n \beta_n \log n} = o(1)$.

Then, for any $0 < a \leq b < 1$, we have

(i) $\lim_{n\to\infty} \max_{a \leq x \leq b} \left|\widehat{\phi}(x) - \phi(x)\right| = 0,\ a.s.$

(ii) $\lim_{n\to\infty} \max_{a \leq x \leq b} \left|M_{RDKE}(\theta^*(x, \widehat{\phi}(x)), x, \widehat{\phi}(x)) - C(x)\right| = 0,\ a.s.$

(iii) $\lim_{n\to\infty} \max_{a \leq x \leq b} \left|\theta^*(x, \widehat{\phi}(x)) - \arctan(\phi'(x))\right| = 0,\ a.s.$

The RDKE procedure can be generalized to handle the case with multiple JLCs. Suppose that the regression function f has the expression

$$f(x, y) = g(x, y) + \sum_{j=1}^{p} C_j(x) I_{y > \phi_j(x)}, \text{ for } (x, y) \in [0,1] \times [0,1],$$

where g is a continuous function in the entire design space, p is the number of JLCs, $0 < \phi_1(x) < \phi_2(x) < \cdots < \phi_p(x) < 1$ for $x \in [0,1]$ are p univariate functions of the JLCs, and $\{C_j(x) \neq 0, \text{ for all } x \in [0,1], j = 1, 2, \cdots, p\}$ are the corresponding jump magnitude functions. For simplicity of presentation, we assume that $p = 2$. In such a case, we define

$$\widehat{\phi}_1(x) = \arg \max_{b_n \leq y \leq 1-b_n} |M_{RDKE}(\theta^*(x,y), x, y)|$$

and

$$\widehat{\phi}_2(x) = \arg \max_{b_n \leq y \leq 1-b_n, |\widehat{\phi}_1(x)-y|>b_n} |M_{RDKE}(\theta^*(x,y), x, y)|.$$

Let $\widehat{\phi}_{(1)}(x) \leq \widehat{\phi}_{(2)}(x)$, for $x \in [b_n, 1-b_n]$, denote the order statistics of $\widehat{\phi}_1(x)$ and $\widehat{\phi}_2(x)$. Then the RDKEs of $\phi_1(x), \phi_2(x), C_1(x)$ and $C_2(x)$ are defined by $\widehat{\phi}_{(1)}(x)$, $\widehat{\phi}_{(2)}(x)$, $M_{RDKE}(\theta^*(x, \widehat{\phi}_{(1)}(x)), x, \widehat{\phi}_{(1)}(x))$ and $M_{RDKE}(\theta^*(x, \widehat{\phi}_{(2)}(x)), x, \widehat{\phi}_{(2)}(x))$, respectively. Qiu showed that these estimators have similar properties to those stated in Theorem 4.1 under some regularity conditions.

In the case when there is a single JLC, Γ, in the design space and the absolute jump magnitudes at Γ have a lower bound, Müller and Song (1994a) suggested a similar jump detection criterion to M_{RDKE}. At a given point (x, y) in the design space, their jump detection criterion $\Delta(\theta, x, y)$ is defined by a difference of two averages of the observations located in two one-sided square-shaped neighborhoods of (x, y). The two neighborhoods are separated along the direction θ and are a certain distance apart. To estimate the true JLC Γ, they assume that there is a sufficiently rich class of candidate JLCs, Ξ, which includes Γ as a member. Similar to jump detection by the RDKE procedure, the value of $\sup_{\theta \in [-\pi/2, \pi/2]} |\Delta(\theta, x, y)|$ should be small if the point (x, y) is a certain distance away from the true JLC, and relatively large if (x, y) is on the true JLC. So for a candidate JLC $\Gamma' \in \Xi$, if the value of $\inf_{(x,y) \in \Gamma'} \sup_{\theta \in [-\pi/2, \pi/2]} |\Delta(\theta, x, y)|$ is large, then the likelihood that Γ equals Γ' is high. Therefore, Müller and Song suggested searching all the candidate JLCs in Ξ; their estimator $\widehat{\Gamma}$ of Γ is then defined by the maximizer of

$$\sup_{\Gamma' \in \Xi} \left[\inf_{(x,y) \in \Gamma'} \left(\sup_{\theta \in [-\pi/2, \pi/2]} |\Delta(\theta, x, y)| \right) \right]. \tag{4.3}$$

This estimator is proved to be almost surely consistent under some regularity conditions.

In applications, however, the sufficiently rich class of candidate JLCs Ξ is difficult to define, and the infimums/supremums involved in expression (4.3) are difficult to compute. To overcome these difficulties, Müller and Song (1994b) suggested a cube splitting algorithm for estimating the single JLC Γ when all design points are equally spaced in the design space. This algorithm is based on the idea that a coarse approximation to the edge can be first obtained from big blocks of aggregated design points. Then, the approximation is improved iteratively as the blocks are split into smaller and smaller sub-blocks.

Carlstein and Krishnamoorthy (1992) discussed the so-called *boundary estimation* problem in which it is assumed that (1) there is a single boundary curve in the design space, (2) the boundary curve partitions the design space into two connected regions, (3) observations in each of the two regions have a same distribution, and (4) the two distributions in the two regions are different. Carlstein and Krishnamoorthy suggest a methodology for estimating the boundary curve, by searching a prespecified rich class of candidate boundary curves, based on some norm measuring the difference between two empirical distributions constructed in the two regions. This problem is different from the jump detection problem in regression surfaces in that the former is for detecting the boundary curve which separates the design space into two regions with different observation distributions, while the latter is for detecting jumps in the mean function of the observations. However, when the means of the two distributions in the boundary estimation setup are assumed to be different, the mean function of the observations has jumps around the boundary curve, and thus the boundary curve can be regarded as a JLC.

4.2.2 Minimax edge detection

Korostelev and Tsybakov (1993) suggested a minimax estimator of the JLCs in several different cases. In the first case considered, the error terms in model (4.1) are i.i.d. with a normal distribution, and the regression function f has the following expression:

$$f(x, y) = I(y \leq g(x)), \text{ for } (x, y) \in [0, 1] \times [0, 1]; \tag{4.4}$$

g is a continuous univariate function satisfying

(1) $h \leq g(x) \leq 1 - h$ for all $x \in [0, 1]$, where h is a small positive constant; and

(2) for any $x, x' \in [0, 1]$, $|g^{(k)}(x) - g^{(k)}(x')| \leq L|x - x'|^{\beta - k}$ where β and L are two positive constants, $k = [\beta]$, and $[\beta]$ denotes the integer part of β.

Condition (2) says that the kth-order derivative of g is Lipschitz $(\beta - k)$ continuous. The collection of all functions satisfying the two conditions is denoted by $\mathcal{G}_0(\beta, L, h)$. Obviously, the regression function f in equation (4.4) is binary, and it has a single JLC which has an expression $y = g(x)$ for $x \in [0, 1]$.

To estimate the JLC, Korostelev and Tsybakov suggested first dividing the design space $[0, 1] \times [0, 1]$ into M slices along the x direction, where $M = [1/\delta_n]$ and $\delta_n = (\log(n)/n)^{1/(\beta+1)}$, and then estimating the JLC in each slice. Specifically,

$$[0, 1] \times [0, 1] = \bigcup_{j=1}^{M} U_j,$$

where $U_1 = [0, \delta_n) \times [0, 1], U_2 = [\delta_n, 2\delta_n) \times [0, 1], \ldots, U_M = [(M-1)\delta_n, 1] \times [0, 1]$. In slice U_j for $1 \leq j \leq M$, the JLC is approximated by the following polynomial function of order k:

$$g_j(x) = \theta_0 + \theta_1(x - (j-1)\delta_n) + \ldots + \theta_k(x - (j-1)\delta_n)^k,$$

where the polynomial coefficients $\{\theta_\ell, \ell = 0, 1, \ldots, k\}$ are estimated by their maximum likelihood estimators $\{\widehat{\theta}_\ell, \ell = 0, 1, \ldots, k\}$. Namely, $\{\widehat{\theta}_\ell, \ell = 0, 1, \ldots, k\}$ are solutions to $\{\theta_\ell, \ell = 0, 1, \ldots, k\}$ of the following maximization problem:

$$\max_{\theta_\ell, \ell=0,1,\ldots,k} \sum l_j(\theta_0, \theta_1, \ldots, \theta_k),$$

where l_j is the log-likelihood function constructed from the observations in U_j. It can be checked that l_j is proportional to

$$\sum_{x_i \in [(j-1)\delta_n, j\delta_n)} I\left(y_i \leq g_j(x_i)\right)(Z_i - 1/2).$$

Then the univariate function g of the JLC can be estimated by the following piecewise polynomial function

$$\begin{aligned} \widehat{g}(x) = \widehat{g}_j(x) &= \widehat{\theta}_0 + \widehat{\theta}_1(x - (j-1)\delta_n) + \ldots + \widehat{\theta}_k(x - (j-1)\delta_n)^k, \\ &\quad \text{for } x \in [(j-1)\delta_n, j\delta_n) \text{ and } j = 1, 2, \ldots, M. \end{aligned} \tag{4.5}$$

Korostelev and Tsybakov proved that, if the design points in the model (4.1) are uniformly distributed in the design space, $\beta \geq 1$, and some other regularity conditions are satisfied, then

$$\lim_{C \to \infty} \lim_{n \to \infty} \sup_{g \in \mathcal{G}_0(\beta, L, h)} P_g\left(\psi_n^{-1} \sup_{0 \leq x \leq 1} |\widehat{g}(x) - g(x)| \geq C\right) = 0; \tag{4.6}$$

here, $\psi_n = (\log(n)/n)^{\beta/(\beta+1)}$ and P_g is the probability measure under the assumption that $\{(x, g(x)), x \in [0, 1]\}$ is the true JLC. This result provides uniform consistency in probability for the JLC estimator $\widehat{g}$.

The second case considered by Korostelev and Tsybakov is more general, where f has the following expression: for $(x, y) \in [0, 1] \times [0, 1]$,

$$f(x, y) = f_1(x, y) I(y \leq g(x)) + f_2(x, y) I(y > g(x)). \tag{4.7}$$

The two continuous functions f_1 and f_2 satisfy the condition that:

$$\sup_{\{(x,y): y \leq g(x)\}} f_1(x, y) \leq a < b \leq \inf_{\{(x,y): y > g(x)\}} f_2(x, y),$$

and a and b are two constants. Korostelev and Tsybakov pointed out that case (4.7) could be easily converted to case (4.4) by the transformation

$$Z_i' = I\left(Z_i > \frac{a+b}{2}\right), \quad i = 1, 2, \ldots, n$$

on the observations, if a and b are known. Then the estimator of the JLC could be obtained similarly to that in equation (4.5) from the new observations $\{Z_i', i = 1, 2, \ldots, n\}$. Korostelev and Tsybakov show that the JLC estimator in such a case has the same large-sample property as stated in expression (4.6). If the constants a and b in the above transformation are unknown, then the constant $(a+b)/2$ should be replaced by its consistent estimator when defining the new observations.

Korostelev and Tsybakov made several other generalizations. One such generalization handles the case when the JLC is an arbitrary smooth curve which divides the whole design space into two connected regions. An estimator of the JLC is suggested in such a case and has a similar property to that stated in expression (4.6).

When the regression function f has expression (4.4), Rudemo and Stryhn (1994) suggested estimating the function g by a step function, where the coefficients are estimated by their MLEs. They demonstrated that this approach can also be used for estimating star-shaped JLC.

4.2.3 Jump estimation based on a contrast statistic

O'Sullivan and Qian (1994) considered the jump detection problem when it is assumed that there is a single JLC in the design space $\Omega = [0,1] \times [0,1]$ and that the JLC has the expression

$$\Gamma_\theta = \{\theta(s) = (\theta_1(s), \theta_2(s)) \in \Omega : s \in [0,1]\}.$$

The JLC is assumed to be a smooth, closed, and simple curve in the design space in the sense that

(1) both θ_1 and θ_2 have first-order derivatives on $[0,1]$, $\int_0^1 (\theta_1'(u))^2\, du < \infty$, and $\int_0^1 (\theta_2'(u))^2\, du < \infty$;

(2) $\theta(0) = \theta(1)$; and

(3) Γ_θ has no crossings (i.e., for any $s_1, s_2 \in (0,1]$, if $s_1 \neq s_2$, then $\theta(s_1) \neq \theta(s_2)$).

At a given point $\theta(s)$ on the JLC, the outward normal direction is defined by

$$\eta_\theta(s) = \left(\frac{\theta_2'(s)}{\sqrt{(\theta_1'(s))^2 + (\theta_2'(s))^2}}, -\frac{\theta_1'(s)}{\sqrt{(\theta_1'(s))^2 + (\theta_2'(s))^2}} \right).$$

A point (x,y) in the design space is in the inside region Γ_θ^I of Γ_θ if

$$\left[(x,y) - (\theta_1(s_{(x,y)}), \theta_2(s_{(x,y)}))\right] (\eta_\theta(s_{(x,y)}))' \leq 0,$$

where $\theta(s_{(x,y)})$ denotes the closest point on Γ_θ to (x,y), and $(\eta_\theta(s_{(x,y)}))'$ is the transpose of $\eta_\theta(s_{(x,y)})$. Otherwise, the point (x,y) is in the outside region Γ_θ^O of Γ_θ.

A jump detection criterion is now defined based on the following definitions: for the true regression function f, let

$$I(f) = \int_{\Gamma_\theta^I} f(u)\, du, \ I(1) = \int_{\Gamma_\theta^I} 1\, du;$$

$$\overline{I}(f) = \int_{\Gamma_\theta^O} f(u)\, du, \ \overline{I}(1) = \int_{\Gamma_\theta^O} 1\, du.$$

Then $I(f)/I(1)$ is the average value of f per unit area inside the JLC, and $\overline{I}(f)/\overline{I}(1)$ is the average value of f per unit area outside the JLC. The *contrast* of the JLC is defined by

$$C(\theta) = \left| \frac{I(f)}{I(1)} - \frac{\overline{I}(f)}{\overline{I}(1)} \right|.$$

O'Sullivan and Qian (1994) assumed that the underlying true JLC is the smooth, closed, and simple curve θ_O which minimizes the negative contrast $L(\theta) = -C(\theta)$ among all possible smooth, closed, and simple curves. Namely, θ_O is the minimizer of

$$\min_{\theta \in \Theta} L(\theta),$$

where Θ is the collection of all smooth, closed and simple curves in Ω.

For n observations $\{(x_i, y_i, Z_i), i = 1, 2, \ldots, n\}$ from the regression model (4.1), a natural estimator of $L(\theta)$ is

$$L_n(\theta) = -\left| \frac{I_n(Z)}{I_n(1)} - \frac{\overline{I}_n(Z)}{\overline{I}_n(1)} \right|,$$

where

$$I_n(Z) = \sum_{(x_i,y_i) \in \Gamma_\theta^I} Z_i, \; I_n(1) = \sum_{(x_i,y_i) \in \Gamma_\theta^I} 1;$$

$$\overline{I}_n(Z) = \sum_{(x_i,y_i) \in \Gamma_\theta^O} Z_i, \; \overline{I}_n(1) = \sum_{(x_i,y_i) \in \Gamma_\theta^O} 1.$$

Let τ be an initial estimator of the true JLC and $\delta + \tau$ be an updated estimator of the true JLC. To guarantee the smoothness of the updated estimator, O'Sullivan and Qian used a similar idea to smoothing spline estimation (cf. the expression (2.18)). They suggest choosing $\delta = (\delta_1, \delta_2)$ such that the following objective function reaches the minimum over a function space Δ:

$$L_{n,\lambda}(\delta) = L_n(\delta + \tau) + \lambda J(\delta), \tag{4.8}$$

where

$$J(\delta) = \frac{1}{2} \int_0^1 [(\delta_1''(s))^2 + (\delta_2''(s))^2] \, ds$$

measures the smoothness of δ, $L_n(\delta + \tau)$ measures the likelihood that the true JLC θ_O is $\delta + \tau$, and $\lambda > 0$ is a smoothing parameter. The function space Δ is defined by

$$\Delta = \{\delta = (\delta_1, \delta_2) : \delta_1, \delta_2 \in W_0^2([0,1])\},$$

where

$$\begin{aligned} W_0^2([0,1]) \;\; = \;\; & \{g : g \text{ and } g' \text{ are absolutely continuous on } [0,1], g(0) = g(1), \\ & g'(0) = g'(1), \text{ and } \int_0^1 (g''(u))^2 du < \infty \}. \end{aligned}$$

To approximate the minimizer of function (4.8) numerically, O'Sullivan and Qian assumed that δ_1 and δ_2 can be expressed by the following linear combinations of M basis functions:

$$\delta_1(s) = \sum_{j=1}^{M} \beta_{1j}\phi_j(s), \quad \delta_2(s) = \sum_{j=1}^{M} \beta_{2j}\phi_j(s), \tag{4.9}$$

where $\beta_1 = (\beta_{11}, \beta_{12}, \ldots, \beta_{1M})'$ and $\beta_2 = (\beta_{21}, \beta_{22}, \ldots, \beta_{2M})'$ are coefficient vectors, and the M basis functions are defined by:

$$\phi_j(s) = I\left(s \in [(j-1)/M, j/M)\right), \; j = 1, 2, \ldots, M.$$

Then the objective function in (4.8) can be expressed in terms of $\beta = (\beta_1', \beta_2')'$ as follows:

$$L_{n,\lambda}(\beta) = L_n^*(\beta) + \frac{\lambda}{2}\left\{\beta_1' \sum \beta_1 + \beta_2' \sum \beta_2\right\},$$

where $\sum$ is a $M \times M$ matrix with its (j,k)-th element $\sum_{jk} = \int_0^1 \phi_j''(s)\phi_k''(s)ds$ for $j, k = 1, 2, \ldots, M$, and $L_n^*(\beta)$ denotes $L_n(\delta + \tau)$.

We still run into a problem with minimizing function (4.8), since the basis functions defined above are obviously not differentiable. Consequently, $\delta = (\delta_1, \delta_2)$ as defined by equation (4.9) is not differentiable either, so its derivatives must be estimated. Suppose that $\{s_j = j/M, j = 0, 1, \ldots, M\}$ are $M + 1$ equally spaced points on $[0, 1]$. Then in applications, the first two derivatives of δ at s_j can be approximated by

$$\frac{\delta(s_{j+1}) - \delta(s_{j-1})}{2h}$$

and

$$\frac{\delta(s_{j+1}) - 2\delta(s_j) + \delta(s_{j-1})}{h^2}$$

respectively, where $h = 1/M$. After using these approximations to propose some numerical algorithms for computing the first and second derivatives of $L_{n,\lambda}(\beta)$ with respect to β, O'Sullivan and Qian suggested choosing β by an algorithm based on Newton-Raphson iteration (cf. Cheney and Kincaid 1994, Section 3.2). Then the true JLC can be estimated by $\delta + \tau$, where δ is determined by this algorithm.

4.2.4 Algorithms for tracking the JLCs

Recently, Peter Hall and co-authors proposed several algorithms for tracking the JLCs in the design space (see Hall and Rau (2000, 2002), Hall *et al.* (2001)). Related discussion can also be found in Hall and Raimondo (1997, 1998). In this subsection, we introduce one such tracking algorithm suggested by Hall *et al.* (2001).

Suppose that $\{(x_i, y_i, Z_i), i = 1, 2, \ldots, n\}$ are n observations generated from the model (4.1), and the regression function f has the expression:

$$f(x, y) = g_1(x, y) + g_2(x, y)I(y \le \gamma(x)), \text{ for } (x, y) \in [0, 1] \times [0, 1],$$

where g_1, g_2, and γ are all continuous functions, and g_2 is also bounded away from zero. Then f has a single JLC with the mathematical expression $y = \gamma(x)$ for $x \in [0, 1]$.

At a given point (x, y) in the design space, Hall *et al.* (2001) considered two competing hypotheses: the null hypothesis H_0 that f is locally constant around (x, y); and the alternative hypothesis H_1 that the point (x, y) is on the JLC, the JLC is locally linear around (x, y) with its normal in the direction of the unit vector θ, and f is locally constant on either side of the JLC. When H_0 is true and the error terms $\{\varepsilon_i, i = 1, 2, \ldots, n\}$ in (4.1) are i.i.d. with normal distribution, the kernel-weighted likelihood function around the point (x, y) is defined by

$$L(\mu, \sigma; x, y) = \Pi_{i=1}^{n} \left[\frac{1}{\sqrt{2\pi}\sigma} \exp\left(-\frac{(Z_i - \mu)^2}{2\sigma^2} \right) \right]^{K_i(x,y)}, \tag{4.10}$$

where μ denotes the value of f at (x, y), $K_i(x, y) = K[(x_i - x)/h, (y_i - y)/h]$, K is a smooth, radially symmetric bivariate density kernel function whose support is the unit disc, and h is a bandwidth. Since the support of the kernel function K is the unit disc, only those observations whose design points are h or less Euclidean distance away from the given point (x, y) are actually included in the kernel-weighted likelihood function $L(\mu, \sigma; x, y)$. Similar to the definition of the MLE discussed in Subsection 2.2.4, the kernel-weighted MLEs of μ and σ are defined by the maximizers of $L(\mu, \sigma; x, y)$ with respect to μ and σ. It can be checked that the kernel-weighted MLEs of μ and σ are actually

$$\overline{Z}(x, y) = \frac{1}{n(x, y)} \sum_{i=1}^{n} Z_i K_i(x, y)$$

and

$$\widehat{\sigma}(x, y) = \left(\frac{1}{n(x, y)} \sum_{i=1}^{n} \left(Z_i - \overline{Z}(x, y)\right)^2 K_i(x, y) \right)^{1/2},$$

respectively, where $n(x, y) = \sum_{i=1}^{n} K_i(x, y)$.

Under the alternative hypothesis H_1, the point (x, y) is assumed to be on the JLC, which is locally linear with a normal direction θ at (x, y). The JLC and its normal direction θ at (x, y) divide all design points into two groups:

$$\mathcal{I}_+(x, y, \theta) = \{(x_i, y_i) : (x_i - x, y_i - y) \cdot \theta \geq 0, i = 1, 2, \ldots, n\}$$

and

$$\mathcal{I}_-(x, y, \theta) = \{(x_i, y_i) : (x_i - x, y_i - y) \cdot \theta < 0, i = 1, 2, \ldots, n\}.$$

Then the kernel-weighted likelihood function under H_1 is defined by

$$\begin{aligned} & L(\mu_+, \mu_-, \sigma; x, y) \\ = \; & \Pi_{(x_i, y_i) \in \mathcal{I}_+(x,y,\theta)} \left[\frac{1}{\sqrt{2\pi}\sigma} \exp\left(-\frac{(Z_i - \mu_+)^2}{2\sigma^2} \right) \right]^{K_i(x,y)} \times \\ & \Pi_{(x_i, y_i) \in \mathcal{I}_-(x,y,\theta)} \left[\frac{1}{\sqrt{2\pi}\sigma} \exp\left(-\frac{(Z_i - \mu_-)^2}{2\sigma^2} \right) \right]^{K_i(x,y)}, \end{aligned} \tag{4.11}$$

where μ_+ and μ_- denote the values of f on the two different sides of the JLC around (x, y). It can be checked that the kernel-weighted MLEs of μ_+, μ_- and σ are

$$\overline{Z}_+(x,y,\theta) = \frac{1}{n_+(x,y,\theta)} \sum_{(x_i,y_i)\in\mathcal{I}_+(x,y,\theta)} Z_i K_i(x,y),$$

$$\overline{Z}_-(x,y,\theta) = \frac{1}{n_-(x,y,\theta)} \sum_{(x_i,y_i)\in\mathcal{I}_-(x,y,\theta)} Z_i K_i(x,y),$$

and $\widehat{\sigma}(x, y)$, respectively, where $n_+(x,y,\theta) = \sum_{(x_i,y_i)\in\mathcal{I}_+(x,y,\theta)} K_i(x,y)$ and $n_-(x,y,\theta) = \sum_{(x_i,y_i)\in\mathcal{I}_-(x,y,\theta)} K_i(x,y)$.

The ratio of the maximum likelihood under H_0 to the maximum likelihood under H_1 is often called the *likelihood ratio statistic*. If H_0 is true, then the value of this statistic should be relatively large; otherwise, its value should be small. Therefore, a *likelihood ratio test* can be conducted: H_0 is rejected if the observed value of the likelihood ratio statistic is small.

In applications, it is often more convenient to consider the logarithm of the likelihood ratio statistic, which we do for equations (4.10) and (4.11). It can be checked that the kernel-weighted local log-likelihood ratio is, up to an additive constant, proportional to $-w(x, y, \theta)$, where

$$\begin{aligned} w(x,y,\theta) &= \frac{1}{n(x,y)} \left[\sum_{i=1}^{n} \left(Z_i - \overline{Z}(x,y,\theta)\right)^2 K_i(x,y) \right. \\ &\quad - \sum_{(x_i,y_i)\in\mathcal{I}_+(x,y,\theta)} \left(Z_i - \overline{Z}_+(x,y,\theta)\right)^2 K_i(x,y) \\ &\quad \left. - \sum_{(x_i,y_i)\in\mathcal{I}_-(x,y,\theta)} \left(Z_i - \overline{Z}_-(x,y,\theta)\right)^2 K_i(x,y) \right] \\ &= \frac{n_+(x,y,\theta)n_-(x,y,\theta)}{n^2(x,y)} \left[\overline{Z}_+(x,y,\theta) - \overline{Z}_-(x,y,\theta)\right]^2. \end{aligned}$$

Thus, when the value of $w(x, y, \theta)$ is large, it is likely that the JLC goes through the point (x, y) with a normal direction θ at (x, y). On the other hand, if the value of $w(x, y, \theta)$ for all possible θ is small, then it is unlikely that the point (x, y) is on the JLC. Based on these properties of $w(x, y, \theta)$, Hall *et al.* (2001) suggested the following tracking algorithm for estimating the JLC.

The first step of the tracking algorithm is to determine an initial approximation $(\widehat{x}^{(0)}, \widehat{y}^{(0)})$ to a point on the JLC; there are two approaches for this step. One approach is to take $(\widehat{x}^{(0)}, \widehat{y}^{(0)})$ to be the maximizer of

$$m(x,y) = \sup_{\theta\in[0,\pi)} w(x,y,\theta)$$

over a reasonably small region through which the JLC is expected to pass. The second approach begins by drawing a line in the design space that traverses a region through which we expect the JLC to pass. Then, all the observations in a neighborhood of this line are projected to this line, obtaining a new data set $\{(x_i', y_i', Z_i'), i = 1, 2, \ldots, n'\}$, where (x_i', y_i') for $i = 1, 2, \ldots, n'$ are on the line and n' is the total number of observations in the neighborhood. From this univariate data, $(\widehat{x}^{(0)}, \widehat{y}^{(0)})$ could be defined by a jump point detected by an appropriate univariate jump detection procedure (cf. Chapter 3 for introduction about some univariate jump detection procedures).

For the next step of the tracking algorithm, suppose we are at the jth estimated point; that is, $(\widehat{x}^{(j)}, \widehat{y}^{(j)})$ is the most recent approximation to a point on the JLC. We wish to update this approximation to $(\widehat{x}^{(j+1)}, \widehat{y}^{(j+1)})$. In general, at a given point (x, y), let $\widehat{\theta}(x, y)$ be the maximizer of $w(x, y, \theta)$ with respect to θ over $[0, \pi)$ and let $\widehat{\theta}_\perp(x, y)$ be a unit vector perpendicular to $\widehat{\theta}(x, y)$. Then $\widehat{\theta}_\perp(x, y)$ is uniquely determined except for its sign. So, at the jth estimated point, we can denote $\widehat{\theta}^{(j)} = \widehat{\theta}_\perp(\widehat{x}^{(j)}, \widehat{y}^{(j)})$ and then define

$$(\widehat{x}^{(j+1)}, \widehat{y}^{(j+1)}) = (\widehat{x}^{(j)}, \widehat{y}^{(j)}) + \delta \widehat{\theta}^{(j)},$$

where δ is a small positive step length and the sign of $\widehat{\theta}^{(j)}$ is determined by tracing the JLC in a specific direction (e.g., the clockwise direction). Alternatively, $(\widehat{x}^{(j+1)}, \widehat{y}^{(j+1)})$ can be defined by the maximizer of $m(x, y)$ on the semicircle

$$\left\{(x, y) : \|(x, y) - (\widehat{x}^{(j)}, \widehat{y}^{(j)})\| = \delta \text{ and } [(x, y) - (\widehat{x}^{(j)}, \widehat{y}^{(j)})]\widehat{\theta}^{(j)} \geq 0\right\}.$$

This process continues until the border of the design space is touched. Then the JLC is traced out in the opposite direction in a similar way, starting from $(\widehat{x}^{(0)}, \widehat{y}^{(0)})$.

Hall *et al.* (2001) suggested that this tracking procedure could be exploited for estimating a single JLC with crossings or multiple JLCs in the design space.

4.2.5 Estimation of JLCs by wavelet transformations

Wang (1998) generalized his 1-D jump and sharp cusp detection approach (cf. Subsection 3.2.5) to 2-D cases for estimating jumps and sharp cusps of regression surfaces. Suppose that the 2-D regression model is (1.3), in which the design points are $\{x_i = i/n_1, y_j = j/n_1, i, j = 1, 2, \ldots, n_1\}$ and n_1 is a power of 2; thus, the sample size of the data is $n = n_1^2$. The regression function f is said to have α cusps along a continuous curve θ in the design space $[0, 1] \times [0, 1]$ if for any point $(x, y) \in \theta$, there exists $\beta_0 \in [0, 2\pi)$ such that for any $\beta \in (\beta_0 - \delta, \beta_0 + \delta)$, when h tends to 0,

$$|f(x + h\cos(\beta), y + h\sin(\beta)) - f(x, y)| \geq c|h|^\alpha,$$

where c and δ are two positive constants that depend on θ and f only. If $\alpha = 0$ in the above expression, then f has jumps at the curve θ.

For two univariate Daubechies wavelets ϕ and ψ, three 2-D wavelets are defined as follows:

$$\begin{aligned}
\Psi^{h,\beta}(x, y) &= \phi(x\cos(\beta) - y\sin(\beta))\psi(x\sin(\beta) + y\cos(\beta)), \\
\Psi^{v,\beta}(x, y) &= \psi(x\cos(\beta) - y\sin(\beta))\phi(x\sin(\beta) + y\cos(\beta)), \\
\Psi^{d,\beta}(x, y) &= \psi(x\cos(\beta) - y\sin(\beta))\psi(x\sin(\beta) + y\cos(\beta)).
\end{aligned}$$

The superscripts h, v and d denote "horizontal", "vertical", and "diagonal" directions, and $\beta \in [0, \pi/4]$ is an orientation-turning parameter. Also, for a scale parameter s, we can define:

$$\Psi_s^{\gamma,\beta}(x, y) = s^{-1}\Psi^{\gamma,\beta}(x/s, y/s)$$

for $\gamma = h, v$ and d. Then the wavelet transformations of f along the three directions (i.e., horizontal, vertical, and diagonal) are

$$T^{\gamma,\beta}f(s, x, y) = \int \Psi_s^{\gamma,\beta}(x - u, y - v)f(u, v)\, dudv$$

for $\gamma = h, v$, and d. Finally, set

$$Tf(s, x, y) = \max_{0 \leq \beta \leq \pi/4} \max_{\gamma = h,v,d} |T^{\gamma,\beta}f(s, x, y)|.$$

It can be checked that if both the univariate wavelets ϕ and ψ have compact supports, then the value of $Tf(s, x, y)$ depends only on the value of f in a neighborhood of (x, y) with size proportional to s^2. If f is differentiable around (x, y), then $Tf(s, x, y)$ is of order s^2. If f has α cusps along θ, then the minimum of $Tf(s, x, y)$ along θ converges to zero at a rate no faster than $s^{\alpha+1}$, as s tends to 0.

In reality, the regression function f is unknown; we only observe a dataset $\{(x_i, y_j, Z_{ij}), i, j = 1, 2, \ldots, n_1\}$ generated from the model (1.3) with equally spaced design points. To detect jumps based on this dataset, two processes $Z_{n_1}(x, y)$ and $W_{n_1}(x, y)$ for $x \in [0, 1] \times [0, 1]$ are defined as follows:

$$\begin{aligned} Z_{n_1}(x, y) &= \frac{1}{n_1^2} \sum_{i=1}^{[n_1 x]} \sum_{j=1}^{[n_1 y]} Z_{ij}, \\ W_{n_1}(x, y) &= \frac{1}{n_1^2} \sum_{i=1}^{[n_1 x]} \sum_{j=1}^{[n_1 y]} \varepsilon_{ij}, \end{aligned}$$

where $\{\varepsilon_{ij}, i, j = 1, 2, \ldots, n_1\}$ are the error terms of model (1.3). The wavelet transformations of $Z_{n_1}(x, y)$ and $W_{n_1}(x, y)$ are defined by

$$\begin{aligned} T^{\gamma,\beta}Z_{n_1}(s, x, y) &= \int \Psi_s^{\gamma,\beta}(x - u, y - v)\, Z_{n_1}(du, dv) \\ &= \frac{1}{n_1^2} \sum_{i=1}^{n_1} \sum_{j=1}^{n_1} \Psi_s^{\gamma,\beta}(x - i/n_1, y - j/n_1)\, Z_{ij} \end{aligned}$$

and

$$\begin{aligned} T^{\gamma,\beta}W_{n_1}(s, x, y) &= \int \Psi_s^{\gamma,\beta}(x - u, y - v)\, W_{n_1}(du, dv) \\ &= \frac{1}{n_1^2} \sum_{i=1}^{n_1} \sum_{j=1}^{n_1} \Psi_s^{\gamma,\beta}(x - i/n_1, y - j/n_1)\, \varepsilon_{ij}, \end{aligned}$$

for $\gamma = h, v, d$ and $\beta \in [0, \pi/4]$.

A criterion for detecting sharp cusps can now be defined based on these wavelet transformations. Let

$$TZ_{n_1}(s, x, y) = \max_{0 \leq \beta \leq \pi/4} \max_{\gamma = h,v,d} |T^{\gamma,\beta} Z_{n_1}(s, x, y)|$$

and

$$TW_{n_1}(s, x, y) = \max_{0 \leq \beta \leq \pi/4} \max_{\gamma = h,v,d} |T^{\gamma,\beta} W_{n_1}(s, x, y)|.$$

Wang (1998) pointed out that, for $\gamma = h, v, d$ and $\beta \in [0, \pi/4]$,

$$T^{\gamma,\beta} Z_{n_1}(s, x, y) = T^{\gamma,\beta} f(s, x, y)[1 + o(1)] + T^{\gamma,\beta} W_{n_1}(s, x, y),$$

where $T^{\gamma,\beta} W_{n_1}(s, x, y)$ is a normal random variable with mean 0 and variance of order $1/n_1$. $Tf(s, x, y)$ is of order s^2 for f differentiable at (x, y), and of order $s^{\alpha+1}$ for f with an α sharp cusp at (x, y). Based on these results, Wang suggested choosing scale s_{n_1} to be

$$s_{n_1} \propto \left(\frac{|\log n_1|^{\eta}}{n_1} \right)^{1/(\alpha+1)},$$

where $\eta > 1$ is a constant and "$\propto$" means "proportional to". In such cases, $TW_{n_1}(s_{n_1}, x, y)$ is smaller than $Tf(s_{n_1}, x, y)$ when f has an α sharp cusp at (x, y), and is much larger than $Tf(s_{n_1}, x, y)$ when f is differentiable at (x, y).

For detecting curves with α sharp cusps, Wang defined the following threshold value

$$C_{n_1} = 2\sigma \frac{\sqrt{\log n_1}}{n_1}$$

and pointed out that (i) $TZ_{n_1}(s_{n_1}, x, y)$ is asymptotically bounded by C_{n_1} when f has no sharp cusps around (x, y), and (ii) it is significantly larger than C_{n_1} when f has sharp cusps around (x, y). Suppose that the regression function f has α sharp cusps along q curves $\theta_1, \theta_2, \ldots, \theta_q$ in the design space and f is continuously differentiable elsewhere. Then Wang defined the estimators of $\{\theta_1, \theta_2, \ldots, \theta_q\}$ by

$$\{\widehat{\theta}\} = \left\{ \theta \in \Theta : \min_{(x,y) \in \theta} TZ_{n_1}(s_{n_1}, x, y) \geq C_{n_1} \right\},$$

where Θ is a class of candidate curves which is rich enough to contain all $\theta_1, \theta_2, \ldots, \theta_q$. Wang proved the following result about $\{\widehat{\theta}\}$.

Theorem 4.2 *Suppose that the following three assumptions are all satisfied:*

(1) the curves $\theta_1, \theta_2, \ldots, \theta_q$ are continuous curves in the design space with finite length and satisfy the Lipschitz condition that the arc length between any two points on a given curve is of order of the geometric distance of the two points;

(2) the regression function f has α cusps along the curves $\theta_1, \theta_2, \ldots, \theta_q$ and is continuously differentiable elsewhere in the design space;

(3) the Daubechies wavelets ϕ and ψ are continuously differentiable on their supports.

Then there exists a constant $A > 0$ such that

$$\lim_{n_1 \to \infty} \left[\frac{1}{s_{n_1}} d_H \left(\bigcup \{\widehat{\theta}\}, \bigcup_{j=1}^{q} \theta_j \right) \leq A \right] = 1,$$

where $d_H(A, B)$ denotes the Hausdorff distance between two point sets A and B in the design space as defined by

$$\begin{aligned} d_H(A,B) &= \max \left\{ \sup_{(x,y)\in A} \inf_{(x',y')\in B} \|(x,y) - (x',y')\|, \right. \\ & \left. \sup_{(x,y)\in B} \inf_{(x',y')\in A} \|(x,y) - (x',y')\| \right\} \end{aligned} \tag{4.12}$$

and $\|\cdot\|$ is the Euclidean distance.

4.3 DETECTION OF ARBITRARY JUMPS BY LOCAL SMOOTHING

In the previous section, we have introduced several jump detection procedures under the assumptions that the number of JLCs is known (most of these procedures assume that there is a single JLC in the design space) and the JLCs are smooth. In applications, however, these assumptions are often hard to satisfy. For instance, the image intensity function of an image has jumps at the boundaries of objects, but it is difficult to know the number of boundary curves beforehand, and these boundary curves may not be smooth at some places either. As an example, there are many corners and angles in the synthetic aperture radar image presented in Figure 1.2. In this section, we introduce a jump detection procedure which can detect arbitrary JLCs of regression surfaces.

4.3.1 Treat JLCs as a pointset in the design space

As pointed out by Qiu (2002), most jump detection procedures discussed in the previous section treat the JLCs as curves and also estimate them by curves. Since curves have a *global* nature in the sense that any two points on a given curve are connected through other points on the same curve, description and estimation of curves do not have much flexibility. Because of that, most conditions imposed on the JLCs by the procedures discussed in the previous section are also global in nature, in the sense that these conditions should be satisfied at *all* points on the JLCs. For example, in the case when there is only one "smooth, closed, and simple" JLC in the design space (cf. Subsection 4.2.3 for a description), the JLC can be described by a parameterized expression of a curve (cf. the expression of Γ_θ in Subsection 4.2.3). However, when there are multiple JLCs in the design space or the JLCs have crossings, this kind of mathematical description becomes difficult. We believe that the major reason some jump detection procedures in the statistical literature impose restrictive assumptions on the JLCs is because of the common philosophy that the JLCs need to be treated as and estimated by curves.

A more flexible description of the JLCs is that the points on the JLCs constitute *a pointset* in the design space. If the true JLCs are described in this way, then they can be estimated by another pointset in the same design space. Since points in a pointset need not form curves, the connection among the points of a pointset is much more flexible than the connection among the points on curves, which makes detection of arbitrary JLCs possible.

The usefulness of a pointset estimator can be emphasized by an analogy with image digitization. As pointed out in Section 1.1, images need to be digitized in both spatial location and brightness measure for computer-aided image processing and analysis. The digitization process essentially uses a 3-D pointset to approximate the surface of the image intensity function. From the practice of image processing, as long as the resolution of a digital image is high and the number of gray levels is large, the visual effect of a digital image should be good enough for human perception. Similarly, when we use a pointset to estimate the true JLCs, as long as the points in the pointset are close enough to the true JLCs and the points are also dense enough, the visual effect of the estimator should be good. This is demonstrated by Figure 4.2, in which the 2-D regression function is assumed to have a single JLC, the line $\{(x, y) : y = 0.5, x \in [0, 1]\}$. In Figure 4.2(a), the set of all dot points represent one estimator of the JLC. Visually, this estimator looks good because all the points in the pointset of the estimator give us a good perception of the true JLC.

However, there are some obvious limitations to estimating the JLCs by a pointset. For instance, when the detected jump points are scattered in the design space in such a way that they look quite different from curves, the set of the detected jump points would not be a good estimator of the JLCs. In the above example, suppose that the dot points in Figure 4.2(b) represent another estimator of the JLC. It can be seen that the detected jumps are widely scattered in this case, and it is hard to perceive the true JLC from them.

4.3.2 Jump detection by local linear estimation

In this subsection, a jump detection procedure suggested by Qiu and Yandell (1997) is introduced; this procedure treats the JLCs as a pointset, as explained above.

We begin by defining a key term in context of a 2-D regression surface. Suppose that the following basic assumptions hold: $n = n_1^2$ observations $\{(x_i, y_j, Z_{ij}), i, j = 1, 2, \ldots, n_1\}$ are generated from the regression model (1.3) in which the design points $\{(x_i, y_j) = (i/n_1, j/n_1), i, j = 1, 2, \ldots, n_1\}$ are equally spaced in the design space $[0, 1] \times [0, 1]$; the error terms are i.i.d. with mean 0 and variance σ^2; and the 2-D regression function f is continuous in the design space except on the JLCs. A point (x, y) on the JLCs is called a *singular point* if it satisfies any one of the following three conditions.

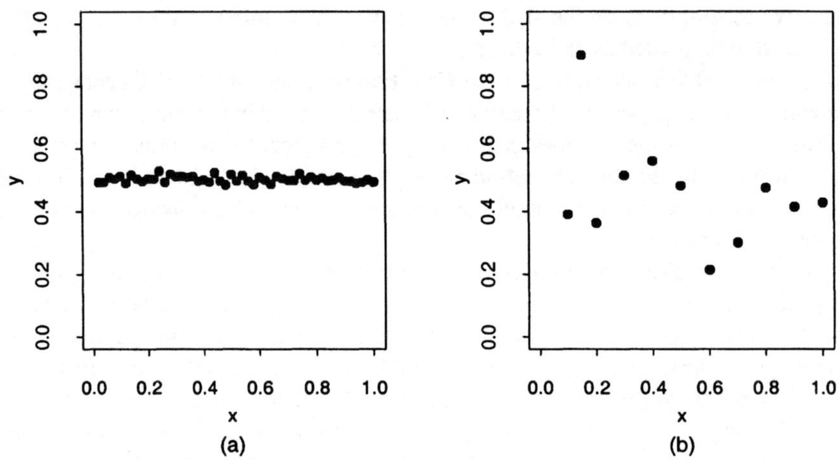

Fig. 4.2 Suppose that the true JLC is the line $\{(x, y) : y = 0.5, x \in [0, 1]\}$. (a): Dot points denote the detected jump positions. They give us a good perception of the true JLC. (b): Dot points denote another set of detected jump positions. They are scattered in the design space and do not estimate the JLC well.

(1) There exists some constant $\eta_0 > 0$ such that, for any $0 < \eta < \eta_0$, the neighborhood of (x, y) with diameter η is divided into more than two connected regions by the JLCs.

(2) There does not exist a constant $\rho_0 > 0$ such that there are two orthogonal lines crossing at (x, y) and two vertical quadrants formed by these two lines belong to two different regions separated by a JLC in the neighborhood of (x, y) with diameter ρ_0.

(3) The jump magnitude of f at (x, y) is zero.

The above conditions define three particular types of singular points of the JLCs. Condition (1) defines all cross points of the JLCs as singular points. Condition (2) corresponds to the corner points of acute angles of the JLCs. Condition (3) refers to the points on the JLCs at which the jump magnitudes are zero (cf. the definition of the JLCs in Section 4.1, by which the JLCs may include finite number of such points). These three types are demonstrated by Figures 4.3(a)–4.3(c), respectively.

Qiu and Yandell's (1997) procedure begins by fitting a local plane at each design point. For any design point (x_i, y_j) for $\ell + 1 \leq i, j \leq n - \ell$, we consider its neighborhood $N(x_i, y_j)$ with window size $k = 2\ell + 1 \ll n$, where ℓ is a nonnegative integer. The neighborhood $N(x_i, y_j)$ is defined by

$$N(x_i, y_j) = \{(x_{i+s}, y_{j+t}),\ s, t = -\ell, -\ell + 1, \ldots, 0, \ldots, \ell - 1, \ell\},$$

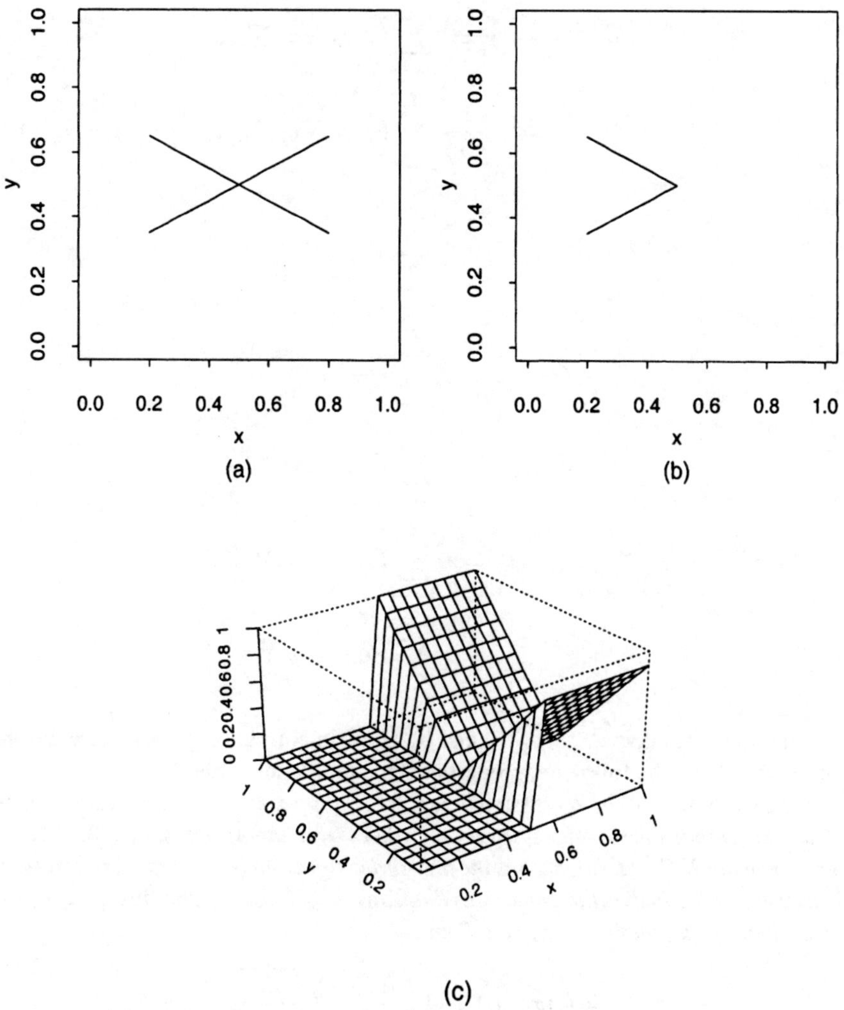

Fig. 4.3 Three types of singular points of the JLCs. (a): Two JLCs cross each other at the singular point (0.5,0.5). (b): A JLC has an acute angle at the singular point (0.5,0.5). (c): Jump magnitude of the true regression surface is zero at the singular point (0.5,0.5).

which is a square including k^2 design points. A least squares (LS) plane is then fitted in this neighborhood. The fitted LS plane is given by

$$\widehat{Z}_{ij}(x,y) = \widehat{\beta}_0^{(i,j)} + \widehat{\beta}_1^{(i,j)}(x - x_i) + \widehat{\beta}_2^{(i,j)}(y - y_j), \text{ for } (x,y) \in N(x_i, y_j),$$

where

$$\widehat{\beta}_0^{(i,j)} = \frac{1}{k^2} Z_{..}$$

$$
\begin{aligned}
\widehat{\beta}_1^{(i,j)} &= \frac{1}{kS_x^2} \sum_{s=-\ell}^{\ell} (x_{i+s} - x_i) Z_{i+s,\cdot} \\
\widehat{\beta}_2^{(i,j)} &= \frac{1}{kS_y^2} \sum_{t=-\ell}^{\ell} (y_{j+t} - y_j) Z_{\cdot,j+t}
\end{aligned} \tag{4.13}
$$

and

$$
\begin{aligned}
Z_{\cdot\cdot} &= \sum_{s,t=-\ell}^{\ell} Z_{i+s,j+t}, \\
Z_{i+s,\cdot} &= \sum_{t=-\ell}^{\ell} Z_{i+s,j+t}, \\
Z_{\cdot,j+t} &= \sum_{s=-\ell}^{\ell} Z_{i+s,j+t}, \\
S_x^2 &= \sum_{s=-\ell}^{\ell} (x_{i+s} - x_i)^2, \\
S_y^2 &= \sum_{t=-\ell}^{\ell} (y_{j+t} - y_j)^2.
\end{aligned}
$$

It can be checked that $\widehat{\beta}_0^{(i,j)}$, $\widehat{\beta}_1^{(i,j)}$, and $\widehat{\beta}_2^{(i,j)}$ are uncorrelated to each other. Furthermore, they have the following properties (Qiu and Yandell 1997).

Theorem 4.3 *In the regression model (1.3), suppose that the regression function f has continuous first-order partial derivatives over the design space $[0,1] \times [0,1]$ except on the JLCs, at which it has the first-order right and left partial derivatives. The window size k satisfies the conditions that $\lim_{n_1 \to \infty} 1/k = 0$ and $\lim_{n_1 \to \infty} k/n_1 = 0$. If there is no jump in $N(x_i, y_j)$, then*

$$
\widehat{\beta}_1^{(i,j)} = f_x'(x_i, y_j) + O\left(\frac{n_1 \sqrt{\log(\log(k))}}{k^2} \right), \quad a.s.
$$

and

$$
\widehat{\beta}_2^{(i,j)} = f_y'(x_i, y_j) + O\left(\frac{n_1 \sqrt{\log(\log(k))}}{k^2} \right), \quad a.s.
$$

If the point (x_i, y_j) is on a JLC and it is not a singular point, then

$$
\widehat{\beta}_1^{(i,j)} = f_x'(\widetilde{x}_i, \widetilde{y}_j) + h_1^{(i,j)} C(i,j) + \gamma_1 C_x(i,j) + O\left(\frac{n_1 \sqrt{\log(\log(k))}}{k^2} \right), \quad a.s.
$$

and

$$
\widehat{\beta}_2^{(i,j)} = f_y'(\widetilde{x}_i, \widetilde{y}_j) + h_2^{(i,j)} C(i,j) + \gamma_2 C_y(i,j) + O\left(\frac{n_1 \sqrt{\log(\log(k))}}{k^2} \right), \quad a.s.,
$$

where $(\tilde{x}_i, \tilde{y}_j)$ *is some point around* (x_i, y_j) *which satisfies (1)* f *has first-order* x *and* y *partial derivatives at* $(\tilde{x}_i, \tilde{y}_j)$*, (2)* $(\tilde{x}_i, \tilde{y}_j)$ *is on the same side of the JLC as* (x_i, y_j)*, and (3) the distance between* $(\tilde{x}_i, \tilde{y}_j)$ *and* (x_i, y_j) *tends to zero faster than* $1/n_1$*.* $C(i,j), C_x(i,j)$*, and* $C_y(i,j)$ *are the jump magnitudes of* f *and its first-order* x *and* y *partial derivatives at* (x_i, y_j)*,* $h_1^{(i,j)}$ *and* $h_2^{(i,j)}$ *are two constants satisfying*

$$\sqrt{\left(h_1^{(i,j)}\right)^2 + \left(h_2^{(i,j)}\right)^2} \sim n_1/k,$$

and γ_1 *and* γ_2 *are two constants between* -1 *and 1.*

In Theorem 4.3, the term $O(n_1\sqrt{\log(\log(k))}/k^2)$ in the related expressions is due to noise. From the results in this theorem, it can be seen that the estimated slopes $\widehat{\beta}_1^{(i,j)}$ and $\widehat{\beta}_2^{(i,j)}$ carry information about both the continuity and the jump parts of f. Therefore, to detect jumps based on the estimated slopes, we need to extract information about the jump part of f from them, which can be accomplished in the following way. At a particular design point (x_i, y_j), let the angle formed by $\vec{v}_{ij} = (\widehat{\beta}_1^{(i,j)}, \widehat{\beta}_2^{(i,j)})$ and the positive direction of the x-axis be denoted by $\theta \in [-\pi/4, 7\pi/4]$. Two neighboring design points (x_{N1}, y_{N1}) and (x_{N2}, y_{N2}) are then determined by the following formulas:

- If $\pi/4 \leq \theta < 3\pi/4$ or $5\pi/4 \leq \theta < 7\pi/4$, then

$$\begin{aligned} x_{N1} &= x_i + \frac{k}{n_1 \cdot \tan\theta}, \quad y_{N1} = y_j + \frac{k}{n_1} \\ x_{N2} &= x_i - \frac{k}{n_1 \cdot \tan\theta}, \quad y_{N2} = y_j - \frac{k}{n_1}. \end{aligned} \tag{4.14}$$

- If $-\pi/4 \leq \theta < \pi/4$ or $3\pi/4 \leq \theta < 5\pi/4$, then

$$\begin{aligned} x_{N1} &= x_i + \frac{k}{n_1}, \quad y_{N1} = y_j + \frac{k}{n_1} \cdot \tan\theta \\ x_{N2} &= x_i - \frac{k}{n_1}, \quad y_{N2} = y_j - \frac{k}{n_1} \cdot \tan\theta. \end{aligned} \tag{4.15}$$

If the two points determined by equations (4.14) and (4.15) are not exactly design points, then we just choose the two design points that are closest to them instead.

Figure 4.4 gives an intuitive explanation about the two neighboring design points (x_{N1}, y_{N1}) and (x_{N2}, y_{N2}) defined by equations (4.14) and (4.15). From the plot, it can be seen that the two points (x_{N1}, y_{N1}) and (x_{N2}, y_{N2}) have the following properties: (1) they are two design points on the line through (x_i, y_j) and with slope $\widehat{\beta}_2^{(i,j)}/\widehat{\beta}_1^{(i,j)}$, and (2) they are closest to (x_i, y_j) among the points on that line whose neighborhoods have no overlap with $N(x_i, y_j)$.

These neighboring design points can be used for defining a jump detection criterion. Recall that $\vec{v}_{ij}$ is the gradient vector of the fitted LS plane in $N(x_i, y_j)$, which implies that the underlying regression surface increases most rapidly along a nearby direction. If the point (x_i, y_j) is on a JLC and is also not a singular point, then as a direct conclusion of Theorem 4.3, the JLC should not be in the neighborhoods $N(x_{N1}, y_{N1})$ and $N(x_{N2}, y_{N2})$ when n_1 is large enough. In other words, the points (x_{N1}, y_{N1}) and (x_{N2}, y_{N2}) are on two different sides of the JLC. Thus, Qiu and Yandell (1997) defined the following jump detection criterion:

$$\delta_{ij} = \min\left\{\|\vec{v}_{ij} - \vec{v}_{N1}\|, \|\vec{v}_{ij} - \vec{v}_{N2}\|\right\}, \tag{4.16}$$

where $\vec{v}_{N1} = (\widehat{\beta}_1^{(N1)}, \widehat{\beta}_2^{(N1)})$ and $\vec{v}_{N2} = (\widehat{\beta}_1^{(N2)}, \widehat{\beta}_2^{(N2)})$ are gradient vectors of the fitted LS planes in $N(x_{N1}, y_{N1})$ and $N(x_{N2}, y_{N2})$, respectively, and $\|\cdot\|$ is the Euclidean norm.

The criterion δ_{ij} has the following properties. If there is no jump in the three neighborhoods $N(x_i, y_j)$, $N(x_{N1}, y_{N1})$ and $N(x_{N2}, y_{N2})$, then $\vec{v}_{ij}, \vec{v}_{N1}$, and $\vec{v}_{N2}$ should be close to each other, and thus δ_{ij} is small. If (x_i, y_j) is on a JLC and it is not a singular point, then by Theorem 4.3,

$$\delta_{ij} \approx C(i,j)\sqrt{\left(h_1^{(i,j)}\right)^2 + \left(h_2^{(i,j)}\right)^2} \sim n_1/k,$$

which tends to infinity when n_1 increases. Hence, large values of δ_{ij} indicate possible jumps.

An equivalent and maybe more intuitive explanation of δ_{ij} is as follows. Let $\vec{u}_{ij} = (\widehat{\beta}_1^{(i,j)}, \widehat{\beta}_2^{(i,j)}, -1)$, $\vec{u}_{N1} = (\widehat{\beta}_1^{(N1)}, \widehat{\beta}_2^{(N1)}, -1)$, and $\vec{u}_{N2} = (\widehat{\beta}_1^{(N2)}, \widehat{\beta}_2^{(N2)}, -1)$ be the normal directions of the LS planes in $N(x_i, y_j)$, $N(x_{N1}, y_{N1})$, and $N(x_{N2}, y_{N2})$, respectively. Then δ_{ij} is the minimum length of vectors $\vec{u}_{ij} - \vec{u}_{N1}$ and $\vec{u}_{ij} - \vec{u}_{N2}$. If (x_i, y_j) is on a JLC, then $\vec{u}_{ij}$ has an abrupt change from both $\vec{u}_{N1}$ and $\vec{u}_{N2}$, and hence δ_{ij} is relatively large.

Example 4.1 Suppose that the true regression function f is

$$f(x,y) = -0.5 - y + 3(x-0.5)^2 + I\left(y > -(x-0.5)^2 + 0.5\right),$$

for $(x, y) \in [0,1] \times [0,1]$. Then this function has a single JLC having expression $y = -(x-0.5)^2 + 0.5$ for $x \in [0,1]$ with constant jump magnitude 1. The true regression surface and the JLC are shown in Figures 4.5(a) and 4.5(b), respectively.

Then 10,000 observations $\{Z_{ij}, i, j = 1, 2, \ldots, 100\}$ are generated from the regression model (1.3) with $n_1 = n_2 = 100$ and $\varepsilon_{11} \sim N(0, 0.5^2)$. In a neighborhood of each design point with $k = 7$, a LS plane is fitted and the regression coefficients are calculated by equation (4.13). The gradient vector $\vec{v}_{ij}$ of the fitted LS plane at each design point is shown in Figure 4.6. It could be seen from the plot that $\vec{v}_{ij}$ indicates the jump direction well.

The jump detection criterion δ_{ij} is then calculated by formulas (4.13)–(4.16). It is plotted in Figure 4.7(a) (a 3-D plot) and Figure 4.7(b) (a corresponding image plot). In the image plot, the brightness at each pixel represents the response value – the darker, the bigger. Clearly, the criterion δ_{ij} does have some ability to reveal the jumps.

A large value of δ_{ij} indicates a possible jump at (x_i, y_j); to determine "large" δ_{ij}, we next derive a threshold value for δ_{ij}. For any constant $b > 0$, it is not difficult to

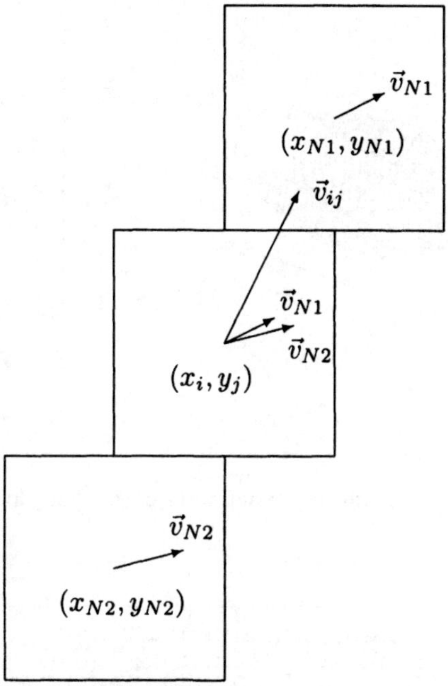

Fig. 4.4 At the design point (x_i, y_j), the jump detection criterion δ_{ij} is defined as the minimum length of the vectors $\vec{v}_{ij} - \vec{v}_{N1}$ and $\vec{v}_{ij} - \vec{v}_{N2}$, where $\vec{v}_{ij}, \vec{v}_{N1}$, and $\vec{v}_{N2}$ are the gradient vectors of the fitted LS planes obtained in $N(x_i, y_j)$, $N(x_{N1}, y_{N1})$, and $N(x_{N2}, y_{N2})$, respectively.

see that

$$
\begin{aligned}
&P(\delta_{ij} > b) \leq P\left(\|\vec{v}_{ij} - \vec{v}_{N1}\| > b\right) \\
&= P\left((\widehat{\beta}_1^{(i,j)} - \widehat{\beta}_1^{(N1)})^2 + (\widehat{\beta}_2^{(i,j)} - \widehat{\beta}_2^{(N1)})^2 > b^2\right) \\
&= E\left\{P\left((\widehat{\beta}_1^{(i,j)} - \widehat{\beta}_1^{(N1)})^2 + (\widehat{\beta}_2^{(i,j)} - \widehat{\beta}_2^{(N1)})^2 > b^2 \middle| \widehat{\beta}_1^{(i,j)}, \widehat{\beta}_2^{(i,j)}\right)\right\}.
\end{aligned}
$$

For fixed $\widehat{\beta}_1^{(i,j)}$ and $\widehat{\beta}_2^{(i,j)}$, $[(\widehat{\beta}_1^{(i,j)} - \widehat{\beta}_1^{(N1)})^2 + (\widehat{\beta}_2^{(i,j)} - \widehat{\beta}_2^{(N1)})^2]/\sigma_{N1}^2$ is approximately χ_2^2 distributed under the assumption that there is no jump in $N(x_i, y_j) \bigcup N(x_{N1}, y_{N1})$, where $\sigma_{N1}^2 = \text{Var}(\widehat{\beta}_1^{(N1)}) = \sigma^2/(kS_x^2)$. Therefore, a natural threshold value for δ_{ij} is

$$
b = \sqrt{\chi^2_{2,1-\alpha_{n_1}} \cdot \frac{\widehat{\sigma}^2}{kS_x^2}} = \widehat{\sigma}\sqrt{\frac{\chi^2_{2,1-\alpha_{n_1}}}{kS_x^2}}, \tag{4.17}
$$

where $\chi^2_{2,1-\alpha_{n_1}}$ is a $1 - \alpha_{n_1}$ quantile of the χ_2^2 distribution and $\widehat{\sigma}$ is a consistent estimator of σ. Then the design points $\{(x_i, y_j) : \delta_{ij} > b, i, j = (3k+1)/2, \ldots, n_1 - (3k-1)/2\}$ could be flagged as jump candidates.

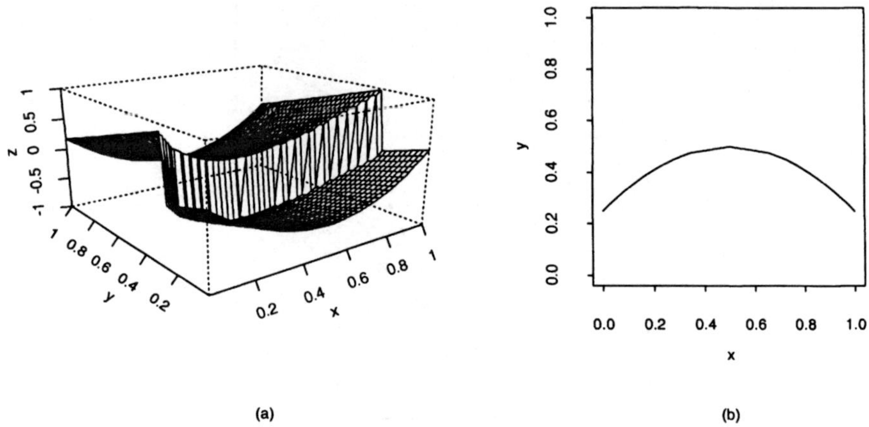

Fig. 4.5 (a): True jump regression surface. (b): Jump location curve.

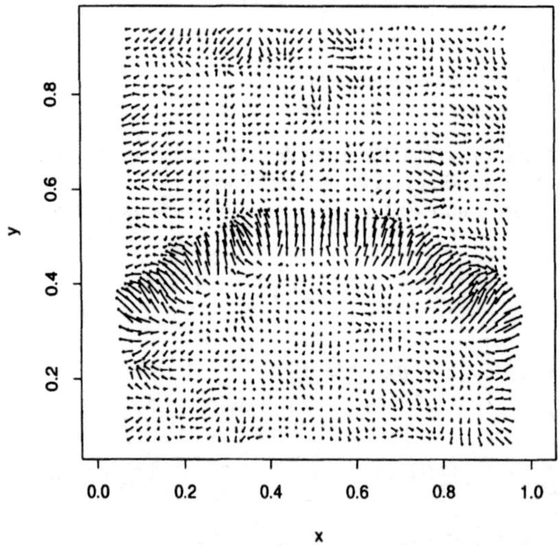

Fig. 4.6 The gradient vector $\vec{v}_{ij}$ of the fitted LS plane at each design point.

The jump detection procedure based on the jump detection criterion δ_{ij} is summarized in the following algorithm.

Jump Detection Algorithm Based on Local LS Estimation

1. At any design point (x_i, y_j) with $\ell + 1 \leq i, j \leq n_1 - \ell$, fit a LS plane in $N(x_i, y_j)$ by formula (4.13).

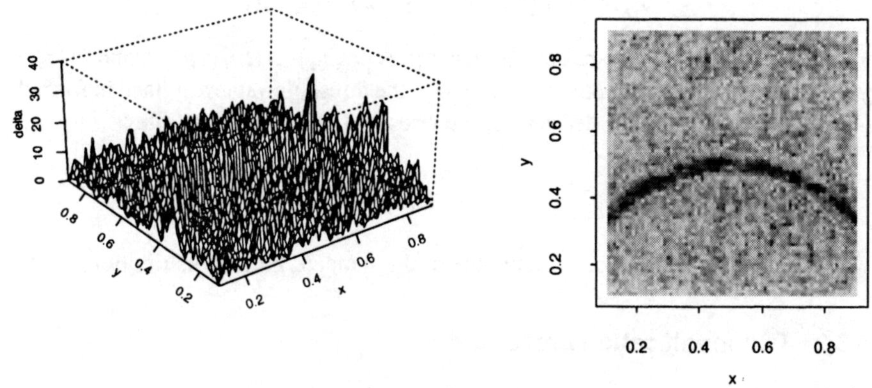

(a) (b)

Fig. 4.7 (a): The jump detection criterion $\{\delta_{ij}\}$ in 3-D plot. (b): The corresponding image plot of (a). The darker the color in the image, the bigger the criterion value.

2. Use formulas (4.14) and (4.15) to determine two neighboring design points of (x_i, y_j), for $(3k+1)/2 \leq i, j \leq n_1 - (3k-1)/2$.

3. Use formula (4.16) to calculate δ_{ij}.

4. Use formula (4.17) to determine the threshold value b.

5. Flag the design point (x_i, y_j) as a jump candidate if $\delta_{ij} > b$.

The detected jumps of the above algorithm have the following large-sample property (Qiu and Yandell 1997).

Theorem 4.4 *Besides the conditions stated in Theorem 4.3, if there are only a finite number of singular points on the JLCs and if* α_{n_1} *in formula (4.17) is chosen such that (i)* $\lim_{n_1 \to \infty} \log(\alpha_{n_1})/\log(\log(k)) = -\infty$ *and (ii)* $\lim_{n_1 \to \infty} \log(\alpha_{n_1})/k^2 = 0$, *then the detected jumps are almost surely consistent in the Hausdorff distance, and the convergence rate could reach* $O(n_1^{-1}\log(n_1))$.

In addition, there are two possibly advantageous revisions of the above algorithm. First, although square-shaped neighborhoods of the design points are used in the algorithm for convenience, other kinds of neighborhoods, such as ellipses and circles, could also be used. Second, because only a few neighborhood points change from one design point to the next, the LS planes fitted can be easily updated; thus the algorithm only requires $O(nk)$ calculations. Taking advantage of the symmetric structures between $\widehat{\beta}_1^{(i,j)}$ and $\widehat{\beta}_2^{(i,j)}$ (cf. equation (4.13)) can also simplify computation.

To conclude this subsection, we would like to point out that variants of the above jump detection algorithm are also possible. For instance, instead of using the criterion

δ_{ij} defined in equation (4.16), we could also use the following criterion:

$$\delta_{ij}^* = \min\left\{|\bar{Z}_{ij} - \bar{Z}_{N1}|, |\bar{Z}_{ij} - \bar{Z}_{N2}|\right\}, \tag{4.18}$$

where $\bar{Z}_{ij}$, $\bar{Z}_{N1}$, and $\bar{Z}_{N2}$ are sample means in $N(x_i, y_j)$, $N(x_{N1}, y_{N1})$, and $N(x_{N2}, y_{N2})$, respectively. Similar to that derived for the threshold value b defined in formula (4.17), a formula could be derived for the threshold value of δ_{ij}^* as follows:

$$b^* = \frac{\sqrt{2}\widehat{\sigma}}{k} Z_{1-\alpha_{n_1}/2},$$

where $Z_{1-\alpha_{n_1}/2}$ is a $1 - \alpha_{n_1}/2$ quantile of the standard normal distribution.

4.3.3 Two modification procedures

Similar to 1-D cases (cf. Subsection 3.3.3), the above jump detection algorithm may flag two kinds of deceptive jump candidates. One kind of deceptive jump candidates consists of those candidates close to the real JLCs; they occur due to the nature of local smoothing. That is, if the point (x_i, y_j) is flagged, then its neighboring design points will be flagged with high probability. This kind of deceptive candidates can make the detected JLCs thick. Another kind of deceptive candidates exists due to the nature of hypothesis testing, on which the threshold value of the jump detection criterion δ_{ij} is based. That is, a point (x_i, y_j) could be flagged as a jump point with probability α_{n_1}, even if it is actually not a jump point. Deceptive candidates of this kind are scattered in the whole design space. To delete these two kinds of deceptive jump candidates, Qiu and Yandell (1997) suggested the following two modification procedures (MPs).

(MP1) (to make the detected JLCs thin) For each $(3k+1)/2 \leq j \leq n_1 - (3k-1)/2$, consider the design points $\{(x_i, y_j) : (3k+1)/2 \leq i \leq n_1 - (3k-1)/2\}$ on line $y = y_j$. Among them, the flagged candidates which satisfy $-\pi/4 \leq \theta < \pi/4$ or $3\pi/4 \leq \theta < 5\pi/4$ are denoted as $\{(x_{i_s}, y_j) : 1 \leq s \leq n_1^*\}$, where θ is defined as in equations (4.14) and (4.15). If there are integers $r_1 < r_2$ such that the increments of the sequence $i_{r_1} < i_{r_1+1} < \ldots < i_{r_2}$ are all less than the window width k, but $i_{r_1} - i_{r_1-1} > k$ and $i_{r_2+1} - i_{r_2} > k$, then we say that $\{(x_{i_s}, y_j) : r_1 \leq s \leq r_2\}$ form a *tie*, and we select the middle point $((x_{i_{r_1}} + x_{i_{r_2}})/2, y_j)$ as a new candidate to replace the tie set. That is, reduce the candidate set to one representative, the middle point, from each tie set. Along the y direction, for each $(3k+1)/2 \leq i \leq n_1 - (3k-1)/2$, consider the design points $\{(x_i, y_j) : (3k+1)/2 \leq j \leq n_1 - (3k-1)/2\}$ on line $x = x_i$. We do the same modification as that along the x direction, except that only those candidates satisfying $\pi/4 \leq \theta < 3\pi/4$ or $5\pi/4 \leq \theta < 7\pi/4$ are considered this time.

(MP2) (to delete scattered jump candidates) For any jump candidate (x_i, y_j), if the number of other jump candidates in $N(x_i, y_j)$ is less than $(k-1)/2$, then we delete (x_i, y_j) from the set of jump candidates.

In the first modification procedure (MP1), along the x direction, the condition $-\pi/4 \leq \theta < \pi/4$ or $3\pi/4 \leq \theta < 5\pi/4$ implies that the possible JLC forms a less than or equal to $\pi/4$ acute angle with the y-axis at the candidate point. This avoids canceling the true JLCs that parallel the x-axis. Compared to the 1-D modification procedure (3.18) in Subsection 3.3.3, (MP1) is obviously the generalization in 2-D cases.

Example 4.2 Let us continue our discussion about the jump detection problem discussed in Example 4.1. Suppose that the jump detection criterion δ_{ij} presented in Figure 4.7 is used for detecting jumps and we choose $\alpha_{n_1} = 0.001$ in formula (4.17). Then the flagged jump candidates are plotted in Figure 4.8(b) by black points. It can be noticed that the detected JLC is quite thick and that there are some scattered jump candidates also. We then use the modification procedures (MP1) and (MP2), successively, to modify the set of jump candidates. The results are plotted in Figures 4.8(c) and 4.8(d), respectively. As a comparison, we plot the real JLC in Figure 4.8(a). Clearly, the modification procedures improve the visual effect of the detected jumps. But we also notice that (i) there are some breaks here or there in the detected JLC displayed in Figure 4.8(d), and (ii) the detected JLC is not thin enough at some places. These phenomena imply that our MPs still have room for improvement.

4.4 JUMP DETECTION IN TWO OR MORE GIVEN DIRECTIONS

The RDKE jump detection procedure introduced in Subsection 4.2.1 searches all directions at a given point by using a rotation transformation for detecting a possible jump. Theoretically, this procedure is appealing, but it is computationally intensive and thus inconvenient to use in applications. To simplify computation, Qiu (2002) suggested a jump detection procedure that searches only two or more given directions at a given point for a possible jump. This procedure is introduced below in four subsections. In Subsection 4.4.1, the basic idea is introduced in the case when only two given directions are searched at a given point for detecting a possible jump. The issue of measuring the performance of various jump detection procedures is discussed in Subsection 4.4.2. The connection between this procedure and the Sobel edge detection procedure, which is popular in the image processing literature, is explored in Subsection 4.4.3. Finally, in Subsection 4.4.4, we discuss a generalization to search more than two given directions at any given point for a possible jump.

4.4.1 Jump detection in two given directions

Our goal is to estimate the JLCs based on n observations $\{(x_i, y_i, Z_i), i = 1, 2, \ldots, n\}$ generated from model (4.1). The regression function f is assumed to be continuous in the design space $[0, 1] \times [0, 1]$ except on some JLCs.

We first define a simpler version of the jump detection criterion introduced in Subsection 4.2.1. Let K_1^* and K_2^* be two one-sided kernel functions, as defined

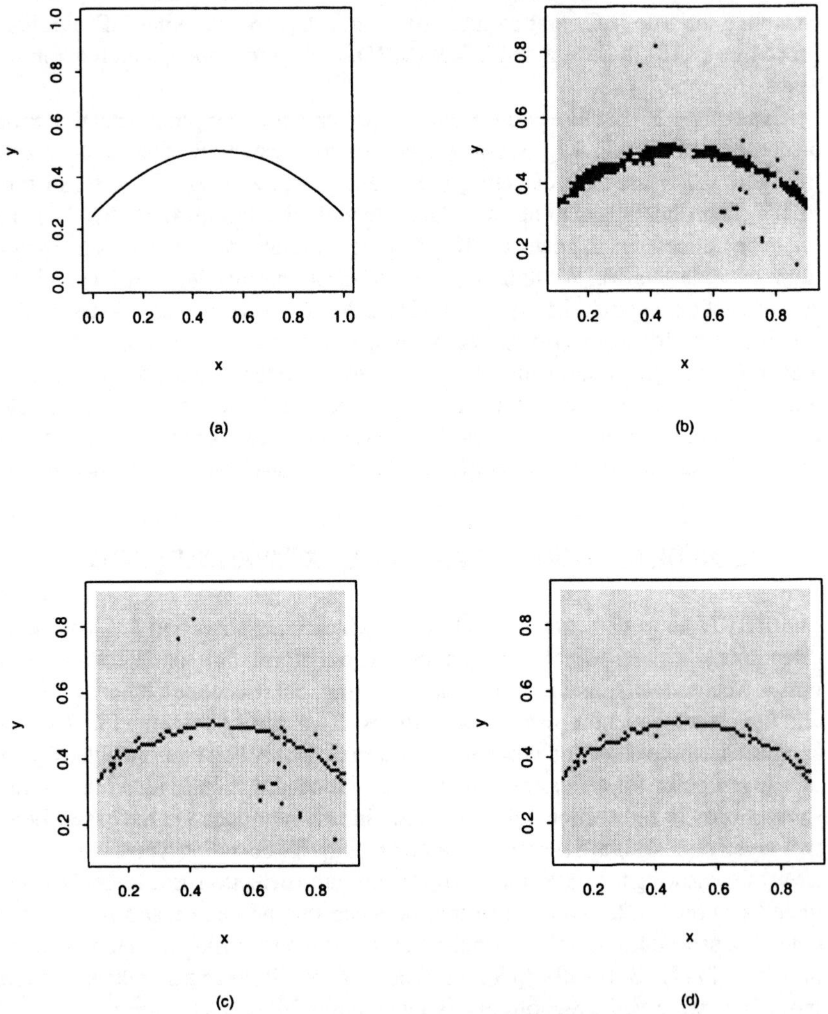

Fig. 4.8 (a): Real jump location curve. (b): Detected jump candidates by the criterion δ_{ij} defined in formula (4.16). (c): Modified jump candidates from those in plot (b) by the modification procedure (MP1). (d): Modified jump candidates from those in plot (c) by the modification procedure (MP2).

previously. Then let

$$M_n^{(1)}(x,y) = \frac{1}{nh_np_n}\sum_{i=1}^{n} Z_i\left[K_1^*\left(\frac{x_i-x}{h_n},\frac{y_i-y}{p_n}\right) - K_2^*\left(\frac{x_i-x}{h_n},\frac{y_i-y}{p_n}\right)\right] \tag{4.19}$$

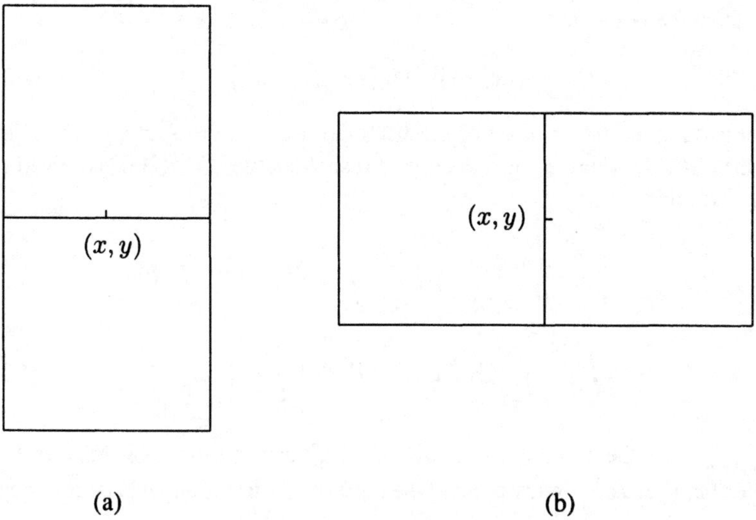

Fig. 4.9 (a): Upper- and lower-sided neighborhoods of (x, y) for constructing $M_n^{(1)}(x, y)$. (b): Left- and right-sided neighborhoods of (x, y) for constructing $M_n^{(2)}(x, y)$.

and

$$M_n^{(2)}(x, y) = \frac{1}{nh_n p_n} \sum_{i=1}^{n} Z_i \left[K_1^* \left(\frac{y_i - y}{h_n}, \frac{x_i - x}{p_n} \right) - K_2^* \left(\frac{y_i - y}{h_n}, \frac{x_i - x}{p_n} \right) \right], \tag{4.20}$$

where h_n and p_n are two bandwidths, $(x, y) \in [b_n, 1 - b_n] \times [b_n, 1 - b_n]$, and $b_n = \sqrt{h_n^2/4 + p_n^2}$. Compared with the jump detection criterion M_{RDKE} defined by equation (4.2), it can be checked that, for $(x, y) \in [b_n, 1 - b_n] \times [b_n, 1 - b_n]$,

$$M_n^{(1)}(x, y) = M_{RDKE}(0, x, y); \; M_n^{(2)}(x, y) = M_{RDKE}(\pi/2, x, y).$$

Therefore, $M_n^{(1)}(x, y)$ is a difference of two weighted averages of the observations in the upper- and lower-sided neighborhoods of (x, y), respectively, and $M_n^{(2)}(x, y)$ is a difference of two weighted averages of the observations in the left- and right-sided neighborhoods of (x, y), respectively. The related one-sided neighborhoods used in $M_n^{(1)}(x, y)$ and $M_n^{(2)}(x, y)$ are displayed in Figures 4.9(a) and 4.9(b).

Intuitively, the quantity $M_n^{(1)}(x, y)$ alone can be used for detecting jumps that are parallel to the x-axis, and the quantity $M_n^{(2)}(x, y)$ alone can be used for detecting jumps that are parallel to the y-axis. By using both of them, it is possible to detect most jumps. To this end, we define

$$M_n(x, y) = \max \left\{ |M_n^{(1)}(x, y)|, |M_n^{(2)}(x, y)| \right\} \tag{4.21}$$

as a jump detection criterion, so a large value of $M_n(x, y)$ indicates a possible jump at (x, y). Then the point set

$$\widehat{D}_n = \{(x_i, y_i) : M_n(x_i, y_i) > u_n\} \tag{4.22}$$

is used for estimating the point set of the JLCs defined by $D = \{(x, y) : (x, y)$ is a point on the JLCs $\}$, where u_n is a positive threshold value. Qiu (2002) derived the following formula for u_n:

$$u_n = \left\{ \frac{\sigma^2 \chi^2_{1,1-\alpha_n/2}}{n h_n p_n} \left[\int_{-1/2}^{1/2} \int_0^1 (K_1^*(u,v))^2 \, dudv + \int_{-1/2}^{1/2} \int_{-1}^{0} (K_2^*(u,v))^2 \, dudv \right] \right\}^{1/2}, \tag{4.23}$$

where $\chi^2_{1,1-\alpha_n/2}$ is the $1 - \alpha_n/2$ quantile of the χ_1^2 distribution. For this threshold value, when (x, y) is a continuity point of the regression surface and n is large enough,

$$P\left(M_n(x, y) > u_n\right) \leq \alpha_n,$$

where α_n is a significance level.

Example 4.3 Suppose that the regression function is $f(x, y) = -2(x - 0.5)^2 - 2(y - 0.5)^2 + 1$ if $(x - 0.5)^2 + (y - 0.5)^2 < 0.25^2$, and $f(x, y) = -2(x - 0.5)^2 - 2(y - 0.5)^2$ otherwise. Then f has a unique JLC which is a circle with the expression $(x - 0.5)^2 + (y - 0.5)^2 = 0.25^2$. The true regression surface is displayed in Figure 4.10(a). 100^2 observations are generated from regression model (1.3) with $\sigma = 0$ (i.e., the observations are noise free). Using the above jump detection procedure, K_1^*, K_2^*, h_n, and p_n are chosen to be $K_1^*(x, y) = \frac{12}{11}(1 - x^2) \cdot \frac{12}{11}(1 - (y - 0.5)^2)$ for $(x, y) \in [-1/2, 1/2] \times [0, 1]$, $K_2^*(x, y) = K_1^*(x, -y)$ for $(x, y) \in [-1/2, 1/2] \times [-1, 0]$, $h_n = 0.05$, and $p_n = 0.05$. Then the absolute values of $M_n^{(1)}$, $M_n^{(2)}$, and M_n are displayed in Figures 4.10(b)–4.10(d), respectively, by three images. The darker the color of the image, the larger the corresponding value. It can be seen that $M_n^{(1)}$ cannot reveal the portions of the JLC that are parallel to the y-axis well, and $M_n^{(2)}$ cannot reveal the portions of the JLC that are parallel to the x-axis well. As a comparison, the jump detection criterion M_n reveals the entire JLC well.

Proper bandwidth selection is vital to this method. We notice that, if the two bandwidths are not chosen carefully, then neither one of $M_n^{(1)}(x, y)$ and $M_n^{(2)}(x, y)$ can detect an arbitrary JLC well, which is demonstrated by Figure 4.11(a). In this plot, the point (x, y) is assumed to be on a JLC which is the diagonal line in the central square. It can be seen that if h_n is chosen large compared to p_n, then both $M_n^{(1)}(x, y)$ and $M_n^{(2)}(x, y)$ would be small because observations on both sides of the JLC are averaged in computing the one-sided kernel estimators; consequently, the jump effect is partially eliminated. On the other hand, when the ratio h_n/p_n of the two bandwidths gets smaller, namely, the neighborhoods get relatively narrower, both $M_n^{(1)}(x, y)$ and $M_n^{(2)}(x, y)$ become more capable of detecting the JLC, because

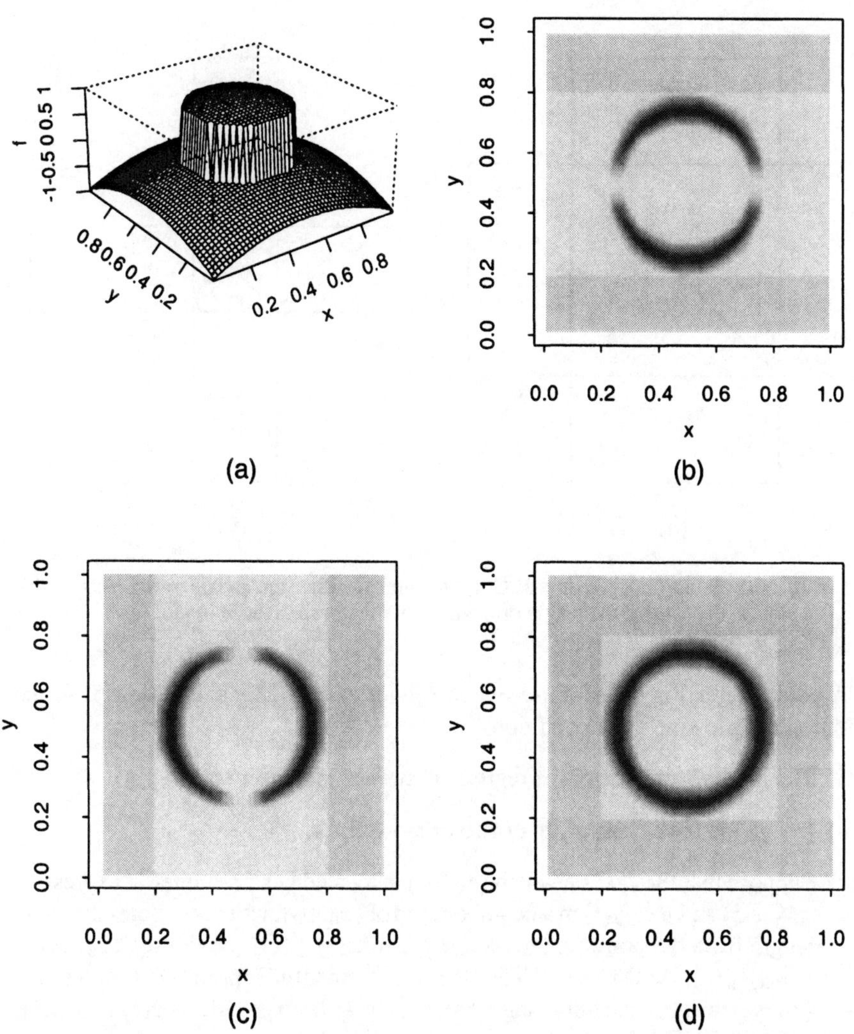

Fig. 4.10 (a): True regression surface. (b): Quantity $|M_n^{(1)}|$. (c): Quantity $|M_n^{(2)}|$. (d): Jump detection criterion M_n. In the three images, darker color represents larger value.

less observations on the two different sides of the JLC are averaged in computing the one-sided kernel estimators, as demonstrated by Figure 4.11(b). Therefore, to detect an arbitrary JLC efficiently, the bandwidths h_n and p_n should satisfy the following assumption (A 4.2):

(A 4.2)

$$\lim_{n \to \infty} h_n / p_n = 0.$$

That is, the neighborhoods become narrower and narrower as n increases.

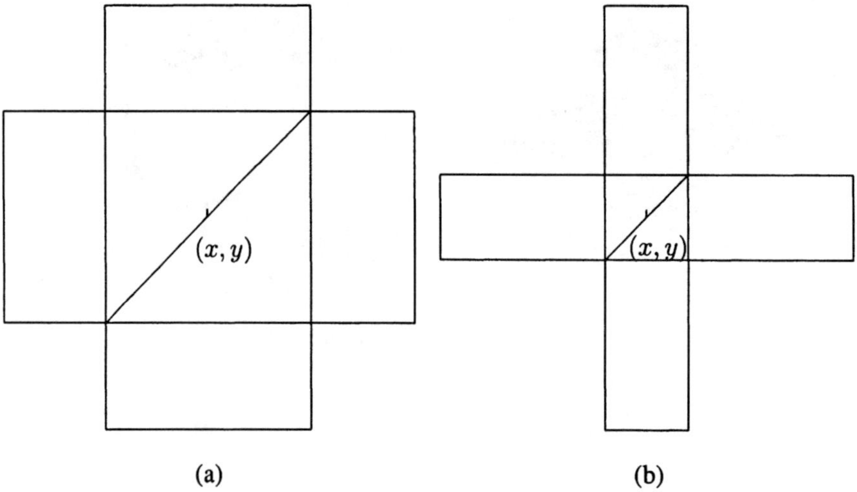

Fig. 4.11 (a): Point (x, y) is on a JLC not parallel to either the x-axis or the y-axis. (b): Same as plot (a) except that the neighborhoods are narrower than those in plot (a).

A point (x_0, y_0) is called a *nonsingular point* of the JLCs if it is on a JLC and satisfies the following three conditions.

(i) The jump magnitude of the regression surface is nonzero at (x_0, y_0).

(ii) (x_0, y_0) is not a cross point of two or more JLCs.

(iii) Suppose that the JLC on which (x_0, y_0) is located has a parametric expression $x = x(t)$ and $y = y(t)$ in a neighborhood of (x_0, y_0), where t denotes the curve length from the point (x_0, y_0) to the point $(x(t), y(t))$ (obviously, $(x_0, y_0) = (x(t_0), y(t_0))$ with $t_0 = 0$). Then the JLC satisfies the Lipschitz (1) condition at (x_0, y_0) in one of the following senses: (1) it can be expressed as $x(t) = \phi(y(t))$ in some neighborhood $N(t_0)$ of t_0, and there exists a constant $\rho_1 > 0$ such that

$$|x(t_1) - x(t_2)| \leq \rho_1 |y(t_1) - y(t_2)|, \text{ for any } t_1, t_2 \in N(t_0);$$

or (2) it has an expression $y(t) = \psi(x(t))$ in $N(t_0)$, and there exists a constant $\rho_2 > 0$ such that

$$|y(t_1) - y(t_2)| \leq \rho_2 |x(t_1) - x(t_2)|, \text{ for any } t_1, t_2 \in N(t_0); \tag{4.24}$$

here, ϕ and ψ are two univariate functions.

Condition (iii) above is essentially the Lipschitz (1) condition expressed in a symmetric way with respect to the $x-$ and $y-$axes. If (x_0, y_0) is on a JLC and it is not a nonsingular point, then it is called a *singular point* of the JLCs.

Notice that this definition of nonsingular points of the JLCs is different from the one given in Subsection 4.3.2. Therefore, the definition of nonsingular points of the JLCs depends on the related jump detection procedure.

We now consider how well the criterion M_n can detect these two types of jump points. From the previous definition, if (x_0, y_0) is a nonsingular point of the JLCs, then it is on a single JLC which satisfies the Lipschitz condition around (x_0, y_0) as stated in condition (iii). Without loss of generality, suppose that the JLC satisfies the Lipschitz condition (4.24). In such a case, the two neighborhoods used in constructing $M_n^{(1)}(x_0, y_0)$ (cf. Figure 4.9(a)) are mostly on two different sides of the JLC under the assumption (A 4.2). The part of the upper-sided neighborhood located below the JLC and the part of the lower-sided neighborhood located above the JLC are both negligible. Consequently, $M_n^{(1)}(x_0, y_0)$ is a good estimator of the jump magnitude of the regression surface at (x_0, y_0), and this jump point is likely to be detected by the criterion M_n. On the other hand, if (x_0, y_0) is a singular point of the JLCs with nonzero jump magnitude, then it is either a crosspoint of several JLCs or a point on a single JLC which does not satisfy the Lipschitz condition around (x_0, y_0). In either case, the property of $M_n^{(1)}(x_0, y_0)$ (also $M_n^{(2)}(x_0, y_0)$) mentioned above may not be true. Thus, procedure (4.19)–(4.23) may not be able to detect the singular points. Some of these properties are summarized in the following theorem (Qiu 2002).

Theorem 4.5 *In model (4.1), suppose that the regression function f is Lipschitz (1) continuous in design space $[0,1] \times [0,1]$ except on the JLCs; the design points satisfy the assumption (A 4.1); and $E|\varepsilon_1|^p < \infty$ for some $p \geq 2$. The two one-sided kernel functions K_1^* and K_2^* are assumed to be density functions and Lipschitz (1) continuous in their supports. The bandwidths h_n and p_n are assumed to satisfy the assumption (A 4.2) and the conditions that*

(i) $\frac{n^\nu}{\beta_n \log n}\left[p_n + \frac{1}{n^\lambda h_n p_n} + \frac{1}{\sqrt{n h_n^2 p_n}}\right] = o(1)$;

(ii) $\frac{n^{2\nu}}{n h_n p_n \beta_n} = O(1)$; *and*

(iii) $\frac{n^{\nu+1/p-1}}{\beta_n h_n p_n \log n} = o(1)$,

where ν is a positive number and $\{\beta_n\}$ is a series of numbers satisfying the condition that $\lim_{n\to\infty} \beta_n = \infty$. The significance level α_n is assumed to satisfy the conditions that (i) $\chi^2_{1,1-\alpha_n/2}/(n h_n p_n) = o(1)$ and (ii) $(\beta_n \log n)^2 n h_n p_n/(n^{2\nu}\chi^2_{1,1-\alpha_n/2}) = O(1)$. Then

$$\lim_{n\to\infty} d_H\left(\widehat{D}_n \bigcap \Omega_\rho, D \bigcap \Omega_\rho\right) = 0, \ a.s.,$$

and the convergence rate is $O(p_n)$, where d_H denotes the Hausdorff distance (cf. expression (4.12)), $\Omega_\rho = \{(x,y) : (x,y) \in [\rho, 1-\rho] \times [\rho, 1-\rho], \sqrt{(x-x^*)^2 + (y-y^*)^2} \geq \rho$, where (x^*, y^*) is some singular point of the JLCs$\}$, and $0 < \rho < 1/2$ is any small constant.

Theorem 4.5 says that, excluding any small areas around the singular points of the JLCs and the border of the design space, the set of detected jumps converges almost surely to the set of true jumps in Hausdorff distance. In particular, by checking the conditions in the above theorem, the convergence rate of $\lim_{n\to\infty} d(\widehat{D}_n \bigcap \Omega_\rho, D \bigcap \Omega_\rho) = 0$, *a.s.*, could reach $o(n^{-1/6+\tau})$ for any $0 < \tau < 1/6$ if h_n and p_n are chosen appropriately. It should be noticed that this result only requires the regression surface to be Lipschitz (1) continuous in the design space except on the JLCs, which is reasonable for most applications. Theorem 4.5 also implies that regions around the singular points of the JLCs and the border of the design space are the places where procedure (4.19)–(4.23) might fail to detect jumps properly.

Example 4.4 (Qiu 2002) Suppose that the regression function is $f(x,y) = \frac{1}{4}(1-x)y + [1+0.2\sin(2\pi x)]I(y \geq 0.6\sin(\pi x)+0.2)$ for $(x,y) \in [0,1] \times [0,1]$. This function has a unique JLC with the expression $\phi(x) = 0.6\sin(\pi x) + 0.2$ and corresponding jump magnitude function $1 + 0.2\sin(2\pi x)$. Observations are generated from model (1.3) at $n = n_1^2$ equally spaced design points $\{(x_i, y_j) = (i/n_1, j/n_1), i, j = 1, 2, \cdots, n_1\}$, where $n_1 = 100$ and $\varepsilon_{11} \sim N(0, 0.5^2)$. Then the true regression surface and a set of observations are displayed in Figures 4.12(a) and 4.12(b), respectively. In procedure (4.19)–(4.23), suppose that K_1^*, K_2^*, h_n, and p_n are chosen to be the same as those in Example 4.3 and the significance level is fixed at $\alpha_n = 0.01$. The detected jumps from the data shown in Figure 4.12(b) are displayed in Figure 4.13(b). It can be noticed that there are two kinds of deceptive jump candidates, as we noticed in Figure 4.8. The first kind consists of those points scattered in the design space due to the nature of hypothesis testing on which the threshold value is based. The second kind of deceptive jump candidates consists of those points near the true JLC, which are due to the nature of local smoothing. Figures 4.13(c) and 4.13(d) show the results after the two modification procedures discussed in Subsection 4.3.3 are applied sequentially to the results in Figure 4.13(b). For convenience of comparison, the true JLC is plotted in Figure 4.13(a).

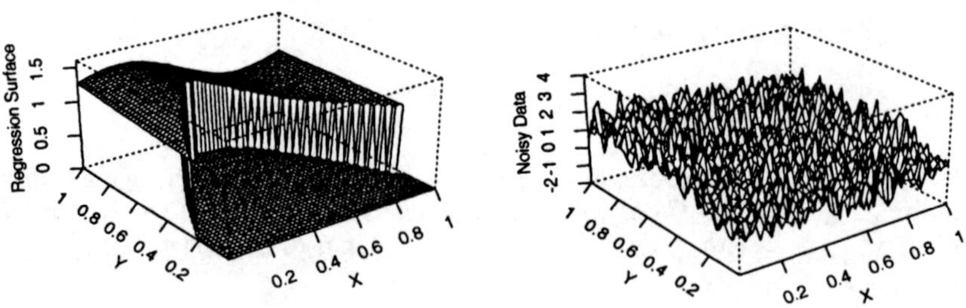

(a) (b)

Fig. 4.12 (a): True regression surface. (b): Noisy version of the regression surface with $\sigma = 0.5$.

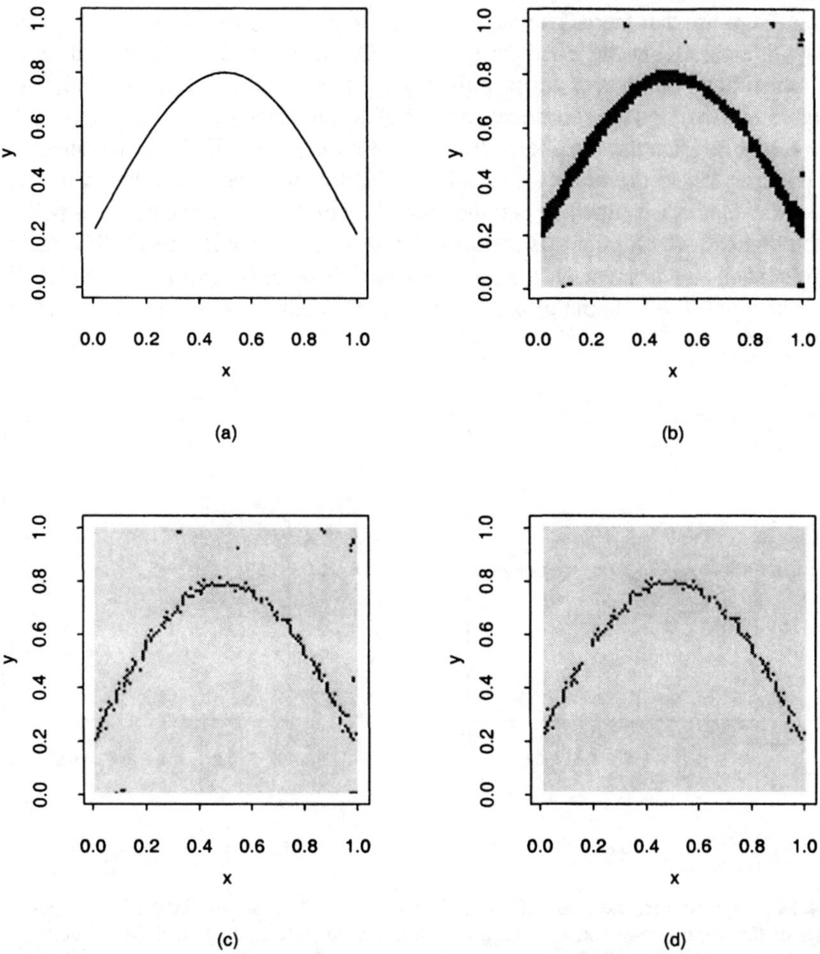

Fig. 4.13 (a): True jump location curve. (b): Detected jump points by procedure (4.19)–(4.23). (c): Modified version of plot (b) by the first modification procedure (MP1) (to make the detected JLC thinner). (d): Modified version of plot (c) by the second modification procedure (MP2) (to delete some scattered jump candidates). In this example, $n_1 = 100, \sigma = 0.5, h_n = p_n = 0.05$, and $\alpha_n = 0.01$.

4.4.2 Measuring the performance of jump detection procedures

Theoretically speaking, the Hausdorff distance defined by expression (4.12) in Subsection 4.2.5 can be used for measuring the distance between the point set of detected jumps and the point set of true jumps. However, in reality, this distance is not easy to compute. There are several infimums/supremums involved, which are computationally expensive.

Another limitation of the Hausdorff distance can be illustrated by the following example. Suppose that there is a unique JLC in the design space $[0,1] \times [0,1]$. This JLC is a line parallel to the x-axis at $y = 0.5$. Consider two sets of detected jump points: the first set consists of design points on the line $y = 0.5$ and a point $(0.8, 0.1)$ (cf. Figure 4.14(a)), and the second set consists of design points on the line $y = 0.2$ (cf. Figure 4.14(b)). Then the Hausdorff distance between the true JLC and the detected jump points is 0.4 in the first case and 0.3 in the second case. Thus, based on the Hausdorff distance, we conclude that the second set of detected jump points is better than the first set, which might be the opposite to what we would expect. The main reason for such an occurrence is that the Hausdorff distance is sensitive to individual points (the point $(0.8, 0.1)$ in this case), due to the supremums/infimums involved in its definition.

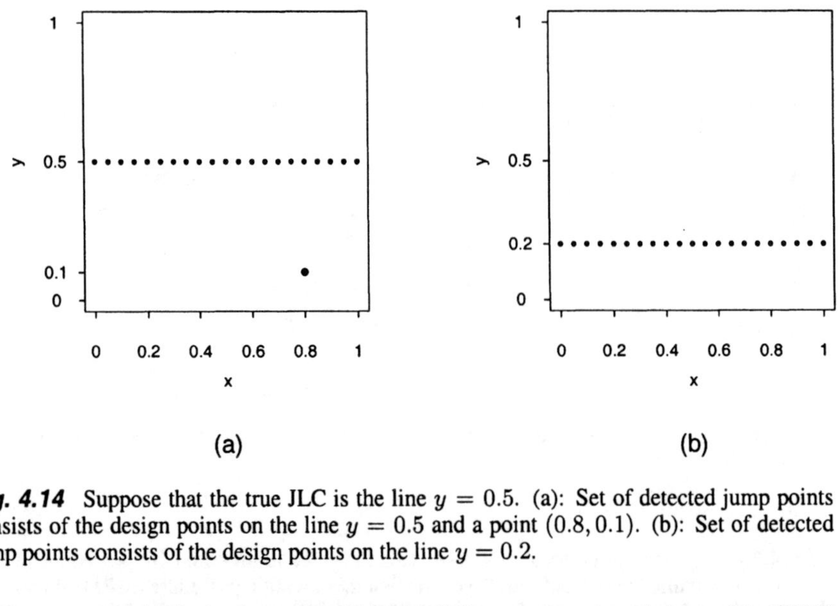

Fig. 4.14 Suppose that the true JLC is the line $y = 0.5$. (a): Set of detected jump points consists of the design points on the line $y = 0.5$ and a point $(0.8, 0.1)$. (b): Set of detected jump points consists of the design points on the line $y = 0.2$.

To overcome the limitations of the Hausdorff distance mentioned above, Qiu (2002) suggested an alternative performance measure defined by

$$d^*\left(\widehat{D}_n, D\right) = w\frac{|\widehat{D}_n \backslash D|}{|\Omega \backslash D|} + (1-w)\frac{|D \backslash \widehat{D}_n|}{|D|}, \tag{4.25}$$

where $0 \leq w \leq 1$ is a constant, Ω denotes the entire design space, D denotes the pointset of the true jumps, $\widehat{D}_n$ is a pointset estimator of D, $A \backslash B$ denotes the set of points in A but not in B, and $|A|$ denotes the number of design points in the point set A.

Theoretically speaking, $|D|$ could be zero because it could happen that none of the design points is exactly on the true JLCs. In applications, we can use $D^* = \{(x_i, y_i) : d((x_i, y_i), D) \leq 1/\sqrt{2n}\}$ to replace D for the purpose of calculating d^*, where $d(\cdot, \cdot)$

denotes the Euclidean distance. That is, all design points whose Euclidean distances to the true JLCs are less than or equal to $1/\sqrt{2n}$ (i.e., half of the distance between two diagonally neighboring design points, if all design points are equally spaced in the design space) are considered in D.

In equation (4.25), $|\widehat{D}_n \backslash D|$ is the number of false jump detections and $|\Omega \backslash D|$ is the total number of continuity points in the design space. Therefore, $|\widehat{D}_n \backslash D|/|\Omega \backslash D|$ is the proportion of the detected continuity points to all continuity points. Similarly, $|D \backslash \widehat{D}_n|/|D|$ is the proportion of the true jump points missed by the jump detection procedure. The measure d^* is their weighted average. The weight w represents the relative importance of the first proportion, and it needs to be specified by the user. In the case that the two proportions are equally important, w can be simply chosen to be 0.5. In the two extreme cases that the JLCs are completely detected with no false jump detection and that all the continuity points are detected with all true jumps missed, d^* equals 0 and 1, respectively. In a general case, d^* is between 0 and 1. The larger its value, the better the jump detection procedure performs, and vice versa.

In the definition of d^*, we use proportions instead of supremum/infimum values to make d^* more robust to individual points. It is not hard to check that d^* is close to zero in the case of Figure 4.14(a) and larger than 0.5 in the case of Figure 4.14(b), which is consistent with what we would expect about the two sets of detected jump points. Compared to the Hausdorff distance, d^* is also quite easy to compute, with computational complexity of $O(n)$. As long as D and $\widehat{D}_n$ are determined, d^* can be calculated immediately.

We should point out that the performance measure d^* is not perfect in applications. In the example of Figure 4.14, if the detected jumps are those design points on the line $y = 0.5 + 1/n_1$, then we would probably expect that the jump detection procedure performs better when n_1 gets larger. However, it can be checked that the value of d^* does not depend on n_1 in such a case. Measuring the jump detection performance properly is an important topic because it is related directly to comparisons of different jump detection procedures. We hope there will be more future research to address this issue.

4.4.3 Connection to the Sobel edge detector

In the image processing literature, the Sobel operator is a "classic" edge detector (cf. Section 6.2 for related discussion). It is based on two so-called Sobel masks, displayed in Figure 4.15. Like most image processing techniques, the Sobel operator is appropriate to use only in cases that the design points are regularly spaced in the design space. For a given design point (x_i, y_i), a 3×3 neighborhood is considered. Convolution of the first Sobel mask with observations in the neighborhood, which is defined by the summation of the products of the numbers in the mask and the observations at the corresponding locations in the neighborhood, is used to estimate the partial derivative of the image intensity function with respect to y. This estimator is denoted by $\widehat{f}_y^{(i)}$. Similarly, the second mask is used to obtain an estimator of the partial derivative with respect to x, which is denoted by $\widehat{f}_x^{(i)}$. Then the quantity

1	2	1
0	0	0
-1	-2	-1

-1	0	1
-2	0	2
-1	0	1

Fig. 4.15 Two Sobel masks.

$[(\widehat{f}_x^{(i)})^2+(\widehat{f}_y^{(i)})^2]^{1/2}$ is used as an edge detection criterion, with large values indicating possible edges.

The criterion $M_n(x_i, y_i)$ defined in equation (4.21) has several strong similarities to the Sobel edge detector but also has some advantages. The quantity $\widehat{f}_y^{(i)}$ is similar to the $M_n^{(1)}(x_i, y_i)$ defined in equation (4.19) in the sense that $\widehat{f}_y^{(i)}$ is also proportional to a difference of two weighted averages along the y-axis direction. Similarly, the quantity $\widehat{f}_x^{(i)}$ is related to $M_n^{(2)}(x_i, y_i)$. The Sobel edge detector uses the Euclidean length of the estimated gradient vector as its edge detection criterion. In equation (4.21), we suggest using $M_n(x_i, y_i)$, which is the maximum of $|M_n^{(1)}(x_i, y_i)|$ and $|M_n^{(2)}(x_i, y_i)|$, to detect jumps based on the following consideration. When the point (x_i, y_i) is on a JLC, the jump structure of the regression surface contaminates some of the four kernel averages used in constructing $M_n(x_i, y_i)$ (cf. equations (4.19)-(4.21)) as estimators of the regression surface at (x_i, y_i). For instance, if the JLC is parallel to the x-axis, then the two kernel averages used in constructing $M_n^{(1)}(x_i, y_i)$ estimate the surface well, but the two kernel averages used in constructing $M_n^{(2)}(x_i, y_i)$ do not provide much helpful information for jump detection. In such a case, if $M_n^{(2)}(x_i, y_i)$ is included in the jump detection criterion as the Sobel edge detector does, it can only make the jump detection criterion noisier and thus have negative impact on jump detection. This negative impact is mostly eliminated in criterion (4.21) because, most likely, $M_n(x_i, y_i)$ equals $M_n^{(1)}(x_i, y_i)$ in such a case.

The conventional Sobel edge detector uses windows of size 3×3 only. With this small size, its ability to remove the noise and enhance the jumps is limited. This limitation is lifted by procedure (4.19)–(4.23) because larger window sizes are allowed in this procedure. The next example shows that larger windows do improve the ability of procedure (4.19)–(4.23) for jump detection.

Example 4.5 In Example 4.4, suppose that n_1 can vary among 100, 200, 300, and 400 and σ can vary among 0.25, 0.5, 0.75, and 1. The two kernel functions K_1^* and K_2^* are chosen to be the same as those in Example 4.3. The two bandwidths h_n and p_n are chosen to be k_1/n_1 and k_2/n_1, respectively, where k_1 and k_2 are two odd integer numbers. For each combination of n_1 and σ, the best pair of window

Table 4.1 For each combination of n_1 and σ, the best pair of window sizes (k_1, k_2) and the corresponding d^* value (in parentheses) are presented.

	n_1			
σ	100	200	300	400
0.25	3,3 (0.00961)	3,3 (0.00469)	3,3 (0.00301)	3,3 (0.00233)
0.50	5,5 (0.01142)	5,5 (0.00564)	5,5 (0.00370)	5,5 (0.00280)
0.75	7,7 (0.01422)	7,7 (0.00701)	7,7 (0.00450)	7,7 (0.00321)
1.0	9,9 (0.01880)	9,9 (0.00784)	9,9 (0.00540)	9,9 (0.00430)

sizes (k_1, k_2) based on the performance measure d^* (cf. equation (4.25)) and 100 replications is presented in Table 4.1, along with the averaged d^* value.

From Table 4.1, it can be seen that (1) d^* gets smaller when n_1 is chosen larger, which reflects the consistency of the detected jump points; (2) the window sizes should be chosen larger if the data is noisier (namely, σ is bigger); (3) the window sizes should be quite stable when the sample size changes. Theorem 4.5 tells us that the convergence rate of the detected jumps is $O(p_n)$. Therefore, the window sizes k_1 and k_2 should be stable to achieve the fastest convergence rate when the sample size increases. These simulation results confirm this conclusion.

One interesting note: from Table 4.1, it seems that the two bandwidths h_n and p_n should be chosen close to each other, which contradicts assumption (A 4.2). Qiu (2002) pointed out with numerical examples that (i) the major reason why h_n and p_n are close to each other in Example 4.5 is that the JLC is smooth in this case, and (ii) when the JLC has large curvatures at some places, h_n should be chosen smaller than p_n in order to detect jumps at such places successfully. That is, assumption (A 4.2) is necessary for procedure (4.19)–(4.23) to detect all nonsingular points of the JLCs successfully, at which the JLCs may have quite different curvatures.

4.4.4 Jump detection in more than two given directions

The jump detection procedure (4.19)–(4.23) searches the x-axis and y-axis directions only at each design point for a possible jump. This idea can be generalized by searching more than two directions as follows. Let $0 \leq \theta_1 \leq \theta_2 \leq \cdots \leq \theta_m < \pi$ be m given directions in $[0, \pi)$, where $m \geq 2$ is an integer. These directions are ordered counterclockwise with the positive y-axis direction corresponding to $\theta = 0$ (cf. Figure 4.9(a)). In applications, we suggest using equally spaced angles in $[0, \pi)$, namely $\theta_j = (j-1)\pi/m$ for $j = 1, 2, \cdots, m$, unless users have some prior information about the jump directions. At a given point (x, y), we define

$$M_{n,m}(x,y) = \max\left\{|M_{RDKE}(\theta_j, x, y)|, j = 1, 2, \cdots, m\right\}, \quad (4.26)$$

where $M_{RDKE}(\theta_j, x, y)$ is defined by equation (4.2) and thus is a difference of two averages of the observations in two one-sided neighborhoods along the direction of $(\cos(\theta_j), \sin(\theta_j))$. Similar to the threshold value defined by equation (4.23), Qiu (2002) derived the following formula for choosing the threshold value of $M_{n,m}(x, y)$:

$$u_{n,m} = \left\{\frac{\sigma^2 \chi^2_{1,1-\alpha_n/m}}{n h_n p_n}\left[\int_{-1/2}^{1/2}\int_0^1 (K_2(u,v))^2\, dudv + \int_{-1/2}^{1/2}\int_{-1}^{0} (K_1(u,v))^2\, dudv\right]\right\}^{1/2}. \quad (4.27)$$

It should be pointed out that both formulas (4.23) and (4.27) are derived based on the Bonferroni correction, which are quite conservative in the sense that the threshold values based on these formulas tend to be larger than what we would expect from the significance level α_n, especially when m is large. To compensate for this, in applications, we suggest choosing α_n a little larger than its regular value. More accurate threshold formulas might be possible by studying the correlation among $\{M_{RDKE}(\theta_j, x, y), j = 1, 2, \cdots, m\}$.

If the generalized procedure (4.26)–(4.27) is used for jump detection, then the definition of nonsingular points of the JLCs discussed in Subsection 4.4.1 should be changed accordingly. More specifically, condition (iii) in the original definition needs to be modified into the following condition (iii)':

(iii)' $\cdots$ Then there exists a neighborhood $N(t_0)$ of t_0 and an angle $\theta \in \{\theta_i\}_{i=1}^m$ such that in this neighborhood the JLC can be expressed by $y(t) = \phi^*(x(t))$ for some Lipschitz (1) continuous function ϕ^*, after the coordinate system is rotated counterclockwise by θ.

It can be checked that the consistency of the detected jumps stated in Theorem 4.5 is still true in this case after M_n and u_n defined by equations (4.21) and (4.23) are replaced by $M_{n,m}$ and $u_{n,m}$ defined by equations (4.26) and (4.27), and after condition (iii) in the original definition of nonsingular points is changed to the above condition (iii)'.

When $m = 2$, $\theta_1 = 0$, and $\theta_2 = \pi/2$, equation (4.26) is identical to equation (4.21). On the other hand, when m is large enough such that $\{\theta_i\}_{j=1}^m$ are dense in $[0, \pi)$, the generalized jump detection procedure (4.26)–(4.27) is almost equivalent to the RDKE procedure discussed in Subsection 4.2.1. It is apparent that the value of m is related directly to the amount of computation required: the larger the value of m is, the more computation the procedure requires. In the previous paragraph, we have also explained that m is related to the conditions on the JLCs required by the procedure. That is, if m is chosen larger, then the conditions for a point on the JLCs to be a nonsingular point would be weaker, and consequently more jump points can be detected by the procedure. So m is related to the tradeoff between the computational complexity of the procedure and its power to detect jumps.

Example 4.6 In the setup of Example 4.5, suppose that $n_1 = 100, \sigma = 0.5$, and $(k_1, k_2) = (5, 5)$. In procedure (4.26)–(4.27), the kernel functions are chosen to be the same as those in Example 4.3 and the value of m can change among 2, 4, 6, 8, 10, 12, 14, 16, 18, and 20. The amount of computation each simulation requires is roughly proportional to the value of m. The averaged values of the performance measure d^* based on 100 replications are plotted in Figure 4.16. This plot shows that jump detection is improved by searching more directions at each design point, at the expense of more computing time when m increases.

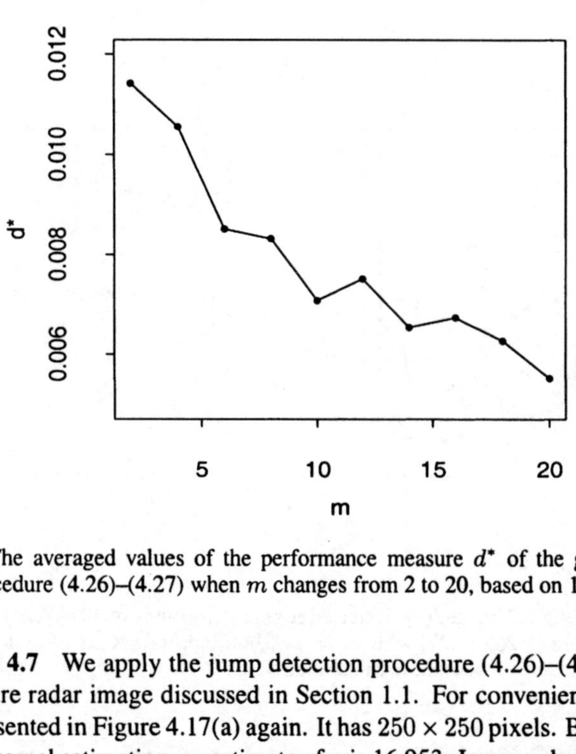

Fig. 4.16 The averaged values of the performance measure d^* of the generalized jump detection procedure (4.26)–(4.27) when m changes from 2 to 20, based on 100 replications.

Example 4.7 We apply the jump detection procedure (4.26)–(4.27) to the synthetic aperture radar image discussed in Section 1.1. For convenience, the original image is presented in Figure 4.17(a) again. It has 250×250 pixels. By a preliminary local linear kernel estimation, an estimate of σ is 16.953. In procedure (4.26)–(4.27), the parameters are chosen to be $(k_1, k_2) = (5, 5)$ and $\alpha_n = 0.01$, as in Example 4.4. The value of m is either 2 or 4. When $m = 2$, procedure (4.26)–(4.27) is the same as procedure (4.19)–(4.23). The detected jumps are presented in Figures 4.17(b) and 4.17(c), respectively, for $m = 2$ and 4. The results in Figure 4.17(c) are then modified by the two modification procedures discussed in Subsection 4.3.3, and the results are plotted in Figure 4.17(d). We can see that most edges are detected well by the procedure. If we compare Figures 4.17(b) and 4.17(c) carefully, we can see that the procedure with $m = 4$ detects edges slightly better than the procedure with $m = 2$ (see e.g., the middle-left parts of the plots).

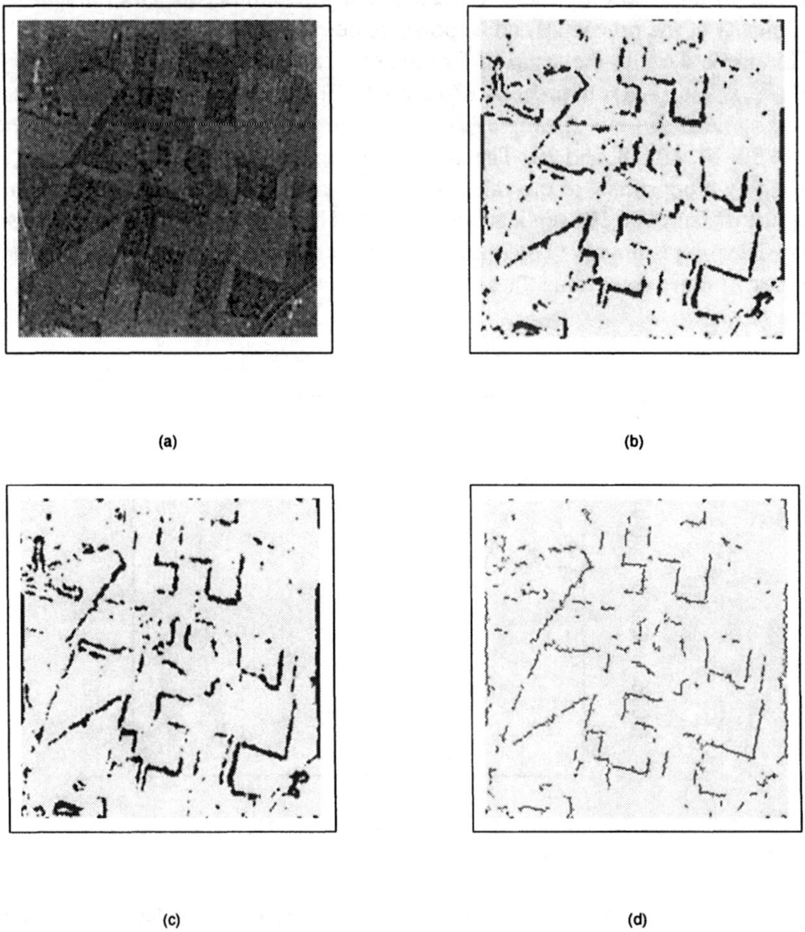

Fig. 4.17 (a): Original image. (b): Detected edges by procedure (4.19)–(4.23). (c): Detected edges by procedure (4.26)–(4.27) with $m = 4$. (d): Modified results of plot (c) by the two modification procedures discussed in Subsection 4.3.3.

4.5 SOME DISCUSSIONS

The 2-D jump detection procedures discussed in Section 4.2 result in smooth curve estimators of the JLCs, but they have some major restrictions. To begin, they require the assumption that there is only one JLC in the design space and the JLC satisfies some smoothness conditions. Although some of these procedures can be generalized for detecting jumps when there are multiple JLCs in the design space, the number of JLCs still needs to be known. The estimators are usually obtained by searching some populations of candidate JLCs, often involving several minimization/maximization processes. These jump detection procedures have some good theoretical properties, but they also have some limitations for practical use. One major limitation is the restrictive assumptions required on the JLCs, such as assuming that the number of JLCs is known; such information is typically tough to determine in applications. Another major limitation lies behind the necessary minimization/maximization processes, which are often computationally expensive.

To make the assumptions on the model more flexible, several jump detection procedures in the statistical literature treat the JLCs as a pointset in the design space. The pointset of the true JLCs is then estimated by another pointset in the same design space. Since points in a pointset need not form curves, pointsets are more flexible to handle than curves. In Section 4.3, we have demonstrated this approach by introducing a jump detection procedure suggested by Qiu and Yandell (1997). From the introduction there, it can be seen that arbitrary JLCs can be detected by such a procedure with relatively simple computation. However, to estimate the JLCs by a pointset, there would be two types of deceptive jump points existing in the pointset of detected jump points. Deceptive jump points of one type are located nearby the true JLCs, making the detected JLCs thick. They are mostly due to the nature of local smoothing. The second type of deceptive jump points consists of points that are scattered in the whole design space. They are mainly caused by the nature of the hypothesis tests on which the threshold values of the related jump detection criteria are based. To remove these two types of deceptive jump points, two modification procedures have been discussed in Subsection 4.3.3.

Theoretically speaking, the distance between the pointset of the true JLCs and the pointset of the detected JLCs can be measured by the Hausdorff distance. However, in applications, this distance is not convenient to compute, and it is also sensitive to individual points of the two pointsets. To overcome these limitations, Qiu (2002) suggested a new measure for evaluating the performance of the jump detection procedures. This measure is a weighted average of the proportion of the detected continuity points among all continuity points and the proportion of the missed jump points among all jump points.

To simplify the computation of the RDKE jump detection procedure discussed in Subsection 4.2.1, Qiu (2002) also suggested searching two or more given directions at a given point for a possible jump point. This jump detection procedure can be regarded as a generalization of the Sobel edge detector commonly used in the image processing literature. It has been shown that the detected jumps by this procedure converge to the true jumps if the two bandwidths used satisfy the assumption (A 4.2) and if other regularity conditions are also satisfied.

There are many research problems on jump detection in regression surfaces that should be addressed in the future research. As pointed out in Sections 4.1 and 4.3, the true JLCs are some curves in the design space. To avoid restrictive assumptions on the model and expensive computation as well, they are regarded as a pointset and estimated by another pointset in the design space. However, the pointset of the detected jumps has some obvious limitations as an estimator of the JLCs even after the detected jumps are modified by the two modification procedures (MP1) and (MP2) discussed in Subsection 4.3.3. For example, the detected JLCs would have some breaks here or there (cf. Figure 4.17(d)), and sometimes they even look quite different from curves (cf. Figure 4.2(b)). To overcome these limitations, one possible approach is to suggest some better modification procedures. Another approach is to replace the pointset of the detected jumps by some curves which go through the center of the detected jumps. The tracking algorithm discussed in Subsection 4.2.4 might also be helpful in this regard.

As mentioned at the end of Subsection 4.4.2, it is important to measure the performance of various jump detection procedures properly in order to compare them in a fair manner. The performance measure d^* discussed in Subsection 4.4.2 overcomes some limitations of the Hausdorff distance, but it has its own weakness. Therefore, much future research is needed to propose better performance measures.

The jump detection procedures discussed in Sections 4.3 and 4.4 cannot detect jumps appropriately around the singular points of the JLCs and the border of the design space. Some special treatments are needed at these places; at this point, no readily usable procedures have been developed.

From our numerical studies (cf. Table 4.1), the optimal bandwidths for jump detection depend mainly on the noise level of the data. They should be chosen to be quite stable when the sample size changes. More empirical guidelines on selecting the bandwidths and other procedure parameters, as well as theoretical study on bandwidth selection, are needed.

Problems

4.1 For the RDKE procedure discussed in Subsection 4.2.1, do the following problems.

4.1.1. Verify that the supports of the rotational kernel functions $K_j(\theta, \cdot, \cdot)$ for $j = 1, 2$ can be obtained by rotating the supports of the two one-sided kernel functions K_1^* and K_2^* by an angle θ counterclockwise about the origin, as demonstrated by Figure 4.1.

4.1.2. Check that assumption (A 4.1) is satisfied if the n design points are equally spaced in the design space $[0, 1] \times [0, 1]$ in $\sqrt{n}$ rows and $\sqrt{n}$ columns (assuming that $\sqrt{n}$ is a positive integer).

4.1.3. Suppose that the true JLC is a closed curve in the design space, which can be expressed by two univariate functions $\phi_1(x)$ and $\phi_2(x)$ for $x \in [A, B] \subseteq [0, 1]$ with $\phi_1(x) < \phi_2(x)$ for $x \in (A, B)$, $\phi_1(A) = \phi_2(A)$ and $\phi_1(B) = \phi_2(B)$, and that the absolute jump magnitudes at the JLC have a known lower bound $C_0 > 0$. Discuss how to estimate the JLC by the jump detection criterion M_{RDKE} defined in equation (4.2).

4.1.4. The file microarray.dat included in the ftp site of this book (see front cover for address) provides a dataset of one small portion of a microarray image. The portion includes 48×48 pixels. It is known that the image intensity function has one "smooth, closed, and simple" JLC (cf. Subsection 4.2.3 for a definition) in the interior of the portion. Use the estimation method suggested in the previous part to estimate the JLC.

4.2 Suppose that the error terms in model (4.1) are i.i.d. with a normal distribution and the regression function f has expression (4.4). Verify that the log-likelihood function constructed from the observations in U_j is, up to an additive constant, proportional to

$$\sum_{x_i \in [(j-1)\delta_n, j\delta_n)} I\left(y_i \leq g(x_i)\right)(Z_i - 1/2),$$

where U_j and δ_n are defined in Subsection 4.2.2.

4.3 Suppose that $\{(x_i, y_i, Z_i), i = 1, 2, \ldots, n\}$ are n observations generated from model (4.1) in which the regression function f has expression (4.7).

4.3.1. If the two constants a and b involved in the regression function f are both known, verify that $\{(x_i, y_i, Z_i'), i = 1, 2, \ldots, n\}$ can be regarded as n observations from model (4.1) in which the regression function f has expression (4.4), where $\{Z_i', i = 1, 2, \ldots, n\}$ are obtained from the transformation

$$Z_i' = I\left(Z_i > \frac{a+b}{2}\right), \quad i = 1, 2, \ldots, n.$$

4.3.2. If both a and b are unknown, suggest a consistent estimator of $(a+b)/2$ used in the above transformation.

4.4 Sketch a "smooth, closed, and simple" curve as defined in Subsection 4.2.3. In the plot, please indicate the inside region Γ_θ^I, the outside region Γ_θ^O, and the outward normal direction at a given point on the curve.

4.5 Based on equations (4.10) and (4.11), verify that the kernel-weighted local log-likelihood ratio at (x, y) is, up to an additive constant, proportional to $-w(x, y, \theta)$, where $w(x, y, \theta)$ is defined in Subsection 4.2.4.

4.6 In Subsection 2.2.4, the least squares (LS) estimation procedure is introduced in 1-D linear regression cases. In 2-D cases, suppose that the linear regression model is

$$Z_{ij} = \beta_0 + \beta_1(x_i - x^*) + \beta_2(y_j - y^*), \text{ for } i, j = 1, 2, \ldots, n_1,$$

where β_0, β_1 and β_2 are the regression coefficients and (x^*, y^*) is a given point in the design space. Then the regression function is actually a plane defined in the design space, and (β_1, β_2) is the gradient direction of this plane. The LS estimators of β_0, β_1, and β_2 are defined by the solution to b_0, b_1, and b_2 of the following minimization problem:

$$\min_{b_0, b_1, b_2 \in R} \sum_{i=1}^{n_1} \sum_{j=1}^{n_1} \left\{Z_{ij} - [b_0 + b_1(x_i - x^*) + b_2(y_j - y^*)]\right\}^2.$$

In Subsection 4.3.2, the 2-D LS estimation of a linear regression model is used in a local fashion. In the neighborhood $N(x_i, y_j)$ of a given design point (x_i, y_j), verify that the LS estimators of the regression coefficients of the local plane in $N(x_i, y_j)$ are given by equation (4.13).

4.7 Equations (4.14) and (4.15) define two neighboring design points of a given design point (x_i, y_j). Verify that the two neighboring design points defined by these two equations are on the line passing (x_i, y_j) along the direction $\vec{v}_{ij}$, as displayed by Figure 4.4.

4.8 Check that, when n_1 is large enough,

$$P(\delta^*_{ij} > b^*) \leq \alpha_{n_1},$$

where δ^*_{ij} and b^* are defined at the end of Subsection 4.3.2.

4.9 Reproduce Figures 4.7 and 4.8.

4.10 For the jump detection procedure discussed in Subsection 4.4.1, do the following problems.

4.10.1. Verify that $M_n^{(1)}$ and $M_n^{(2)}$ defined by equations (4.19) and (4.20) have the following relationship with the jump detection criterion M_{RDKE}:

$$M_n^{(1)}(x, y) = M_{RDKE}(0, x, y); \; M_n^{(2)}(x, y) = M_{RDKE}(\pi/2, x, y).$$

4.10.2. Check that, when n is large enough,

$$P\left(M_n(x, y) > u_n\right) \leq \alpha_n,$$

where M_n and u_n are defined by equations (4.21) and (4.23), respectively.

4.11 In the example of Figure 4.14, verify that:

(i) the Hausdorff distance between the true JLC and the detected jump points is 0.4 in the case of Figure 4.14(a) and is 0.3 in the case of Figure 4.14(b).

(ii) the d^* value is about zero in the case of Figure 4.14(a) and is larger than 0.5 in the case of Figure 4.14(b).

4.12 The synthetic aperture radar image shown in Figure 4.17(a) can be downloaded from the ftp site of this book (see the front cover for address). Using this data, reproduce Figure 4.17.

5

Jump-Preserving Surface Estimation By Local Smoothing

5.1 INTRODUCTION

Because the JLCs of a 2-D regression surface often represent outlines of objects (e.g., edges of an image of one or more objects) or places where the structure of the related 2-D process changes (e.g., places where an equitemperature surface in high sky or deep ocean changes dramatically), it is important to locate their positions accurately from observed data. Various jump detection techniques for this purpose have been discussed in the previous chapter.

On the other hand, in some applications, our ultimate goal is to reconstruct the regression surface from noisy data or to recover the true surface by removing noise; this is also the major objective of the conventional 2-D regression analysis discussed in Chapter 2. However, the conventional 2-D smoothing procedures are appropriate for reconstructing continuous surfaces only. If they are used directly for estimating jump surfaces, then many jumps would be blurred by them at the same time as noise is removed. This fact has been demonstrated in 1-D cases in Chapter 3, and it is also true in 2-D cases. In this chapter, we introduce some recent smoothing techniques which aim to accomplish the goal of estimating the regression surface with jumps preserved.

In 1-D cases, as mentioned in Subsection 3.2.1, there are two possible approaches to estimating a regression curve after its possible jumps are detected. One approach is to estimate the regression curve as usual in design subintervals separated by the detected jumps. The second approach has two steps: the estimated jump part of the regression function is first removed from the observations, and then the continuity part of the regression function is estimated as usual from the resulting new data obtained in the first step.

We now wish to generalize these two approaches to 2-D cases: in order to use the first approach, the design subspaces separated by the estimated JLCs should be well defined; in order to use the second approach, the jump part of the 2-D regression function and its estimator should have explicit mathematical expressions. However, these requirements can only be satisfied in some simple cases, such as the case when the number of JLCs is known and the JLCs can be expressed by univariate functions (cf. the discussion about multiple JLCs in Subsection 4.2.1). In general cases with arbitrary JLCs, neither one of these two approaches can be used easily.

When estimating arbitrary JLCs, the major strategy discussed in Chapter 4 is to treat the JLCs as a pointset in the design space and then to estimate the JLCs by another pointset in the same design space. By this strategy, the detected jumps are some set of points instead of curves in the design space, making "the two different sides" of a detected jump point undefined. Therefore, it is still a challenging problem to reconstruct the surface with possible jumps preserved, even after the JLCs have been estimated by a pointset.

Recently, several jump-preserving surface reconstruction procedures have been proposed in the statistical literature. Some of these procedures have been discussed in 1-D cases in Chapter 3; for instance, the sigma filter and the M smoother discussed by Chu *et al.* (1998) and introduced in Subsection 3.4.3 can be generalized without much difficulty for estimating regression surfaces with possible jumps preserved.

In this chapter, we focus on new procedures that have not been previously discussed. In Section 5.2, a three-stage local smoothing surface reconstruction procedure is introduced. This procedure preserves the jumps by fitting a principal component line through the detected jump points in a neighborhood of a given point. In Sections 5.3 and 5.4, two other local smoothing, jump-preserving surface reconstruction procedures are introduced; these preserve the jumps automatically without detecting the jumps explicitly. The first procedure preserves the jumps by introducing a threshold parameter, and the second one preserves the jumps by combining several local surface estimators in an appropriate way. A so-called "adaptive weighting smoothing" procedure is introduced in Section 5.5. Finally, some concluding remarks, including future research areas, are provided in Section 5.6.

5.2 A THREE-STAGE PROCEDURE

Qiu (1998) suggested a three-stage procedure for reconstructing regression surfaces with possible jumps preserved. In the first stage, jump candidate points are detected using a jump detection criterion. In the second stage, a local principal component line is fitted through the detected jump points in a neighborhood of a given design point. If there is a single JLC in the neighborhood, then this principal component line provides a first-order approximation to the JLC in that neighborhood. In the third stage, observations on the same side of the line as the given point are combined using a weighted average procedure to estimate the surface at the given point. If there are no jump candidate points in the neighborhood, then all observations in that neighborhood are used in surface estimation. If, however, a JLC exists in the neighborhood, then only those observations on one specific side of the principal component line are used. Thus blurring is automatically avoided around the jump locations. The three stages of this surface reconstruction procedure are discussed in some detail in the following three subsections.

5.2.1 Jump detection

The first stage of Qiu's procedure is to estimate the JLCs by a pointset in the design space. In Chapter 4, we discussed a number of jump detection procedures. Some of them estimate the JLCs by curves, while the others treat the JLCs as a pointset and estimate them by another pointset. Since curves in the design space can also be regarded as a pointset (e.g., for practical purposes, they can be replaced by those design points whose Euclidean distances to the curves are less than certain pre-specified value; see D^* defined in Subsection 4.4.2 for an example), generally speaking, any existing jump detection procedure could be used in the first stage of the three-stage procedure.

In this section, we assume that $n = n_1^2$ observations $\{(x_i, y_j, Z_{ij}), i, j = 1, 2, \ldots, n_1\}$ have been generated from model (1.3) in which the design points $\{(x_i, y_j) = (i/n_1, j/n_1), i, j = 1, 2, \ldots, n_1\}$ are equally spaced in the design space $[0, 1] \times [0, 1]$; the error terms $\{\varepsilon_{ij}, i, j = 1, 2, \ldots, n_1\}$ are independent and identically distributed (i.i.d.) with mean zero and unknown variance σ^2; and the 2-D regression function f is continuous in the design space except at the JLCs, which are defined in Section 4.1.

For simplicity, Qiu used the criterion δ_{ij}^*, defined by equation (4.18) in the neighborhood $N(x_i, y_j)$ with window size $k = 2\ell + 1$, for jump detection at any given design point (x_i, y_j), where $\ell + 1 \leq i, j \leq n_1 - \ell$ and $\ell \ll n_1$ is a positive integer. The detected jumps have properties similar to those stated in Theorem 4.4.

For the purpose of surface estimation by the three-stage procedure, the detected jumps obtained in the first stage do not need to be modified by any procedures such as those discussed in Subsection 4.3.3. This is valid as long as the window size used in the first stage is chosen smaller than the window size used in the third stage of the procedure, because the detected jumps are only used in the intermediate steps of the entire surface estimation procedure. If users do choose to modify the detected jumps before surface estimation, the results should still be similar to those without jump modifications.

5.2.2 First-order approximation to the JLCs

After the first stage, the jump candidate points are identified; surface estimation is now the primary goal. In regions with jump candidate points, special treatment is necessary before this estimation takes place. This treatment is outlined below as the second stage.

At any given design point (x_i, y_j), we consider its neighborhood $N^*(x_i, y_j) = \{(x_{i+s}, y_{j+t}) : s, t = -\ell^*, \ldots, 0, \ldots, \ell^*\}$ with window size $k^* = 2\ell^* + 1$, where the values of k^* and ℓ^* can differ from the k and ℓ used in the first stage. The jump candidate points in that neighborhood are denoted by $\{(w_r, v_r), r = 1, 2, \ldots, m\}$. We use (W, V) to denote a vector variable taking values over $\{(w_r, v_r), r = 1, 2, \ldots, m\}$, $\bar{w}, \bar{v}, \sigma_{ww}$ and σ_{vv} to denote the sample means and variances of $\{w_r, r = 1, 2, \ldots, m\}$ and $\{v_r, r = 1, 2, \ldots, m\}$, and σ_{wv} to denote their sample covariance.

As demonstrated by Figure 4.8(b), most jump candidate points detected in the first stage of the procedure are located in bands of width $\sqrt{2}k/n_1$, centered at the JLCs, since the jump detection criterion would have large values at these points due to the jump effect. The closer to the JLCs, the denser the jump candidate points tend to be, because more jump information is involved.

We then search for a line $a(W - \bar{w}) + b(V - \bar{v}) = 0$ with $a^2 + b^2 = 1$ such that

$$\text{Var}\left[a(W - \bar{w}) + b(V - \bar{v})\right] = \min_{\tilde{a}^2+\tilde{b}^2=1} \text{Var}\left[\tilde{a}(W - \bar{w}) + \tilde{b}(V - \bar{v})\right], \quad (5.1)$$

where "Var" is taken with respect to the random vector (W, V), which is assumed to have a Uniform distribution over $\{(w_r, v_r), r = 1, 2, \ldots, m\}$. An extreme case is that $\text{Var}(a(W - \bar{w}) + b(V - \bar{v})) = 0$, in which case all jump candidate points are on the line $a(W - \bar{w}) + b(V - \bar{v}) = 0$.

The vector (a, b) determined by equation (5.1) is actually the second principal component (PC), the normalized eigenvector of the smaller eigenvalue of the covariance matrix of (W, V). The line $a(W - \bar{w}) + b(V - \bar{v}) = 0$ is the usual PC line that indicates the direction in which the jump candidate points have the biggest dispersion, as demonstrated in Figure 5.1.

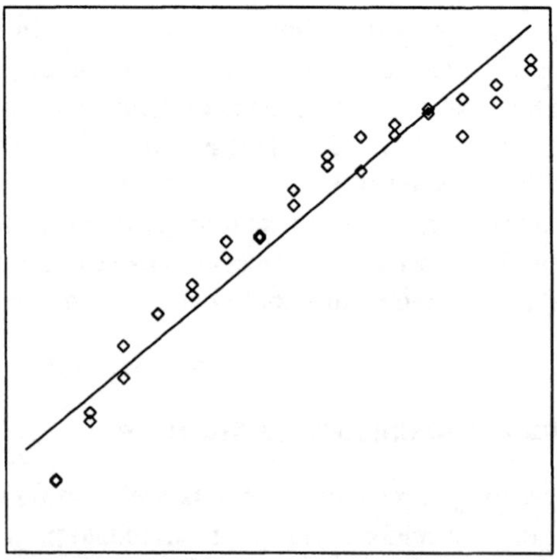

Fig. 5.1 The little diamonds denote the detected jump candidate points in a neighborhood of a given design point. The solid line denotes the local principal component line, which goes through the center of the detected jump candidate points.

After some calculations, the local PC line $a(W-\bar{w})+b(V-\bar{v})=0$ turns out to be

$$\sigma_{wv}(W-\bar{w})+(\lambda_1-\sigma_{ww})(V-\bar{v})=0, \tag{5.2}$$

where

$$\lambda_1=\frac{1}{2}\left(\sigma_{ww}+\sigma_{vv}-\sqrt{(\sigma_{ww}-\sigma_{vv})^2+4\sigma_{wv}^2}\right)$$

is the smaller eigenvalue of the covariance matrix of (W,V). It is not hard to check that $\lambda_1-\sigma_{ww}\leq 0$. Hence, the PC line has an increasing trend when $\{w_r, r=1,2,\ldots,m\}$ and $\{v_r, r=1,2,\ldots,m\}$ are positively correlated (i.e., $\sigma_{wv}>0$), and vice versa. This is intuitively reasonable.

Theorem 5.1 (Qiu 1998) *Suppose that the 2-D regression function f has continuous first-order partial derivatives over the design space $[0,1]\times[0,1]$ except on the JLCs at which the first-order partial derivatives have limits; the error terms satisfy the condition that $E|\varepsilon_{11}|^p<\infty$ for some $p>4$; the significance level α_{n_1} used in defining the threshold value of δ_{ij}^* (cf. Subsection 4.3.2) satisfies the conditions that $\lim_{n_1\to\infty}(\log n_1)^{-1}Z_{1-\alpha_{n_1}/2}=\infty$ and $\lim_{n_1\to\infty} n_1^{-2/p}Z_{1-\alpha_{n_1}/2}=0$; and the window sizes k and k^* satisfy the conditions that*

(i) $0<\lim_{n_1\to\infty} n_1^{-2/p}k<\infty$,

(ii) $\lim_{n_1\to\infty} k^*/n_1=0$,

(iii) $\lim_{n_1\to\infty} k/k^*=0$.

If (x_i,y_j) is the closest design point to a given nonsingular point P (by the definition of singular points in Subsection 4.3.2) on a JLC at which the JLC has a unique tangent line, then the PC line in $N^*(x_i,y_j)$ converges to the tangent line at P both pointwise (i.e., for a given point on the tangent line, its closest point on the PC line converges to it when n_1 tends to infinity) and in direction with probability 1. The convergence rate can reach $O(n_1^{-1+2/p})$.

The above result establishes convergence properties for the first-order approximation to the JLCs of the local PC lines. Second- or higher-order approximations (i.e., approximations by local quadratic or higher-order polynomials) might also be possible, but, from our numerical studies, second- or higher-order approximations have much larger variability than the first-order approximation. Therefore, we suggest using the first-order approximation in applications.

5.2.3 Estimation of jump regression surfaces

In the third stage, surface estimation can finally take place. In regions without any jump candidate points, the regression surface can be estimated as usual by various local smoothing procedures, such as the kernel smoothing procedures discussed in Chapter 2. However, in regions with jump candidate points, we only want to use points on the same side of the JLC as the given point. We can combine these two methods into one estimator, as described below.

In the neighborhood $N^*(x_i, y_j)$, observations on the same side of the fitted local PC line as the design point (x_i, y_j) are combined using a smoothing procedure to fit the regression surface at (x_i, y_j). The subscripts of the design points of these observations are determined by the following set:

$$S(x_i, y_j) = \begin{cases} \{(s,t) : \sigma_{wv}(x_{i+s} - \bar{w}) + (\lambda_1 - \sigma_{ww})(y_{j+t} - \bar{v}) \geq 0, \\ \qquad s, t = -\ell^*, \cdots, 0, \cdots, \ell^*\}, \\ \qquad \text{if } \sigma_{wv}(x_i - \bar{w}) + (\lambda_1 - \sigma_{ww})(y_j - \bar{v}) \geq 0; \\ \{(s,t) : \sigma_{wv}(x_{i+s} - \bar{w}) + (\lambda_1 - \sigma_{ww})(y_{j+t} - \bar{v}) < 0, \\ \qquad s, t = -\ell^*, \cdots, 0, \cdots, \ell^*\}, \\ \qquad \text{otherwise.} \end{cases}$$

Different smoothing procedures can be used here to average the observations. For simplicity of presentation, Qiu used the Nadaraya-Watson kernel smoothing procedure (cf. Subsection 2.3.3 for introduction). Let K be a bivariate kernel function defined on $[-L/2, L/2] \times [-L/2, L/2]$ with $L > 0$ a constant. Then the surface estimator at (x_i, y_j) of the three-stage procedure is defined by

$$\widehat{f}(x_i, y_j) = \frac{\sum_{s,t} K\left(\frac{Ls}{k^*}, \frac{Lt}{k^*}\right) Z_{i+s,j+t} I\left(\{(s,t) \in S(x_i, y_j)\}\right)}{\sum_{s,t} K\left(\frac{Ls}{k^*}, \frac{Lt}{k^*}\right) I\left(\{(s,t) \in S(x_i, y_j)\}\right)}. \tag{5.3}$$

We now examine how this estimator behaves in the continuity regions. When there are no jump candidate points in $N^*(x_i, y_j)$, it can be checked that σ_{wv} and $\lambda_1 - \sigma_{ww}$ in (5.2) are both zero. In the estimation procedure (5.3), all observations in the neighborhood are actually used in the surface estimation. If the number of jump candidate points is very small, such that no JLC is possible in the neighborhood (e.g., in the case that several jump candidate points exist in a corner of the neighborhood), then only a small amount of observations are dropped from the surface estimation. Although the three-stage procedure works well in these cases, Qiu still suggested making a judgment to ensure that a JLC is possible in the neighborhood (e.g., by the criterion that the number of jump candidate points is more than, say, $k^*/2$) before finding the PC line. The computation involved in each of such judgments is much less than the amount involved in finding the PC line and can thus save considerable computation time, since most design points would not be jump candidate points.

It is also worth mentioning that the first two stages (namely, the stages to detect the jump candidate points and define the local PC lines) of the three-stage procedure can be updated from one design point to the next because only a few points change; this will further reduce computation time.

The following result basically says that the surface estimator defined in formula (5.3) behaves as well as the conventional kernel estimator in the continuity regions (cf. Stone (1982) for optimal convergence rates of nonparametric regression estimators). In the neighborhood of JLCs, however, it sacrifices some of the convergence rate because only part of the observations in $N^*(x_i, y_j)$ are used in estimating $f(x_i, y_j)$.

Theorem 5.2 (Qiu 1998) *Besides the conditions in Theorem 5.1, suppose that f has continuous second-order partial derivatives in the design space, except on the*

JLCs at which its second-order partial derivatives have limits, and the kernel function K is nonnegative and satisfies the following conditions:

(i) $K(x,y) = 0$ *if* $(x,y) \notin [-L/2, L/2] \times [-L/2, L/2]$;

(ii) $\int_{-L/2}^{L/2} \int_{-L/2}^{L/2} K(u,v)\, dudv = 1$;

(iii) $\int_{-L/2}^{L/2} uK(u,v)\, dudv = \int_{-L/2}^{L/2} vK(u,v)\, dudv = 0$.

Then the surface estimator $\widehat{f}$ defined by formula (5.3) is uniformly L_2 consistent in

$$\mathcal{D}_\varepsilon^* = [\varepsilon, 1-\varepsilon] \times [\varepsilon, 1-\varepsilon] \backslash O(JLCs, \varepsilon),$$

where $O(JLCs, \varepsilon) = \{(x,y) : \|(x,y) - (x_*, y_*)\| < \varepsilon, (x_*, y_*)$ *is any point on the JLCs* $\}$, *and* $0 < \varepsilon < 0.5$ *is any small constant. The convergence rate could reach* $O(n_1^{-4/3})$ *if* $k^* \sim n_1^{2/3}$. *The surface estimator is L_2 consistent pointwise in* $O(JLCs, \varepsilon) \backslash \{(x,y) : \|(x,y) - (x^*, y^*)\| < \varepsilon, (x^*, y^*)$ *is some singular point of the* $JLCs\}$. *If* $k^* \sim n_1^{1/2}$, *then the pointwise convergence rate could reach* $O(n_1^{-1})$.

Example 5.1 Suppose that the true regression function is $f(x,y) = x + I(\{(y > 0.375 + 0.25x)$ or $(x > 0.5)\})$ for $(x,y) \in [0,1] \times [0,1]$, which has a unique JLC $\{(x,y) : y = 0.375 + 0.25x$ and $0 \leq x \leq 0.5\} \bigcup \{(x,y) : x = 0.5$ and $0 \leq y \leq 0.5\}$ with constant jump magnitude 1. Also, according to the definition of singular points in Subsection 4.3.2, the point $(0.5, 0.5)$ is a singular point of the JLC. This true regression surface is displayed in Figure 5.2(a). Observations are generated from model (1.3) at equally spaced design points in design space $[0,1] \times [0,1]$, with errors from $N(0, \sigma^2)$ distribution, with $\sigma = 0.5$ and $n_1 = n_2 = 100$. One set of data points are presented in Figure 5.2(b).

We then apply the three-stage procedure for surface reconstruction with $(k, k^*) = (9, 15)$ and $K(x,y) = \frac{9}{16}(1-x^2)(1-y^2)$ for $(x,y) \in [-1,1] \times [-1,1]$. Like most local smoothing procedures, the three-stage procedure has the boundary problem; i.e., its surface estimator around the border regions of the design space is not well defined. In this example, we use the symmetric "padding" method to overcome this difficulty. That is, we expand the design space in such a way that an "observation" in the expanded area takes the same value of the corresponding observation in the original design space; "corresponding" means that the two design points are symmetric about the border of the design space. Figure 5.2(c) shows the gradient directions $\{\vec{v}_{ij}\}$ of the fitted local LS planes as discussed in Subsection 4.3.2. It is apparent that $\{\vec{v}_{ij}\}$ reveal the JLC well even when the gradient direction of the true surface disagrees with the jump direction (cf. the left part of the JLC). The jump detection criterion δ_{ij}^* is shown in Figure 5.2(d), where the darker the color, the larger the value of δ_{ij}^*. Figure 5.2(d) shows that it is quite reasonable to use the "padding" method to handle the "boundary problem" in this case. The averaged surface fit of the three-stage procedure based on 100 replications and $\alpha_{n_1} = 0.001$ is presented in Figure 5.2(e). It can be seen that the procedure works well except at the places around the singular point. For instance, consider the cross section at $x = 0.25$: the 2.5 and 97.5 percentiles of 100 replications of the surface fit are plotted in Figure 5.2(f), as shown by the lower and upper dashed curves, respectively.

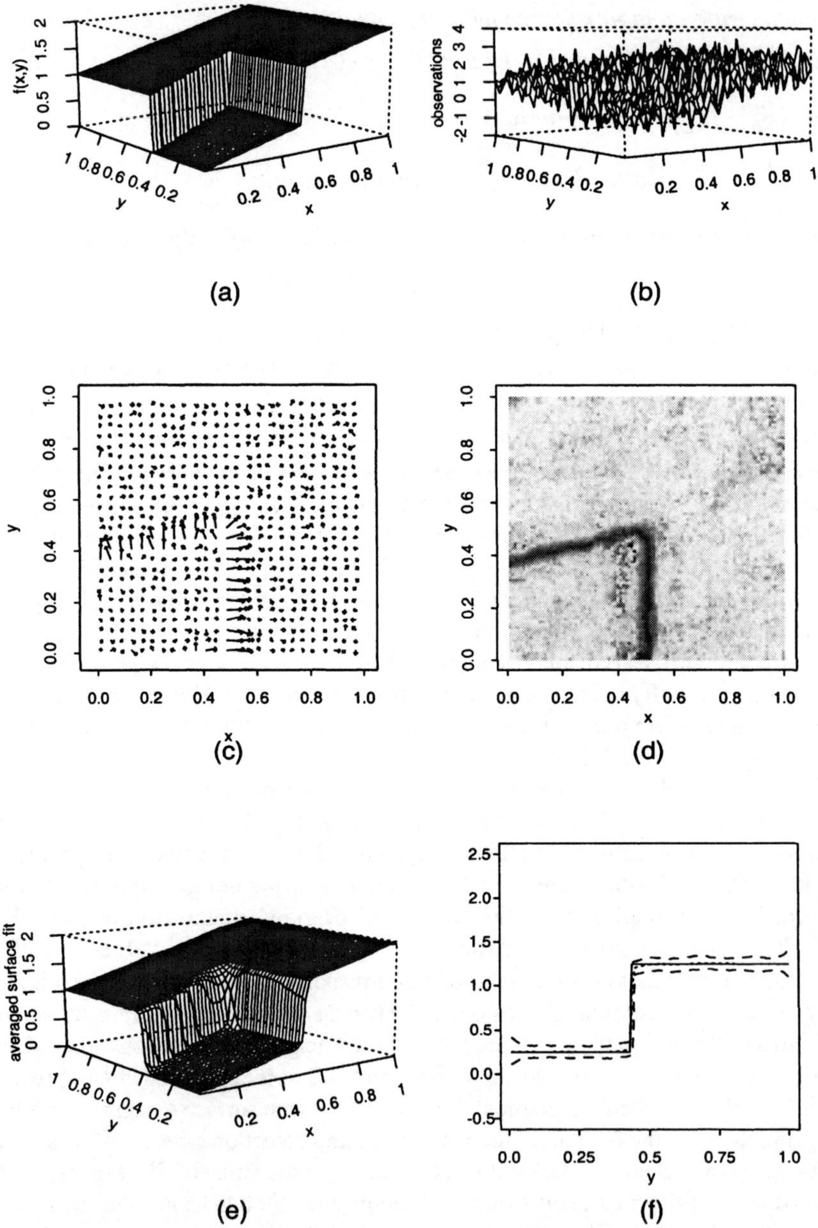

Fig. 5.2 (a): Original regression surface. (b): Observations. (c): Gradient directions $\{\vec{v}_{ij}\}$ of the fitted local LS planes (cf. Subsection 4.3.2). (d): Jump detection criterion $\{\delta^*_{ij}\}$. (e): Averaged surface fit based on 100 replications. (f): 2.5 and 97.5 percentiles of the 100 replications of the surface fit in the cross section of $x = 0.25$. In plot (f), the dashed curves represent the two percentiles, the solid curve is the cross section of the true regression surface, and the dotted curve is the averaged fit.

Example 5.2 In this example, we apply the three-stage procedure to the global topographical elevation data. Figure 5.3(a) shows this data contaminated by i.i.d. noise with $N(0, 1, 500^2)$ distribution. The darker the color, the higher is the elevation. The resolution of the data is 1 degree × 1 degree in longitude and latitude. Hence, the sample size is 181 × 360 =65,160. The elevation measurement is relative to the sea surface level, ranging from −8,964.0 meters to 6,096.0 meters in this data set. Because the number of JLCs cannot be counted accurately in this case, some jump detection procedures based on the assumption that the number of JLCs is known (such as those suggested by Korostelev and Tsybakov (1993), Müller and Song (1994a), O'Sullivan and Qian (1994), and Qiu (1997) (cf. Section 4.2 for introduction)) are not appropriate for detecting jumps in this case.

Theoretically speaking, the topographical elevation changes gradually and abrupt jumps can hardly be found. However, given the low resolution of the data, the jumps are quite obvious, particularly between land and sea. With the jump detection criterion δ^*_{ij} defined by equation (4.18) and its threshold value, both of which are discussed in Subsection 4.3.2, the detected jump locations are presented in Figure 5.3(e). Its modified version by the two procedures discussed in Subsection 4.3.3 is shown in Figure 5.3(f). In this step, α_{n_1} is set to be 0.001 as in the previous example, and the window width k is chosen to be 5 for good visual impression.

We then apply the three-stage jump-preserving surface fitting procedure to this dataset. The fitted surface with $k^* = 15$ and the same kernel function as that used in the previous example is presented in Figure 5.3(b). In the procedure, the detected jumps without jump modifications (i.e., those in Figure 5.3(e)) are used for fitting local PC lines. We can see that the three-stage procedure performs quite well in preserving the jumps, but some angles of the JLCs seem to be smoothed away. This phenomenon might be explained by these two reasons: (1) the angles are close to some singular points of the JLCs, and (2) the *first* order approximation to the JLCs of the local PC lines behaves this way.

As a comparison, we plot the fitted surfaces by the conventional Nadaraya-Watson kernel smoothing method and the local median smoothing method in Figures 5.3(c) and 5.3(d), respectively. The same window sizes and kernel function are used in the three methods. The MSE values of the fitted surfaces by the three-stage procedure, the conventional Nadaraya-Watson kernel smoothing procedure, and the local median smoothing procedure are 5.979×10^5, 6.195×10^5, and 8.684×10^5, respectively.

An interesting side note: for this data set, there is no "boundary problem" since the data are from a globe. Longitude 0 degree west is the same location as that of longitude 0 degree east, although they are at two ends in the plot. In the Arctic Circle and the Antarctic Circle, neighborhood relationship can also be carefully defined.

5.3 SURFACE RECONSTRUCTION WITH THRESHOLDING

The three-stage surface reconstruction procedure introduced in the previous section preserves the jumps by detecting the possible jump locations first and then approximating the JLCs with local PC lines. Although this procedure is theoretically ap-

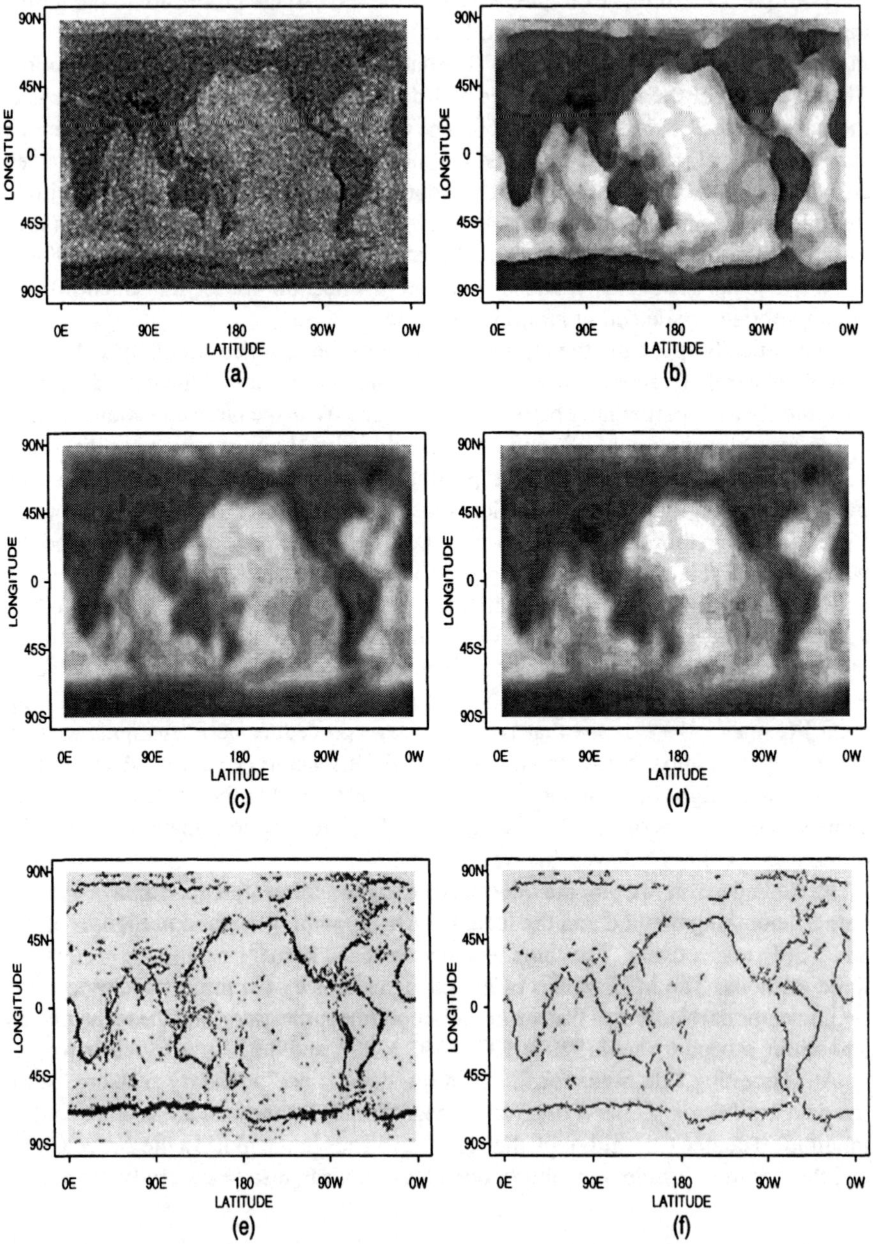

Fig. 5.3 (a): Global topographical elevation data contaminated by i.i.d. noise with $N(0, 1, 500^2)$ distribution. (b): Fitted surface by the three-stage procedure. (c): Fitted surface by the conventional Nadaraya-Watson kernel smoothing method. (d): Fitted surface by the local median smoothing method. (e): Detected jump positions by the jump detection criterion δ^*_{ij} discussed in Subsection 4.3.2. (f): Modified version of the image in plot (e) by the two modification procedures discussed in Subsection 4.3.3.

pealing with relatively simple computation due to its local smoothing nature, it is still quite complicated to use, compared to most conventional local smoothing procedures discussed in Chapter 2. Such conventional surface estimators usually have explicit mathematical formulas (cf. the multivariate Nadaraya-Watson kernel estimator discussed in Subsection 2.3.3).

Generally speaking, there are two major benefits for a curve/surface estimator to have an explicit mathematical formula. One is that such an estimator is often easy to use and simple to compute; as long as the procedure parameters involved are chosen properly, the estimator can be computed immediately. The second major benefit is that it is usually easier for us to study the statistical properties of such an estimator. As a contrast, the surface estimator of the three-stage procedure can be obtained only after the possible jump locations are detected and the local PC lines are fitted, which are both quite technical stages for general users.

A major goal of this author's recent research is to reconstruct the surface with possible jumps preserved, by means of a procedure which is relatively easy to use like most conventional local smoothing procedures. In this section, we introduce such a recent surface reconstruction procedure suggested by Qiu (2004). This procedure reconstructs the surface by local piecewisely linear kernel smoothing and can preserve possible jumps without detecting them explicitly. Its surface estimator has an explicit formula like most conventional local smoothing surface estimators; thus, it is quite convenient to use. The basic idea and the form of the surface estimator for this procedure are introduced in Subsection 5.3.1. Selection of its two procedure parameters is discussed in Subsection 5.3.2.

5.3.1 Surface reconstruction by local piecewisely linear kernel smoothing

The regression model in this section is set up similarly to that in Section 5.2. Suppose that $\{(x_i, y_j, Z_{ij}), i, j = 1, 2, \ldots, n_1\}$ are observations from the regression model (1.3) in which the design points $\{(x_i, y_j) = (i/n_1, j/n_1), i, j = 1, 2, \ldots, n_1\}$ are equally spaced in the design space $[0,1] \times [0,1]$, $\{\varepsilon_{ij}\}$ are i.i.d. random errors with mean 0 and variance σ^2, $n = n_1^2$ is the sample size, and f is an unknown nonparametric regression function that is continuous in the entire design space except on the JLCs.

It can be checked that estimated surfaces by most conventional local smoothing procedures are not statistically consistent at the JLCs. For example, the local linear kernel (LK) estimator $\widehat{f}_{LK}(x, y)$ of $f(x, y)$, for $(x, y) \in [0,1] \times [0,1]$, is defined by the solution to α of the following minimization problem (see e.g., Fan and Gijbels 1996; Ruppert and Wand 1994):

$$\min_{\alpha, \beta_1, \beta_2} \sum_{i=1}^{n_1} \sum_{j=1}^{n_1} \{Z_{ij} - \alpha - \beta_1(x_i - x) - \beta_2(y_j - y)\}^2 K\left(\frac{x_i - x}{h_n}, \frac{y_j - y}{p_n}\right), \tag{5.4}$$

where K is a bivariate kernel function with support $[-1/2, 1/2] \times [-1/2, 1/2]$, and h_n and p_n are two bandwidth parameters. It can be checked that $\widehat{f}_{LK}(x, y)$ does not converge to $f(x, y)$ if the point (x, y) is on a JLC with nonzero jump magnitude.

The reason why most conventional surface estimators (such as the LK estimator $\widehat{f}_{LK}$) are not statistically consistent around the JLCs is obvious: a *continuous* function, which is a linear function in the case of expression (5.4), has been used by these procedures for estimating a *jump* function either locally or globally, as noted in Subsection 3.4.2 for 1-D cases.

In the particular case of local linear kernel surface estimation, a natural idea to overcome the above-mentioned limitation of $\widehat{f}_{LK}$ is to fit a piecewisely linear function in expression (5.4). This piecewisely linear function may have jumps along the boundaries between the four quadrants $Q_{11}(x, y), Q_{12}(x, y), Q_{21}(x, y)$ and $Q_{22}(x, y)$ of the neighborhood $[x - h_n/2, x + h_n/2] \times [y - p_n/2, y + p_n/2]$ of a given point (x, y), as shown in Figure 5.4. The fitted function then provides four estimators of the surface at (x, y), which correspond to four pieces of the fitted planes in $Q_{11}(x, y), Q_{12}(x, y), Q_{21}(x, y)$, and $Q_{22}(x, y)$, respectively.

We now consider how these four estimators behave. If the point (x, y) is on a single JLC and the JLC has the right and left tangent lines at (x, y) (a JLC with parametric expressions $x = x(t)$ and $y = y(t)$ has a right (left) tangent line at a given point if the right (left) derivatives of $x(t)$ and $y(t)$ both exist at that point), then at least one of the four estimators should estimate $f(x, y)$ well, as demonstrated by Figure 5.5. In Figure 5.5(a), the right and left tangent lines of the JLC at (x, y) are located in a single quadrant. In such a case, the three estimators constructed in the other three quadrants should all estimate $f(x, y)$ well. In Figure 5.5(b), the right and left tangent lines are located in two different quadrants which are next to each other. In this case, the two estimators constructed in the other two quadrants should estimate $f(x, y)$ well. In the last situation shown by Figure 5.5(c), the two tangent lines are in two opposite quadrants. Here, only one estimator constructed in one of the other two quadrants estimates $f(x, y)$ well. That is, in all three cases, at least one of the four estimators estimates $f(x, y)$ well. The best one of the four, which can be determined by comparing the weighted residual sums of squares of the four fitted planes in the four quadrants, can be used as the estimated value of $f(x, y)$. On the other hand, if there are no jumps in the neighborhood of (x, y), then all four estimators estimate $f(x, y)$ well, and their average should provide a reasonable estimator of $f(x, y)$. In applications, however, since it is often unknown whether the point (x, y) is a jump point, a data-based mechanism is needed for making such a judgment. To this end, a threshold parameter is introduced in the procedure by Qiu (2004), and the point (x, y) is regarded as a jump point if the range of the four estimators exceeds the threshold value.

From the above intuitive introduction to the procedure by Qiu, we mathematically define the four estimators in a neighborhood $N_n(x, y) = [x - h_n/2, x + h_n/2] \times [y - p_n/2, y + p_n/2]$ of a given point $(x, y) \in [0, 1] \times [0, 1]$ by the following minimization

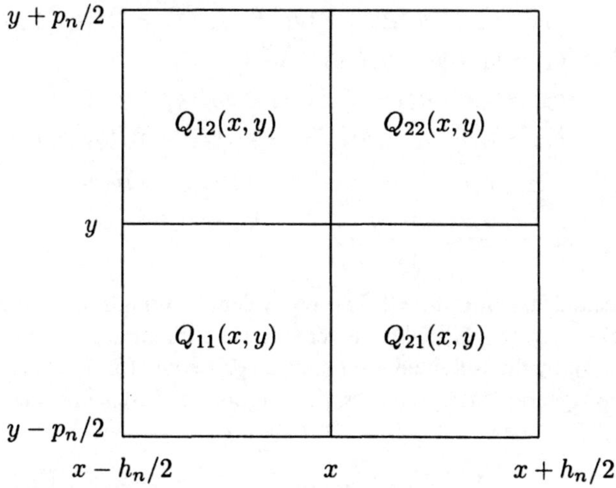

Fig. 5.4 The neighborhood of a given point (x, y) consists of four quadrants: $Q_{11}(x, y), Q_{12}(x, y), Q_{21}(x, y)$, and $Q_{22}(x, y)$.

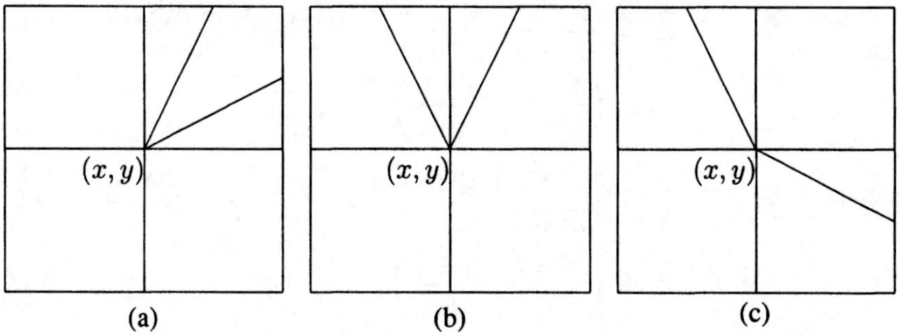

Fig. 5.5 (a): Right and left tangent lines of the JLC at (x, y) located in a single quadrant. (b): Two tangent lines in two different quadrants which are next to each other. (c): Two tangent lines in two opposite quadrants.

problem:

$$\min_{a_{s_1 s_2}, b_{s_1 s_2}, c_{s_1 s_2};\ s_1, s_2 = 1,2}$$

$$\sum_{i=1}^{n_1} \sum_{j=1}^{n_1} \{Z_{ij} - [a_{11} + b_{11}(x_i - x) + c_{11}(y_j - y) +$$

$$(a_{21} - a_{11})I(x_i \geq x) + (b_{21} - b_{11})(x_i - x)I(x_i \geq x) +$$

$$
\begin{aligned}
&(c_{21}-c_{11})(y_j-y)I(x_i \geq x)+\\
&(a_{12}-a_{11})I(y_j \geq y)+(b_{12}-b_{11})(x_i-x)I(y_j \geq y)+\\
&(c_{12}-c_{11})(y_j-y)I(y_j \geq y)+\\
&(a_{22}-a_{21}-a_{12}+a_{11})I(x_i \geq x)I(y_j \geq y)+\\
&(b_{22}-b_{21}-b_{12}+b_{11})(x_i-x)I(x_i \geq x)I(y_j \geq y)+\\
&(c_{22}-c_{21}-c_{12}+c_{11})(y_j-y)I(x_i \geq x)I(y_j \geq y)]\}^2\\
&K\left(\frac{x_i-x}{h_n},\frac{y_j-y}{p_n}\right). \qquad (5.5)
\end{aligned}
$$

It can be checked that procedure (5.5) is equivalent to fitting a linear function $a_{s_1s_2} + b_{s_1s_2}(u-x) + c_{s_1s_2}(v-y)$, which is regarded as a function of (u,v) in $Q_{s_1s_2}(x,y)$ (cf. Figure 5.4), by the weighted least squares procedure (5.4) for $s_1, s_2 = 1, 2$. By some routine algebraic manipulations, the solution of (5.5) is given by the following equations:

$$
\begin{aligned}
\widehat{a}_{s_1s_2}(x,y) &= \frac{1}{\Delta_{s_1s_2}}\sum_{i=1}^{n_1}\sum_{j=1}^{n_1}\left\{A^{(1)}_{s_1s_2}+A^{(2)}_{s_1s_2}(x_i-x)+A^{(3)}_{s_1s_2}(y_j-y)\right\}\\
&\quad Z_{ij}K\left(\frac{x_i-x}{h_n},\frac{y_j-y}{p_n}\right);\\
\widehat{b}_{s_1s_2}(x,y) &= \frac{1}{\Delta_{s_1s_2}}\sum_{i=1}^{n_1}\sum_{j=1}^{n_1}\left\{A^{(2)}_{s_1s_2}+A^{(4)}_{s_1s_2}(x_i-x)+A^{(5)}_{s_1s_2}(y_j-y)\right\}\\
&\quad Z_{ij}K\left(\frac{x_i-x}{h_n},\frac{y_j-y}{p_n}\right);\\
\widehat{c}_{s_1s_2}(x,y) &= \frac{1}{\Delta_{s_1s_2}}\sum_{i=1}^{n_1}\sum_{j=1}^{n_1}\left\{A^{(3)}_{s_1s_2}+A^{(5)}_{s_1s_2}(x_i-x)+A^{(6)}_{s_1s_2}(y_j-y)\right\}\\
&\quad Z_{ij}K\left(\frac{x_i-x}{h_n},\frac{y_j-y}{p_n}\right) \qquad (5.6)
\end{aligned}
$$

for $s_1, s_2 = 1, 2$, where

$$
\begin{aligned}
\Delta_{s_1s_2} &= \begin{vmatrix} B^{(00)}_{s_1s_2} & B^{(10)}_{s_1s_2} & B^{(01)}_{s_1s_2}\\ B^{(10)}_{s_1s_2} & B^{(20)}_{s_1s_2} & B^{(11)}_{s_1s_2}\\ B^{(01)}_{s_1s_2} & B^{(11)}_{s_1s_2} & B^{(02)}_{s_1s_2}\end{vmatrix},\\
A^{(1)}_{s_1s_2} &= B^{(20)}_{s_1s_2}B^{(02)}_{s_1s_2} - B^{(11)}_{s_1s_2}B^{(11)}_{s_1s_2},\\
A^{(2)}_{s_1s_2} &= B^{(01)}_{s_1s_2}B^{(11)}_{s_1s_2} - B^{(10)}_{s_1s_2}B^{(02)}_{s_1s_2},\\
A^{(3)}_{s_1s_2} &= B^{(10)}_{s_1s_2}B^{(11)}_{s_1s_2} - B^{(01)}_{s_1s_2}B^{(20)}_{s_1s_2},\\
A^{(4)}_{s_1s_2} &= B^{(00)}_{s_1s_2}B^{(02)}_{s_1s_2} - B^{(01)}_{s_1s_2}B^{(01)}_{s_1s_2},\\
A^{(5)}_{s_1s_2} &= B^{(01)}_{s_1s_2}B^{(10)}_{s_1s_2} - B^{(00)}_{s_1s_2}B^{(11)}_{s_1s_2},\\
A^{(6)}_{s_1s_2} &= B^{(00)}_{s_1s_2}B^{(20)}_{s_1s_2} - B^{(10)}_{s_1s_2}B^{(10)}_{s_1s_2},
\end{aligned}
$$

$$B_{s_1 s_2}^{(r_1 r_2)} = \sum_{i=1}^{n_1} \sum_{j=1}^{n_1} (x_i - x)^{r_1} (y_j - y)^{r_2} K_{s_1 s_2} \left(\frac{x_i - x}{h_n}, \frac{y_j - y}{p_n} \right),$$

and $K_{s_1 s_2}(x/h_n, y/p_n) = K(x/h_n, y/p_n)$ if $(x, y) \in Q_{s_1 s_2}(0,0)$ and 0 otherwise, for $s_1, s_2 = 1, 2$ and $r_1, r_2 = 0, 1, 2$.

For simplicity of presentation, a point (x, y) is called a *nonsingular point* of the JLCs if it is on a single JLC, the JLC has the left and right tangent lines at (x, y), and the jump magnitude at (x, y) is nonzero. All other points on the JLCs are called *singular points*. Obviously, a point (x, y) is a singular point if it is a cross point of several JLCs, or if it is on a single JLC but the JLC does not have the left or right tangent line at (x, y), or if the jump magnitude at (x, y) is zero. Again, this definition of nonsingular points is specifically for the local piecewisely linear kernel smoothing procedure discussed in this section and may be different from the definitions of nonsingular points discussed in other places (e.g., in Subsections 4.3.2 and 4.4.1).

The quantities $\widehat{a}_{11}(x, y), \widehat{a}_{12}(x, y), \widehat{a}_{21}(x, y)$, and $\widehat{a}_{22}(x, y)$ defined in equation (5.6) are the four previously mentioned estimators of $f(x, y)$; we now consider the selection of the *best* estimator. If there are no jumps in $N_n(x, y)$, then all of them should estimate $f(x, y)$ well, and it is reasonable to use their average for estimating $f(x, y)$. If the point (x, y) is a nonsingular point of the JLCs, then at least one of the four estimators estimates $f(x, y)$ well, as explained previously (cf. Figure 5.5). A natural idea to estimate $f(x, y)$ in the latter case is to choose one of $\widehat{a}_{11}(x, y), \widehat{a}_{12}(x, y), \widehat{a}_{21}(x, y)$ and $\widehat{a}_{22}(x, y)$ based on the weighted residual sums of squares (RSS) of the four pieces of fitted planes in the four quadrants of $N_n(x, y)$. In order to tell whether there are jumps in $N_n(x, y)$, a threshold parameter u_n is introduced below, and we conclude that there are no jumps in $N_n(x, y)$ if

$$\text{range}\left(\widehat{a}_{11}(x, y), \widehat{a}_{12}(x, y), \widehat{a}_{21}(x, y), \widehat{a}_{22}(x, y)\right) \leq u_n, \tag{5.7}$$

where range$(\widehat{a}_{11}(x, y), \widehat{a}_{12}(x, y), \widehat{a}_{21}(x, y), \widehat{a}_{22}(x, y))$ denotes the difference between the maximum and minimum of $\widehat{a}_{11}(x, y), \widehat{a}_{12}(x, y), \widehat{a}_{21}(x, y)$, and $\widehat{a}_{22}(x, y)$. Finally, the piecewisely linear kernel (PLK) estimator of $f(x, y)$ is defined by

$$\begin{aligned} & \widehat{f}_{PLK}(x, y) \\ = & \begin{cases} \left(\widehat{a}_{11}(x, y) + \widehat{a}_{12}(x, y) + \widehat{a}_{21}(x, y) + \widehat{a}_{22}(x, y)\right)/4, & \text{if (5.7) holds} \\ \tilde{a}(x, y), & \text{otherwise,} \end{cases} \end{aligned} \tag{5.8}$$

where $\tilde{a}(x, y)$ is the one of $\widehat{a}_{11}(x, y), \widehat{a}_{12}(x, y), \widehat{a}_{21}(x, y)$, and $\widehat{a}_{22}(x, y)$, which satisfies

$$\begin{aligned} & RSS(\tilde{a}(x, y)) \\ = & \min\left\{RSS(\widehat{a}_{11}(x, y)), RSS(\widehat{a}_{12}(x, y)), RSS(\widehat{a}_{21}(x, y)), RSS(\widehat{a}_{22}(x, y))\right\} \end{aligned}$$

and $RSS(\widehat{a}_{s_1 s_2}(x, y))$ is the weighted residual sum of squares defined by

$$RSS(\widehat{a}_{s_1 s_2}(x, y)) = \sum_{i=1}^{n_1} \sum_{j=1}^{n_1} \Big\{ Z_{ij} - \Big[\widehat{a}_{s_1 s_2}(x, y) + \widehat{b}_{s_1 s_2}(x, y)(x_i - x) +$$

$$\widehat{c}_{s_1 s_2}(x,y)(y_j - y)]\}^2 K_{s_1 s_2}\left(\frac{x_i - x}{h_n}, \frac{y_j - y}{p_n}\right), \quad (5.9)$$

for $s_1, s_2 = 1, 2$. If there are two or more of $\widehat{a}_{11}(x,y), \widehat{a}_{12}(x,y), \widehat{a}_{21}(x,y)$, and $\widehat{a}_{22}(x,y)$, which have the smallest RSS value, then $\tilde{a}(x,y)$ is defined by their simple average.

The entire jump-preserving surface reconstruction procedure is summarized as follows.

- At a given point (x,y), compute the four estimators $\widehat{a}_{11}(x,y), \widehat{a}_{12}(x,y), \widehat{a}_{21}(x,y)$, and $\widehat{a}_{22}(x,y)$ by equation (5.6).

- Compute the range of the four estimators and then compare this range to the threshold value u_n (cf. equations (5.7) and (5.10) below).

- If the range is smaller than or equal to u_n, then use the simple average of the four estimators as the surface estimator at (x,y).

- If the range is larger than u_n, then compute $RSS(\widehat{a}_{11}(x,y))$, $RSS(\widehat{a}_{12}(x,y))$, $RSS(\widehat{a}_{21}(x,y))$, and $RSS(\widehat{a}_{22}(x,y))$ by equation (5.9) and search for $\tilde{a}(x,y)$ (the one of $\widehat{a}_{11}(x,y), \widehat{a}_{12}(x,y), \widehat{a}_{21}(x,y)$, and $\widehat{a}_{22}(x,y)$ with the smallest RSS value). The surface estimator at (x,y) is then defined to be $\tilde{a}(x,y)$.

By using the above procedure, the regression surface is estimated with jumps automatically preserved. Unlike the two-stage or three-stage procedures such as those suggested by Müller and Song (1994a), O'Sullivan and Qian (1994), and Qiu (1997, 1998) (cf. Sections 4.2 and 5.2 for introduction), explicit jump detection is avoided in the present procedure, which has several benefits. First, this procedure is easy to use because its surface estimator has an explicit mathematical formula (cf. equation (5.8)) like most conventional surface estimators. The second major benefit is that no restrictive assumptions on the JLCs are required by the present procedure, and, consequently, it can find more applications. In addition, Qiu (2004) proved that the reconstructed surface by equation (5.8) is uniformly almost surely consistent in the design space, excluding an arbitrarily small region around the border of the design space and around the singular points of the JLCs.

As a comparison, some existing procedures (e.g., O'Sullivan and Qian 1994; Qiu 1997) use smooth curves to estimate the JLCs before the surface is reconstructed. These procedures also assume that the number of JLCs is known (e.g., some procedures assume that only one JLC exists in the design space) and the JLCs are smooth.

We now make several technical remarks about the surface reconstruction procedure (5.8). First, in procedure (5.8), there are several alternative ways to define $\widehat{f}_{PLK}(x,y)$ when expression (5.7) holds. For instance, $\widehat{f}_{PLK}(x,y)$ can be defined by the weighted average

$$\frac{\sum_{s_1=1}^{2}\sum_{s_2=1}^{2} \widehat{a}_{s_1 s_2}(x,y)/RSS(\widehat{a}_{s_1 s_2}(x,y))}{\sum_{s_1=1}^{2}\sum_{s_2=1}^{2} 1/RSS(\widehat{a}_{s_1 s_2}(x,y))}.$$

However, here we need to compute $RSS(\widehat{a}_{11}(x,y)), RSS(\widehat{a}_{12}(x,y)), RSS(\widehat{a}_{21}(x,y))$, and $RSS(\widehat{a}_{22}(x,y))$, which is not always necessary by using the definition (5.8). A second alternative is that $\widehat{f}_{PLK}(x, y)$ can be defined by the conventional local linear kernel estimator constructed in the entire neighborhood $N_n(x, y)$. In this case, we need to compute the conventional local linear kernel estimator of $f(x, y)$ in the entire neighborhood $N_n(x, y)$ after $\widehat{a}_{11}(x, y)$, $\widehat{a}_{12}(x, y)$, $\widehat{a}_{21}(x, y)$, and $\widehat{a}_{22}(x, y)$ are computed. Therefore, these two alternative approaches require some extra computation compared to the approach used in procedure (5.8). If the extra computation is not an issue for a specific application, then the alternative approaches might be more reasonable to use.

Second, from Figures 5.5(a) and 5.5(b), it seems that a potential improvement of procedure (5.8) is to use more than one of $\widehat{a}_{11}(x, y)$, $\widehat{a}_{12}(x, y)$, $\widehat{a}_{21}(x, y)$, and $\widehat{a}_{22}(x, y)$ for estimating $f(x, y)$ when expression (5.7) holds. For example, in the case of Figure 5.5(a), $\widehat{f}_{PLK}(x, y)$ could be defined by the average of $\widehat{a}_{11}(x, y)$, $\widehat{a}_{12}(x, y)$, and $\widehat{a}_{21}(x, y)$. However, it is a challenging task to distinguish the three cases demonstrated by Figure 5.5 based on data; this idea needs much future research.

Third, at the beginning of this subsection, it is assumed that the design points are equally spaced in the design space, which is true in many applications, including the image reconstruction problem. On the other hand, from the construction of $\widehat{f}_{PLK}$, it can be seen that this procedure can also work well when the design points are not equally spaced. In such cases, if the design points have some homogeneity (cf. assumption (A 4.1) in Subsection 4.2.1), then constant bandwidths may still be appropriate to use. If the design points have no homogeneity at all, then variable bandwidths might be more appropriate: in regions where the design points are sparse, the bandwidths need to be relatively large to include enough observations for estimating f; in regions where the design points are dense, the bandwidths should be relatively small.

Fourth, like most local smoothing estimators, the estimator $\widehat{f}_{PLK}$ is only defined in $[h_n/2, 1 - h_n/2] \times [p_n/2, 1 - p_n/2]$. It is not well defined in the border regions of the design space. In the numerical example presented in the next subsection, the symmetric "padding" method is used to overcome this difficulty, as we did in Subsection 5.2.3.

5.3.2 Selection of procedure parameters

We first examine the extreme values for u_n, in order to obtain some intuition about its role in the estimation procedure (5.8). At a given point (x, y) in the design space, if the threshold parameter u_n is chosen to be positive infinity, then the estimator $\widehat{f}_{PLK}(x, y)$ is close to the conventional local linear kernel estimator $\widehat{f}_{LK}(x, y)$ constructed in the entire neighborhood $N_n(x, y)$. This estimator removes the noise well in the continuity regions of the regression function f, but blurs the jumps around the JLCs at the same time. At the other extreme, if u_n is chosen to be zero in equation (5.8), then $\widehat{f}_{PLK}(x, y)$ equals $\tilde{a}(x, y)$, which preserves the jumps well. However, its ability to remove the noise is limited since it is actually constructed in only a single quadrant of $N_n(x, y)$. Because the main purpose for using the threshold parameter u_n is to make the estimator $\widehat{f}_{PLK}(x,y)$ able both to preserve the jumps around the JLCs and to remove the noise efficiently, it is important to choose this parameter properly.

We now provide some asymptotic results for the four estimators $\widehat{a}_{11}(x,y), \widehat{a}_{12}(x,y), \widehat{a}_{21}(x,y)$, and $\widehat{a}_{22}(x,y)$. From their definitions, these estimators are clearly independent of each other. Also, because they are all linear estimators, they are asymptotically normally distributed when the sample size tends to infinity. Then it can be checked that when there are no jumps in $N_n(x,y)$, their asymptotic means all equal $f(x,y)$ and their asymptotic variances (AV) are

$$AV(\widehat{a}_{s_1s_2}(x,y)) = \frac{\sigma^2}{n_1^2 h_n p_n \widetilde{\Delta}_{s_1s_2}^2} \int_{-1/2}^{1/2}\int_{-1/2}^{1/2} \left(\widetilde{\gamma}_{s_1s_2}^{(1)} + \widetilde{\gamma}_{s_1s_2}^{(2)} u + \widetilde{\gamma}_{s_1s_2}^{(3)} v\right)^2 K_{s_1s_2}^2(u,v)\, dudv,$$

for $s_1, s_2 = 1, 2$, where

$$\begin{aligned}
\widetilde{\Delta}_{s_1s_2} &= \begin{vmatrix} \widetilde{\beta}_{s_1s_2}^{(00)} & \widetilde{\beta}_{s_1s_2}^{(10)} & \widetilde{\beta}_{s_1s_2}^{(01)} \\ \widetilde{\beta}_{s_1s_2}^{(10)} & \widetilde{\beta}_{s_1s_2}^{(20)} & \widetilde{\beta}_{s_1s_2}^{(11)} \\ \widetilde{\beta}_{s_1s_2}^{(01)} & \widetilde{\beta}_{s_1s_2}^{(11)} & \widetilde{\beta}_{s_1s_2}^{(02)} \end{vmatrix}, \\
\widetilde{\gamma}_{s_1s_2}^{(1)} &= \widetilde{\beta}_{s_1s_2}^{(20)}\widetilde{\beta}_{s_1s_2}^{(02)} - \widetilde{\beta}_{s_1s_2}^{(11)}\widetilde{\beta}_{s_1s_2}^{(11)}, \\
\widetilde{\gamma}_{s_1s_2}^{(2)} &= \widetilde{\beta}_{s_1s_2}^{(01)}\widetilde{\beta}_{s_1s_2}^{(11)} - \widetilde{\beta}_{s_1s_2}^{(10)}\widetilde{\beta}_{s_1s_2}^{(02)}, \\
\widetilde{\gamma}_{s_1s_2}^{(3)} &= \widetilde{\beta}_{s_1s_2}^{(10)}\widetilde{\beta}_{s_1s_2}^{(11)} - \widetilde{\beta}_{s_1s_2}^{(01)}\widetilde{\beta}_{s_1s_2}^{(20)},
\end{aligned}$$

and

$$\widetilde{\beta}_{s_1s_2}^{(r_1r_2)} = \int_{-1/2}^{1/2}\int_{-1/2}^{1/2} u^{r_1} v^{r_2} K_{s_1s_2}(u,v)\, dudv$$

for $s_1, s_2 = 1, 2$ and $r_1, r_2 = 0, 1, 2$.

Using these asymptotic results, we derive a formula for u_n. For simplicity, let us assume that the kernel function K is symmetric in the sense that $K(x,y) = K(y,x) = K(-x,y)$ for any $(x,y) \in [-1/2, 1/2] \times [-1/2, 1/2]$. For example, the bivariate Epanechnikov kernel function $K(x,y) = \frac{144}{121}(1-x^2)(1-y^2)I((x,y) \in [-1/2,1/2] \times [-1/2,1/2])$ satisfies this condition. Then, it can be checked that $AV(\widehat{a}_{s_1s_2}(x,y)) = \sigma^2\phi^2/(n_1^2 h_n p_n)$ for $s_1, s_2 = 1, 2$, where

$$\phi^2 = \frac{1}{\widetilde{\Delta}_{11}^2} \int_{-1/2}^{1/2}\int_{-1/2}^{1/2} \left(\widetilde{\gamma}_{11}^{(1)} + \widetilde{\gamma}_{11}^{(2)} u + \widetilde{\gamma}_{11}^{(3)} v\right)^2 K_{11}^2(u,v)\, dudv.$$

If we define

$$\xi_{s_1s_2}(x,y) = \frac{\widehat{a}_{s_1s_2}(x,y) - f(x,y)}{\sigma\phi/(n_1 h_n^{1/2} p_n^{1/2})},$$

for $s_1, s_2 = 1, 2$, then $\xi_{11}(x,y), \xi_{12}(x,y), \xi_{21}(x,y)$ and $\xi_{22}(x,y)$ are asymptotically i.i.d. with a common asymptotic distribution $N(0,1)$. Obviously,

$$\begin{aligned}
& \text{range}\left(\widehat{a}_{11}(x,y), \widehat{a}_{12}(x,y), \widehat{a}_{21}(x,y), \widehat{a}_{22}(x,y)\right) \\
= \; & \text{range}\left(\xi_{11}(x,y), \xi_{12}(x,y), \xi_{21}(x,y), \xi_{22}(x,y)\right) \cdot \left[\sigma\phi/(n_1 h_n^{1/2} p_n^{1/2})\right].
\end{aligned}$$

Table 5.1 Several quantiles of the limiting distribution of range($\xi_{11}(x,y)$, $\xi_{12}(x,y)$, $\xi_{21}(x,y)$, $\xi_{22}(x,y)$).

$1-\alpha_{n_1}$	$R_{1-\alpha_{n_1}}$	$1-\alpha_{n_1}$	$R_{1-\alpha_{n_1}}$
0.5000	1.9746	0.9250	3.3959
0.6000	2.2070	0.9500	3.6253
0.7000	2.4596	0.9750	3.9647
0.8000	2.7785	0.9900	4.3911
0.8500	2.9747	0.9990	5.2724
0.9000	3.2314	0.9999	5.8557

Thus, a natural choice for u_n is

$$u_n = \frac{\widehat{\sigma}\phi R_{1-\alpha_{n_1}}}{n_1 h_n^{1/2} p_n^{1/2}}, \tag{5.10}$$

where $R_{1-\alpha_{n_1}}$ is the $1-\alpha_{n_1}$ quantile of the limiting distribution of range($\xi_{11}(x,y)$, $\xi_{12}(x,y), \xi_{21}(x,y), \xi_{22}(x,y)$), $0 \leq \alpha_{n_1} \leq 1$ is a significance level, and $\widehat{\sigma}$ is a consistent estimator of σ. A formula for the limiting distribution of range($\xi_{11}(x,y), \xi_{12}(x, y), \xi_{21}(x,y), \xi_{22}(x,y)$) can be found in Balakrishnan and Cohen (1991). Based on that formula, several values of $R_{1-\alpha_{n_1}}$ are calculated and presented in Table 5.1.

As discussed at the beginning of this subsection, the threshold parameter u_n controls the trade-off between noise-removing and jump-preservation abilities of procedure (5.8). By formula (5.10) and Table 5.1, the asymptotic probability that a continuity point is treated as a jump point by procedure (5.8) is known. For example, if $R_{1-\alpha_{n_1}}$ is chosen to be 4.3911 in equation (5.10), then, asymptotically speaking, about one percent of the continuity points are treated as jump points. Therefore, the threshold parameter u_n is related to the asymptotic *false jump detection rate*.

We now briefly discuss bandwidth selection of h_n and p_n for procedure (5.8). In Subsections 2.4.4 and 2.5.2, we discussed several existing data-driven bandwidth selection procedures, including the plug-in procedures, the Mallow's C_p criterion, the cross-validation (CV) procedure, and the generalized cross-validation procedure. These procedures can also be used for selecting the bandwidth parameters h_n and p_n here. For example, if the CV procedure is used, then the optimal values of h_n and p_n are chosen by minimizing the following CV score:

$$CV(h_n, p_n) = \frac{1}{n_1^2} \sum_{i=1}^{n_1} \sum_{j=1}^{n_1} \left(Z_{ij} - \widehat{f}_{-i,-j}(x_i, y_j) \right)^2, \tag{5.11}$$

where $\widehat{f}_{-i,-j}(x,y)$ is the "leave-1-out" estimator of $f(x,y)$. Namely, the observation (x_i, y_j, Z_{ij}) is left out in constructing $\widehat{f}_{-i,-j}(x,y)$ for $i, j = 1, 2, \cdots, n_1$.

In applications, if we are not sure which α_{n_1} value should be chosen, then it can be determined by the CV procedure, together with the bandwidth parameters. To this end, let us write the left-hand side of equation (5.11) as $CV(R_{1-\alpha_{n_1}}, h_n, p_n)$. Then the optimal values of $R_{1-\alpha_{n_1}}$, h_n, and p_n can be determined by minimizing $CV(R_{1-\alpha_n}, h_n, p_n)$.

Example 5.3 Suppose that the regression function is $f(x,y) = -4(x-0.5)^2 - 4(y-0.5)^2$ when $0 \leq x < 0.5$ and $0 \leq y < 0.75 - 0.5x$; and $f(x,y) = -4(x-0.5)^2 - 4(y-0.5)^2 + 1$ otherwise. Then this function has a single JLC: $\{(x,y) : y = 0.75 - 0.5x, 0 \leq x \leq 0.5\} \bigcup \{(x,y) : x = 0.5, 0 \leq y \leq 0.5\}$. A set of observations are generated from the model (1.3) at $n = 100^2$ equally spaced design points with $\varepsilon_{11} \sim N(0, \sigma^2)$ and $\sigma = 0.25$. The true regression surface and the observations are presented in Figures 5.6(a) and 5.6(b), respectively.

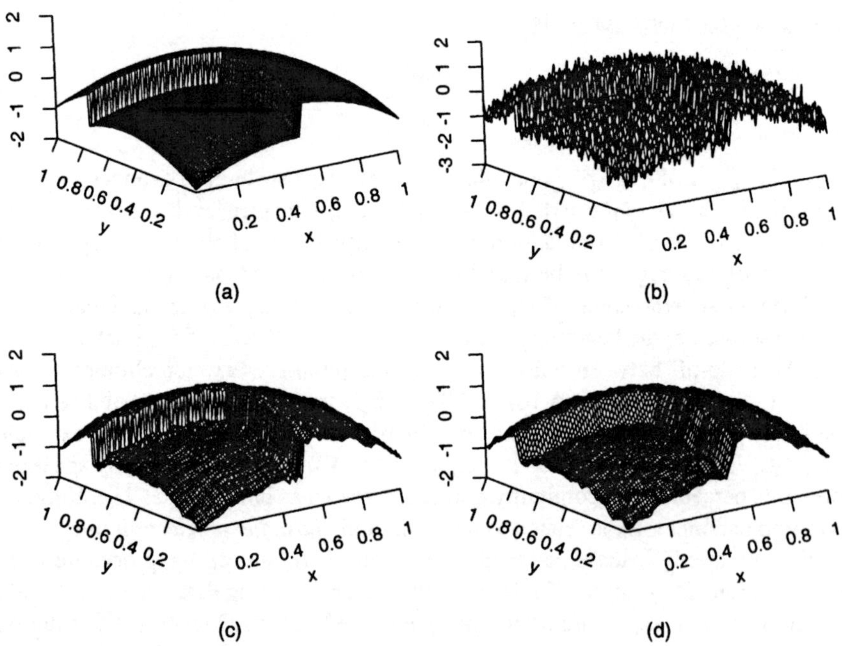

Fig. 5.6 (a): True regression surface. (b): Set of observations with $\varepsilon_{11} \sim N(0, \sigma^2)$, $\sigma = 0.25$, and $n_1 = 100$. (c): Reconstructed surface by procedure (5.8) with $h_n = p_n = 0.16$. (d): Reconstructed surface by the conventional local linear kernel procedure with $h_n = p_n = 0.05$.

The jump-preserving surface reconstruction procedure (5.8) is then applied to the dataset shown in Figure 5.6(b). In the procedure, the kernel function is chosen to be the bivariate Epanechnikov function $K(x,y) = \frac{144}{121}(1-x^2)(1-y^2)I((x,y) \in [-1/2, 1/2] \times [-1/2, 1/2])$. The significance level α_{n_1} is fixed at 0.01. The bandwidths h_n and p_n are assumed to be the same, and, by the CV procedure (5.11), they are chosen to be 0.16. The fitted surface by procedure (5.8) is presented in Figure 5.6(c). As a comparison, the fitted surface by the conventional local linear kernel smoothing procedure (5.4) with the same kernel function is presented in Figure 5.6(d). In the conventional procedure, the two bandwidths are also assumed to be the same, and they are chosen to be 0.05 by the CV procedure. From the plots, it can be seen that the jumps are preserved reasonably well by procedure (5.8), whereas they are blurred by the conventional procedure, as expected. It can also be noticed that procedure (5.8) performs reasonably well in the boundary regions.

5.4 SURFACE RECONSTRUCTION WITH GRADIENT ESTIMATION

The surface reconstruction procedure discussed in the previous section preserves the possible jumps automatically after introducing a threshold parameter. The idea of using a threshold parameter is quite common in the image processing literature (cf. discussions in Chapters 6 and 7); however, the performance of the reconstructed surface depends heavily on this parameter. Therefore, it needs to be chosen properly, which adds much complexity to the surface reconstruction procedure.

In this section, we introduce a local smoothing surface reconstruction procedure suggested by Qiu (2003b) which does not require any procedure parameters besides the bandwidths. Like the piecewisely linear kernel smoothing procedure discussed in the previous section, the surface estimator of the current procedure has an explicit mathematical formula. Thus, the procedure is easy to use and the surface estimator is simple to compute. This procedure is introduced below in four subsections.

5.4.1 Gradient estimation and three possible surface estimators

The gradient estimator is developed in the following context. Suppose that $n = n_1^2$ observations $\{(x_i, y_j, Z_{ij}), i, j = 1, 2, \ldots, n_1\}$ are generated from the regression model (1.3) in which the design points $\{(x_i, y_j) = (i/n_1, j/n_1), i, j = 1, 2, \ldots, n_1\}$ are equally spaced in the design space $[0, 1] \times [0, 1]$, and the regression function f is continuous in the entire design space except on the JLCs. At a given point (x, y) in the design space, we consider a circular neighborhood $N_n(x, y) = \{(u, v) : (u, v) \in [0, 1] \times [0, 1], \sqrt{(u - x)^2 + (v - y)^2} \leq h_n\}$ where h_n is a bandwidth parameter. A major consideration for using circular neighborhoods here is their rotation-invariance property. For some applications in geology, meteorology, and oceanography, a pre-specified coordinate system is unavailable. In such cases, it is desirable to have the estimated surface be coordinate free, and circular neighborhoods would make this possible. For some applications such as image reconstruction, the coordinate system is well defined; therefore, square-shaped or other neighborhoods can also be used.

Next, in $N_n(x,y)$, a local plane is fitted by local linear kernel smoothing:

$$\min_{a,b,c} \sum_{i=1}^{n} \sum_{j=1}^{n} \{Z_{ij} - [a + b(x_i - x) + c(y_j - y)]\}^2 K\left(\frac{x_i - x}{h_n}, \frac{y_j - y}{h_n}\right), \quad (5.12)$$

where K is a radially symmetric, bivariate density kernel function with support $\{(x,y) : x^2+y^2 \leq 1\}$. The solution of expression (5.12) is denoted by $\widehat{a}(x,y)$, $\widehat{b}(x,y)$, and $\widehat{c}(x,y)$. Their formulas are the same as those in equation (5.6) after the kernel functions $K_{s_1 s_2}$ for $s_1, s_2 = 1, 2$ there are replaced by K here. Then $\widehat{a}(x,y)$ is the conventional local linear kernel estimator of $f(x,y)$, and the gradient direction of the fitted plane is $G(x,y) = (\widehat{b}(x,y), \widehat{c}(x,y))$. The weighted residual mean square (WRMS) of the fitted plane is

$$\begin{aligned} e(x,y) &= \sum_{i=1}^{n_1} \sum_{j=1}^{n_1} \left\{ \left\{ Z_{ij} - \left[\widehat{a}(x,y) + \widehat{b}(x,y)(x_i - x) + \widehat{c}(x,y)(y_j - y) \right] \right\}^2 \right. \\ &\quad \left. K\left(\frac{x_i - x}{h_n}, \frac{y_j - y}{h_n}\right) \right\} \Big/ \sum_{i=1}^{n_1} \sum_{j=1}^{n_1} K\left(\frac{x_i - x}{h_n}, \frac{y_j - y}{h_n}\right). \quad (5.13) \end{aligned}$$

The gradient direction $G(x,y)$ is a good estimator of the direction in which f increases fastest around (x,y). Also, if the point (x,y) is on a JLC, then $G(x,y)$ also indicates the orientation of the JLC at (x,y) well, which can be seen from Figure 5.8(c) in Subsection 5.4.3 below.

We now use this gradient direction to obtain two one-sided estimators as follows. The neighborhood $N_n(x,y)$ is first divided into two parts $N_n^{(1)}(x,y)$ and $N_n^{(2)}(x,y)$ by a line which passes (x,y) and is perpendicular to $G(x,y)$, as shown in Figure 5.7. Then in $N_n^{(1)}(x,y)$ and $N_n^{(2)}(x,y)$, we fit two one-sided local planes by local linear kernel smoothing:

$$\min_{a,b,c} \sum_{(x_i,y_j) \in N_n^{(\ell)}(x,y)} \{Z_{ij} - [a + b(x_i - x) + c(y_j - y)]\}^2 K\left(\frac{x_i - x}{h_n}, \frac{y_j - y}{h_n}\right), \quad (5.14)$$

for $\ell = 1, 2$. The solution of expression (5.14) is denoted by $(\widehat{a}^{(\ell)}(x,y), \widehat{b}^{(\ell)}(x,y), \widehat{c}^{(\ell)}(x,y))$ for $\ell = 1, 2$. Their formulas can also be obtained from equation (5.6), after replacing the kernel functions $K_{s_1 s_2}$ for $s_1, s_2 = 1, 2$ there by the kernel function K here and changing the ranges of the summations to $\sum_{(x_i,y_j) \in N_n^{(\ell)}(x,y)}$. The WRMS values of the two fitted one-sided planes are defined to be

$$\begin{aligned} e^{(\ell)}(x,y) &= \sum_{(x_i,y_j) \in N_n^{(\ell)}(x,y)} \left\{ \left\{ Z_{ij} - \left[\widehat{a}^{(\ell)}(x,y) + \widehat{b}^{(\ell)}(x,y)(x_i - x) + \right. \right. \right. \\ &\quad \left. \left. \widehat{c}^{(\ell)}(x,y)(y_j - y) \right] \right\}^2 K\left(\frac{x_i - x}{h_n}, \frac{y_j - y}{h_n}\right) \Big\} \Big/ \\ &\quad \sum_{(x_i,y_j) \in N_n^{(\ell)}(x,y)} K\left(\frac{x_i - x}{h_n}, \frac{y_j - y}{h_n}\right), \quad (5.15) \end{aligned}$$

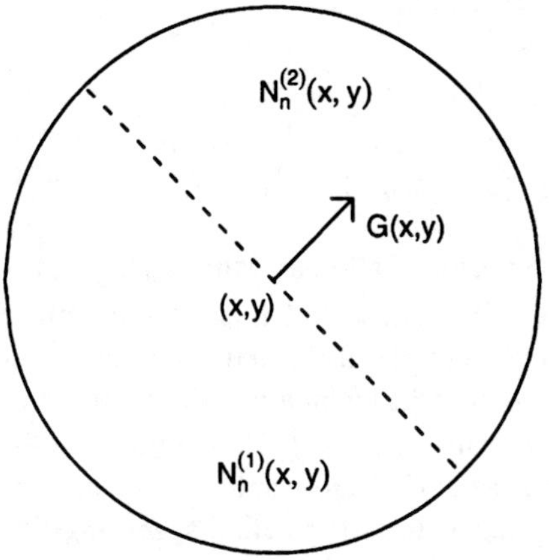

Fig. 5.7 Neighborhood $N_n(x, y)$ of the point (x, y) is divided into two parts $N_n^{(1)}(x, y)$ and $N_n^{(2)}(x, y)$ along the gradient direction $G(x, y)$.

for $\ell = 1$ and 2.

Intuitively, a comparison of the three estimators $\widehat{a}(x, y)$, $\widehat{a}^{(1)}(x, y)$, and $\widehat{a}^{(2)}(x, y)$ can be done to determine the best overall estimator of $f(x, y)$, based on whether $N_n(x, y)$ contains any jump points. When there are no jumps in $N_n(x, y)$, all three estimators are statistically consistent for estimating $f(x, y)$ under some regularity conditions. Then, the conventional estimator $\widehat{a}(x, y)$ is preferred because it averages more observations. When there is a single JLC in $N_n(x, y)$ and the JLC has a unique tangent line at each of its points in the neighborhood, $\widehat{a}(x, y)$ does not estimate $f(x, y)$ well because observations on both sides of the JLC are averaged and the jumps are blurred. Here, one of the two parts of $N_n(x, y)$ should be mostly on one side of the JLC, and the one-sided estimator constructed in that part should be able to estimate the surface well. Thus, in both cases, at least one of the three estimators $\widehat{a}(x, y)$, $\widehat{a}^{(1)}(x, y)$ and $\widehat{a}^{(2)}(x, y)$ estimates $f(x, y)$ well when the sample size is reasonably large.

In applications, however, it is often unknown whether a given point (x, y) is on a JLC, so a data-driven mechanism is needed to choose one of the three estimators for estimating $f(x, y)$. This goal leads to three possible solutions, which are described in the next three subsections.

5.4.2 Choose one of the three estimators based on the WRMS values

To estimate $f(x, y)$ properly, one possibility is to choose among the three estimators according to the corresponding WRMS values defined by equations (5.13) and (5.15). If this method is used, it can be checked that $e(x, y) \geq (e^{(1)}(x, y) + e^{(2)}(x, y))/2$ when n_1 is large enough. Therefore, $\widehat{a}(x, y)$ will not be selected, so the surface estimator is actually defined by

$$\widehat{f}_1(x,y;\underline{Z}) = \begin{cases} \widehat{a}^{(1)}(x,y), & \text{if } e^{(1)}(x,y) < e^{(2)}(x,y) \\ \widehat{a}^{(2)}(x,y), & \text{if } e^{(1)}(x,y) > e^{(2)}(x,y) \\ (\widehat{a}^{(1)}(x,y) + \widehat{a}^{(2)}(x,y))/2, & \text{if } e^{(1)}(x,y) = e^{(2)}(x,y), \end{cases} \tag{5.16}$$

where $\underline{Z}$ denotes the vector of all observations $\{Z_{ij}, i,j = 1,2,\ldots,n_1\}$. Clearly, the estimator $\widehat{f}_1(x,y;\underline{Z})$ is not well defined in the boundary regions of the design space. In all numerical examples presented in this section, the symmetric "padding" method is used for handling the boundary problem, as we did in the previous section.

As explained previously, $\widehat{f}_1(x,y;\underline{Z})$ should preserve the jump well at each point of the JLCs at which the JLCs have a unique tangent line and the jump magnitude is nonzero; such a point is hereafter called a *nonsingular point* of the JLCs. For convenience, other points of the JLCs are called *singular points*. Again, this definition of nonsingular points is valid for the present procedure only and may be different from other definitions of nonsingular points used at other places.

Example 5.4 In Figure 5.8, plot (a) shows the original surface $f(x,y) = -2(x-0.5)^2-2(y-0.5)^2+1$ if $(x-0.5)^2+(y-0.5)^2 < 0.25^2$, and $f(x,y) = -2(x-0.5)^2-2(y-0.5)^2$ otherwise. It has one JLC, which is a circle with constant jump size 1. A data set generated from model (1.3) with $n_1 = 100$ and $\varepsilon_{11} \sim N(0, 0.5^2)$ is presented in plot (b). The conventional procedure (5.12) is then applied to this data set with $h_n = 0.1$ and $K(x,y) = [\exp(-(x^2+y^2)/2) - e^{-0.5}]/(2\pi - 3\pi e^{-0.5})I(x^2+y^2 \leq 1)$. The gradient direction $G(x,y)$ is shown in plot (c); it indicates the orientation of the JLC well. The conventional estimator $\widehat{a}(x,y)$ is presented in plot (d); the noise is mostly removed, but the jumps are also blurred. Next, the procedure (5.16) is applied to the same data with the same window width and kernel function. The WRMS values of $e(x,y)$, $e^{(1)}(x,y)$, and $e^{(2)}(x,y)$ in the cross section of $y = 0.5$ are presented in plot (e). It can be seen that $e(x,y)$ can never be the smallest one among the three WRMS values. The fitted surface by (5.16) is displayed in plot (f). It is expected to be relatively noisy compared to the conventional estimator $\widehat{a}(x,y)$ because it is defined by observations in one of the two parts of $N_n(x,y)$ only; this is seen from Figure 5.8. The estimator $\widehat{f}_1(x,y;\underline{Z})$ preserves the jumps well but is indeed quite noisy.

5.4.3 Choose one of the three estimators based on their estimated variances

For estimating $f(x,y)$, an alternative strategy is to choose one of $\widehat{a}(x,y)$, $\widehat{a}^{(1)}(x,y)$, and $\widehat{a}^{(2)}(x,y)$ based on their variances. Intuitively, the variance of $\widehat{a}(x,y)$ would be smaller than the variances of $\widehat{a}^{(1)}(x,y)$ and $\widehat{a}^{(2)}(x,y)$ in the continuity regions of f, because the former averages more observations. As a matter of fact, when the kernel function K is a constant in its support, it can be checked that

$$\text{Var}(\widehat{a}^{(1)}(x,y)) \approx \text{Var}(\widehat{a}^{(2)}(x,y)) \approx 2\text{Var}(\widehat{a}(x,y)) \approx C^*(nh_n)^{-2}\sigma^2$$

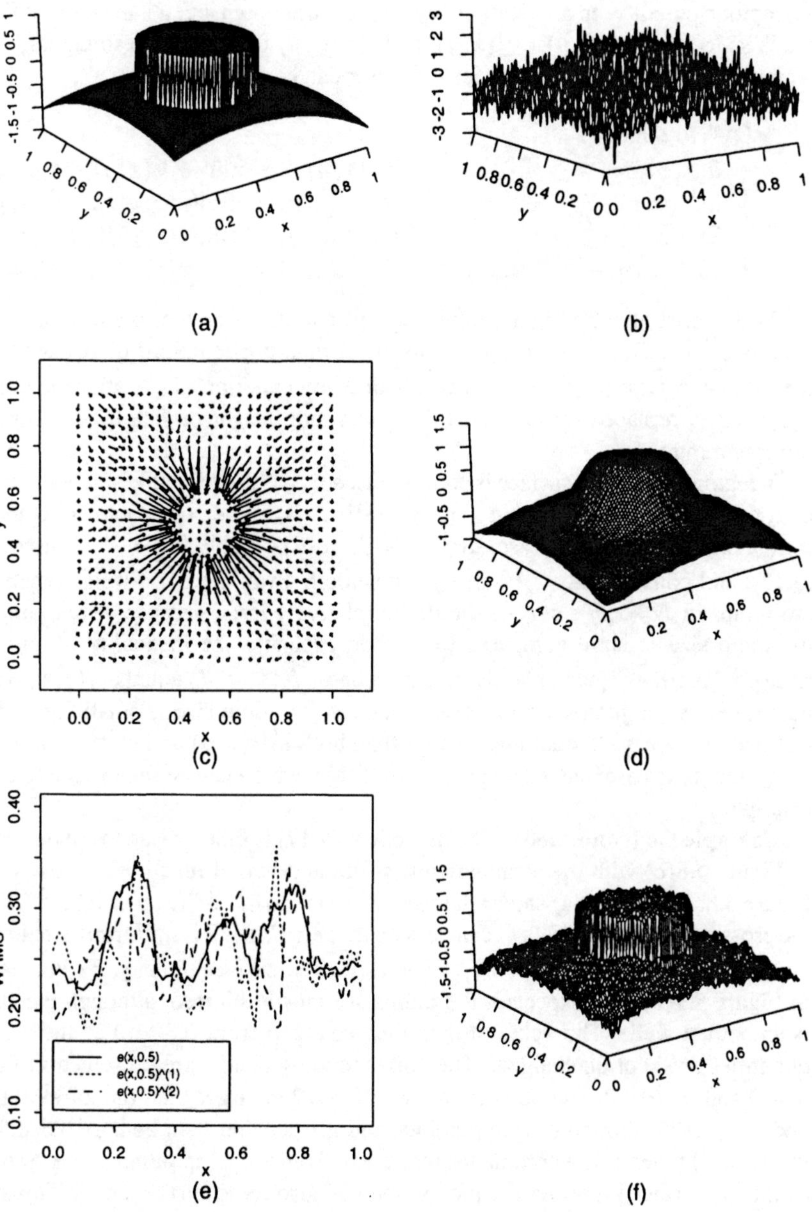

Fig. 5.8 (a): True surface. (b): Observations. (c): Gradient direction G. (d): Conventional surface estimator $\widehat{a}$ with $h_n = 0.1$. (e): WRMS values $e(x, 0.5)$, $e^{(1)}(x, 0.5)$, and $e^{(2)}(x, 0.5)$. (f): Jump-preserving surface estimator $\widehat{f_1}$ with $h_n = 0.1$.

when there are no jumps in $N_n(x, y)$, where C^* is a constant and "$\approx$" denotes asymptotic equality. In applications, σ^2 is often unknown, but it can be estimated by the WRMS values $e(x, y), e^{(1)}(x, y)$, and $e^{(2)}(x, y)$ for the three estimators. Based on these considerations, the surface estimator can be defined by

$$\widehat{f}_2(x, y; \underline{Z}) = \tag{5.17}$$
$$\begin{cases} \widehat{a}(x, y), & \text{if } e(x, y)/2 \leq \min(e^{(1)}(x, y), e^{(2)}(x, y)) \\ \widehat{a}^{(1)}(x, y), & \text{if } e^{(1)}(x, y) < \min(e(x, y)/2, e^{(2)}(x, y)) \\ \widehat{a}^{(2)}(x, y), & \text{if } e^{(2)}(x, y) < \min(e(x, y)/2, e^{(1)}(x, y)) \\ (\widehat{a}^{(1)}(x, y) + \widehat{a}^{(2)}(x, y))/2, & \text{if } e^{(1)}(x, y) = e^{(2)}(x, y) < e(x, y)/2. \end{cases}$$

As demonstrated by Figure 5.8(e), $e(x, y)$ can never be the smallest one among $e(x, y), e^{(1)}(x, y)$, and $e^{(2)}(x, y)$. Thus, by comparing equations (5.16) and (5.17), the estimator $\widehat{f}_1(x, y; \underline{Z})$ can also be obtained by equation (5.17) after the quantity $e(x, y)/2$ is replaced by $e(x, y)$; this observation might be helpful for computer programming.

In regions where the surface is continuous, $e(x, y)/2$ is probably smaller than both $e^{(1)}(x, y)$ and $e^{(2)}(x, y)$ because $e(x, y), e^{(1)}(x, y)$, and $e^{(2)}(x, y)$ are all consistent estimators of σ^2. In this case, $\widehat{a}(x, y)$ will most probably be selected in equation (5.17), and consequently $\widehat{f}_2(x, y; \underline{Z})$ can smooth away the noise well. When there are jumps in $N_n(x, y)$, $e(x, y)$ should be relatively large because of the jumps. If the jump size is large compared to σ, then probably the inequality $e(x, y)/2 > \min(e^{(1)}(x, y), e^{(2)}(x, y))$ holds. In such a case, $\widehat{f}_2(x, y; \underline{Z})$ equals $\widehat{f}_1(x, y; \underline{Z})$, and consequently the jumps are preserved well. If the jump size is small compared to σ, however, $e(x, y)/2$ could be smaller than both $e^{(1)}(x, y)$ and $e^{(2)}(x, y)$. In such cases, the jumps are blurred by procedure (5.17), which can be seen in the following example.

Example 5.4 (continued) The procedure (5.17) is first applied to the data shown in Figure 5.8(b) with the same window width and kernel function as those used in Figure 5.8. Figure 5.9(a) shows the values of $e(x, y)/2, e^{(1)}(x, y)$, and $e^{(2)}(x, y)$ in the cross section of $y = 0.5$. It can be seen that $e(x, 0.5)/2$ is uniformly smaller than both $e^{(1)}(x, 0.5)$ and $e^{(2)}(x, 0.5)$. The estimated surface $\widehat{f}_2(x, y; \underline{Z})$ is presented in Figure 5.9(b). As expected, the jumps are mostly blurred, although the surface is smoothed well. The value of σ is then changed from 0.5 to 0.2, and all other quantities are kept unchanged. The corresponding results are presented in Figures 5.9(c) and 5.9(d). It can be seen that $e(x, 0.5)/2$ is larger than one of $e^{(1)}(x, 0.5)$ and $e^{(2)}(x, 0.5)$ around the jump points, and it is smaller than both $e^{(1)}(x, 0.5)$ and $e^{(2)}(x, 0.5)$ when x is a certain distance away from the jump points. As a result, the estimated surface preserves the jumps well and also removes the noise efficiently.

5.4.4 A two-step procedure

From the previous two subsections, we notice that procedure (5.17) provides a good surface estimator when the ratio of the jump size to σ (which is called *signal-to-noise ratio* (SNR) hereafter) is large; procedure (5.16) is preferred when this ratio is small.

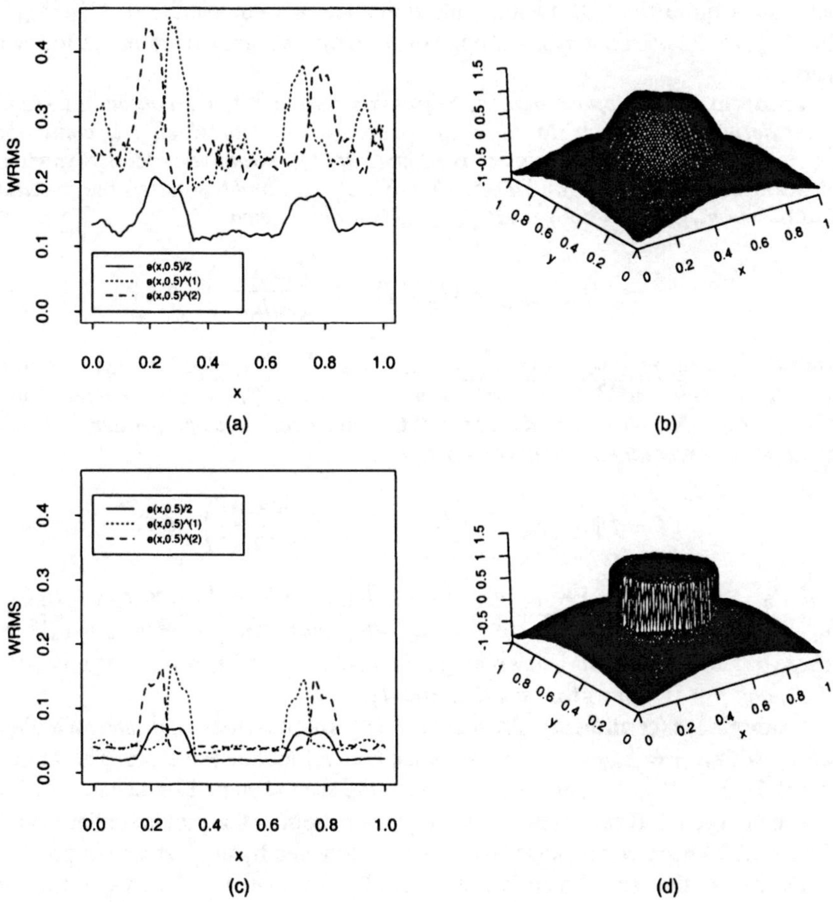

Fig. 5.9 (a): Values of $e(x, 0.5)/2$, $e^{(1)}(x, 0.5)$, and $e^{(2)}(x, 0.5)$ when $\sigma = 0.5$. (b): $\widehat{f}_2$. Plots (c) and (d) are the same as plots (a) and (b) except that $\sigma = 0.2$ in the data.

Based on this observation, Qiu (2003b) suggested the following two-step procedure: first, procedure (5.16) is applied to the original data for decreasing the noise level; second, procedure (5.17) is applied to the estimated surface obtained in the first step to further remove noise. That is, the surface estimator is defined by

$$\widehat{f}(x, y; \underline{Z}) = \widehat{f}_2(x, y; \underline{\widehat{f}}_1), \tag{5.18}$$

where $\underline{\widehat{f}}_1$ denotes the vector of $\{\widehat{f}_1(i/n_1, j/n_1; \underline{Z}), i, j = 1, 2, \ldots, n_1\}$. It should be noticed that the bandwidths used in the two steps of procedure (5.18), which are hereafter denoted by h_{n1} and h_{n2}, could be different.

Because both steps of procedure (5.18) involve local smoothing only, its computation is quite straightforward, with computational complexity of $O(n^2h_{n1}^2) + O(n^2h_{n2}^2)$. This estimator has strong consistency, as summarized in the following theorem.

Theorem 5.3 *Suppose that the regression function f has continuous second-order derivatives in each closed set of the design space in which f is continuous; $E(\varepsilon_{11}^4) < \infty$; the kernel function K is a Lipschitz (1) continuous, radially symmetric, bivariate density function; and the two bandwidths h_{n1} and h_{n2} satisfy the conditions that $h_{n1} \sim h_{n2}$, $h_{n1} = o(1)$, and $1/(n_1h_{n1}) = o(1)$. Then*

$$\|\widehat{f} - f\|_{D_{h_{n1}+h_{n2}}} = O\left(h_{n1}^2\right) + O\left(\frac{\log(n)}{nh_{n1}}\right), \quad a.s.,$$

where $D_{h_{n1}+h_{n2}} = \{(u,v) : (u,v) \in [h_{n1} + h_{n2}, 1 - h_n - h_{n2}] \times [h_{n1} + h_{n2}, 1 - h_n - h_{n2}], \sqrt{(u-u^*)^2 + (v-v^*)^2} \geq h_{n1} + h_{n2}$, and (u^*, v^*) is some point on the JLCs$\}$. If it is further assumed that the JLCs have continuous first-order derivatives at all points except their singular points, then

$$\|\widehat{f} - f\|_{J_{h_{n1}+h_{n2},\rho}} = O\left(h_{n1}\right) + O\left(\frac{\log(n)}{nh_{n1}}\right), \quad a.s.,$$

with $J_{h_{n1}+h_{n2},\rho} = \{(u,v) : (u,v) \in [h_{n1} + h_{n2}, 1 - h_n - h_{n2}] \times [h_{n1} + h_{n2}, 1 - h_n - h_{n2}], \sqrt{(u-u^*)^2 + (v-v^*)^2} < h_{n1} + h_{n2}$ and $\sqrt{(u-u_*)^2 + (v-v_*)^2} \geq \rho$, where (u^*, v^*) is a nonsingular point of the JLCs, (u_*, v_*) is a singular point of the JLCs, and $\rho > 0$ is any given small constant$\}$.

Example 5.4 (continued) Procedure (5.18) is applied to the data shown in Figure 5.8(b), with $h_{n1} = h_{n2} = 0.1$ and the same kernel function as before. The values of $e(x, 0.5)/2$, $e^{(1)}(x, 0.5)$, and $e^{(2)}(x, 0.5)$ of the second step of procedure (5.18) are shown in Figure 5.10(a). It can be seen that the pattern of this plot is similar to that of Figure 5.9(c) because the noise level has been lowered by the first step of procedure (5.18) before the second step is executed. The estimated surface from procedure (5.18) is presented in Figure 5.10(b). We can see that procedure (5.18) performs much better than both procedures (5.16) and (5.17) in this example (cf. Figures 5.8(f) and 5.9(b) for their estimators).

In applications, the window widths h_{n1} and h_{n2} can be determined by minimizing the following cross-validation score

$$CV(h_{n1}, h_{n2}) = \frac{1}{n_1^2} \sum_{i=1}^{n_1} \sum_{j=1}^{n_1} \left(Z_{ij} - \widehat{f}_{-i,-j}(x_i, y_j; \underline{Z})\right)^2, \tag{5.19}$$

where $\widehat{f}_{-i,-j}(x, y; \underline{Z})$ is the "leave-one-out" estimator of $f(x, y)$. Namely, the observation (x_i, y_j, Z_{ij}) is left out in constructing $\widehat{f}_{-i,-j}(x, y; \underline{Z})$.

Example 5.5 In this example, we apply the three surface estimators $\widehat{f}_1$, $\widehat{f}_2$, and $\widehat{f}$ defined in equations (5.16)–(5.18) to a test image frequently used in the image processing literature. The original test image has four gray levels: 20, 75, 150, and 235; and several circular JLCs with three different jump magnitudes: 75, 160,

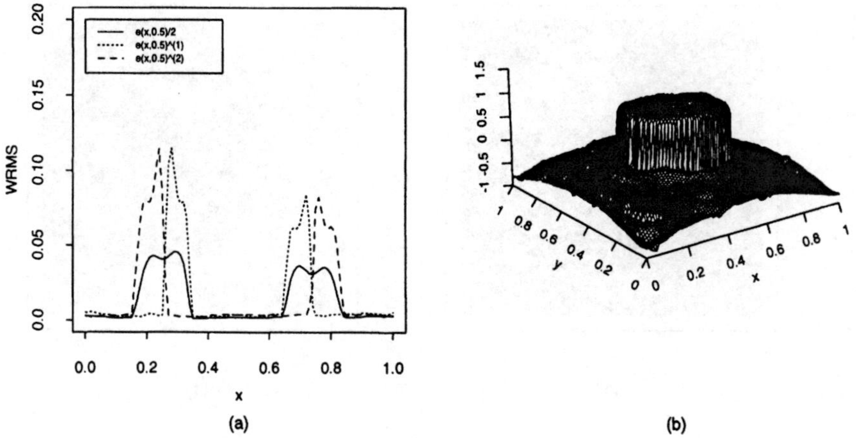

Fig. 5.10 (a): Values of $e(x, 0.5)/2, e^{(1)}(x, 0.5)$, and $e^{(2)}(x, 0.5)$ of the second step of procedure (5.18). (b): Fitted surface of procedure (5.18) with $h_{n1} = h_{n2} = 0.1$.

and 215. Its resolution is 256 × 256. We then add i.i.d. noise with distribution $N(0, 75^2)$ to the original test image; the noisy image is shown in Figure 5.11(a). The three surface estimation procedures are then applied to the noisy image, with bandwidths chosen to be 0.05, 0.02 and (0.02, 0.04), respectively, by the CV procedure (cf. equation (5.19)). Their reconstructed images are presented in Figures 5.11(b)–5.11(d), respectively. From the plots, it can be seen that procedure (5.17) blurs the edges, especially in the region around the smallest circle. The reconstructed images by procedures (5.16) and (5.18) look reasonably good. If we compare them carefully, we can see that the reconstructed image of procedure (5.18) is smoother than that of procedure (5.16). Also, both procedures slightly blur some parts of the edges close to the image border. One possible explanation of this phenomenon is that both procedures use the symmetric padding procedure to handle the boundary regions, and these parts are close to some singular points of the edges artificially created by the symmetric padding procedure.

5.5 SURFACE RECONSTRUCTION BY ADAPTIVE WEIGHTS SMOOTHING

Polzehl and Spokoiny (2000) suggested a so-called *adaptive weights smoothing* (AWS) procedure for nonparametric estimation of regression curves or surfaces. That procedure was further discussed and improved by Polzehl and Spokoiny (2003a, 2003b). It is introduced below in 2-D cases only; the basic idea is introduced in Subsection 5.5.1, and selection of its procedure parameters is discussed in Subsection 5.5.2.

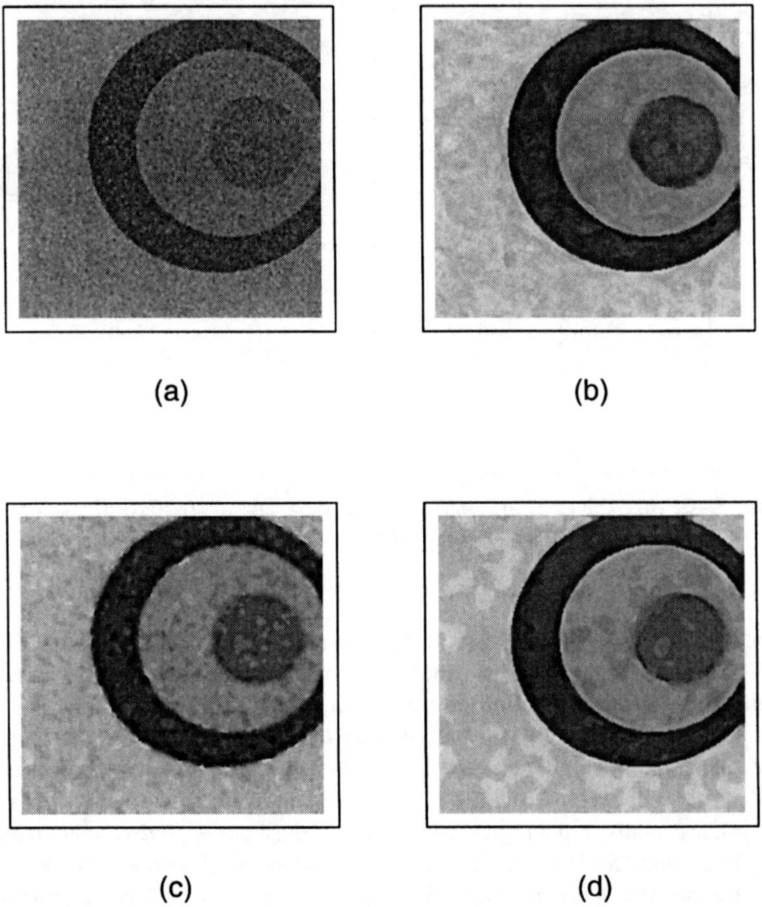

Fig. 5.11 (a): The noisy test image. (b)-(d): Reconstructed images by procedures (5.16)–(5.18), respectively.

5.5.1 Adaptive weights smoothing

The context for adaptive weights smoothing is as follows. Suppose that $\{(x_i, y_i, Z_i), i = 1, 2, \ldots, n\}$ are n observations from the regression model (4.1) in which the regression function f has the following structure:

$$f(x, y) = \sum_{m=1}^{M} a_m I((x, y) \in A_m), \text{ for } (x, y) \in [0, 1] \times [0, 1], \tag{5.20}$$

where $\{a_m, m = 1, 2, \ldots, M\}$ are M constants and $\{A_m, m = 1, 2, \ldots, M\}$ represent a partition of the design space $[0, 1] \times [0, 1]$. By equation (5.20), the regression function f is assumed to be piecewise constant in the design space. The constants $\{a_m, m = 1, 2, \ldots, M\}$, the partition $\{A_m, m = 1, 2, \ldots, M\}$, and the integer M are all assumed unknown.

The major goal of the AWS procedure is to estimate the unknown constants $\{a_m, m = 1, 2, \ldots, M\}$ and the partition $\{A_m, m = 1, 2, \ldots, M\}$, based on n observations. Since each pointset A_m could be small (e.g., A_m could include as few as one design point), expression (5.20) is flexible enough to provide an approximation to the regression surfaces encountered in most applications, including the image intensity surface of a digital image discussed in Section 1.1. On the other hand, if, for a given application, most A_m's in (5.20) contain many points, then the results from the AWS procedure will benefit from this assumption.

The AWS procedure is defined iteratively. An initial estimator of f at a given design point (x_i, y_i) is given by

$$\widehat{f}_0(x_i, y_i) = \frac{1}{N_0(x_i, y_i)} \sum_{(x_j, y_j) \in U_0(x_i, y_i)} Z_j,$$

where $U_0(x_i, y_i)$ is a prespecified neighborhood of (x_i, y_i) and $N_0(x_i, y_i)$ denotes the total number of observations in $U_0(x_i, y_i)$. That is, $\widehat{f}_0(x_i, y_i)$ is defined by the simple average of the observations in $U_0(x_i, y_i)$. Its variance can be estimated by

$$\widehat{s}_0^2(x_i, y_i) = \frac{\widehat{\sigma}^2}{N_0(x_i, y_i)},$$

where $\widehat{\sigma}^2$ denotes a consistent estimator of σ^2. In applications, people can simply define

$$\widehat{f}_0(x_i, y_i) = Z_i;$$

namely, $U_0(x_i, y_i) = \{(x_i, y_i)\}$ and $N_0(x_i, y_i) = 1$ in such a case.

At the k-th iteration with $k \geq 1$, the updated estimator of f at (x_i, y_i) and the estimator of its variance are defined by

$$\widehat{f}_k(x_i, y_i) = \frac{\sum_{(x_j, y_j) \in U_k(x_i, y_i)} w_k((x_i, y_i), (x_j, y_j)) Z_j}{\sum_{(x_j, y_j) \in U_k(x_i, y_i)} w_k((x_i, y_i), (x_j, y_j))}$$

and

$$\widehat{s}_k^2(x_i, y_i) = \frac{\widehat{\sigma}^2 \sum_{(x_j, y_j) \in U_k(x_i, y_i)} w_k^2((x_i, y_i), (x_j, y_j))}{\left\{\sum_{(x_j, y_j) \in U_k(x_i, y_i)} w_k((x_i, y_i), (x_j, y_j))\right\}^2},$$

where $U_k(x_i, y_i)$ is a prespecified neighborhood of (x_i, y_i) and $w_k((x_i, y_i), (x_j, y_j))$ is a weight. The neighborhood $U_k(x_i, y_i)$ is often defined as a circle:

$$U_k(x_i, y_i) = \{(x_j, y_j) : \|(x_i, y_i) - (x_j, y_j)\| \leq h_k\},$$

where h_k is a radius parameter and $\|\cdot\|$ is the Euclidean distance. The weight $w_k((x_i, y_i), (x_j, y_j))$ is defined by

$$w_k((x_i, y_i), (x_j, y_j)) = K_\ell\left(\ell_{ij}^{(k)}\right) K_s\left(s_{ij}^{(k)}\right)$$

where

$$\ell_{ij}^{(k)} = \frac{\|(x_i, y_i) - (x_j, y_j)\|^2}{h_k^2},$$

$$s_{ij}^{(k)} = \left(\frac{\widehat{f}_{k-1}(x_i, y_i) - \widehat{f}_{k-1}(x_j, y_j)}{\lambda \widehat{s}_{k-1}(x_i, y_i)} \right)^2,$$

K_ℓ and K_s are two kernel functions selected to be monotonically nonincreasing on the positive semi-axis, and λ is a positive constant.

Clearly, $\widehat{f}_k(x_i, y_i)$ is defined as a weighted average of the observations in $U_k(x_i, y_i)$; this weight is determined by the kernel functions K_ℓ and K_s, the distance between the two points (x_i, y_i) and (x_j, y_j), and the difference between the estimated values of f at these two points from the previous iteration. The weight is larger when the distance between the two points or the difference between the estimated values of f at the two points (obtained from the previous iteration) is smaller. This is intuitively reasonable because the chance that the two related design points belong to a same pointset in $\{A_m, m = 1, 2, \ldots, M\}$ is larger in such cases.

The radius parameter h_k is determined by the following recursive formula:

$$h_k = C_h h_{k-1}, \text{ for } k \geq 1,$$

where $h_0 = 1/\sqrt{n}$, and C_h is a parameter. The AWS procedure stops if $h_k \geq h_{max}$, where h_{max} is a prespecified maximum radius for the procedure.

Recently, a related pointwise adaptive approach was suggested by Polzehl and Spokoiny (2003c) for surface reconstruction. This new approach estimates the surface at a given design point by the simple average of the observations in a neighborhood of the given design point. This neighborhood is selected from a prespecified family of neighborhoods by a multiple testing procedure.

5.5.2 Selection of procedure parameters

The above AWS procedure has several procedure parameters to determine. Polzehl and Spokoiny (2000) pointed out that the performance of the procedure depends strongly on the choice of the parameters involved, but there is no one universally optimal choice of these parameters. However, they provided some guidelines about the selection of the procedure parameters; these are briefly discussed below.

Polzehl and Spokoiny (2003a) pointed out that the most important parameter of the procedure is λ, which scales the statistical penalty $s_{ij}^{(k)}$. Small values of λ would lead to overpenalization, which may result in unstable performance of the method in the homogeneous situation. Large values of λ could result in loss of adaptivity of the method. A reasonable way to choose λ for a specific application is based on the condition of free extension, which was called "propagation condition" by Polzehl and Spokoiny. By this condition, λ should be chosen such that the impact of the statistical penalty in the computed weights $w_k((x_i, y_i), (x_j, y_j))$ is negligible when the true regression function is a constant. The recommended λ value is the $1 - \alpha$ quantile of the χ^2 distribution with 1 degree of freedom, and the default α is 0.034.

Briefly, the other parameters may be chosen as follows. The parameter h_{max} could be selected by cross-validation or another parameter selection procedure discussed in Chapter 2. The default C_h value is 1.25, and the default kernel function is $K_s(u) = \exp(-u)$. The kernel function K_ℓ could be any compactly supported density function. In the literature, the variance estimator $\widehat{\sigma}^2$ is often defined by

$$\widehat{\sigma}^2 = \frac{1}{n} \sum_{i=1}^{n} \widehat{e}_i^2,$$

where $\widehat{e}_i = Z_i - \widehat{f}(x_i, y_i)$ is the pseudo-residual and $\widehat{f}(x_i, y_i)$ is a preliminary surface estimator constructed by some conventional surface estimation procedures, such as the local linear kernel estimation procedure (5.4). Polzehl and Spokoiny (2000) pointed out that the estimator $\widehat{\sigma} = (t_{0.75} - t_{0.25})/1.35$ is often preferable in applications, where $t_{0.25}$ and $t_{0.75}$ are the first and third quartiles of all pseudo-residuals. In addition, they suggest that a bootstrap-based procedure may be used for choosing some of the parameters mentioned above.

5.6 SOME DISCUSSIONS

In this chapter, we have introduced several recent jump-preserving surface reconstruction procedures. The three-stage procedure introduced in Section 5.2 preserves the jumps by detecting the possible jumps and fitting a local PC line going through the center of the detected jump points, in a neighborhood of any given point. This procedure is appealing theoretically, but it is relatively complicated to use. The surface estimation procedure discussed in Section 5.3 preserves the possible jumps by fitting a local piecewisely linear function in a neighborhood of a given point and by introducing a threshold parameter. The possible jumps are not explicitly detected in this procedure, and its surface estimator has a mathematical formula like most conventional local smoothing surface estimators. Therefore, this procedure is relatively easy to use and relatively simple to compute. To get rid of the threshold parameter, a two-stage surface estimation procedure is discussed in Section 5.4. In the first stage, a local plane is fitted by local linear kernel smoothing in a circular neighborhood of a given point. Then the neighborhood is divided into two parts along the gradient direction of the fitted plane. Two one-sided planes are fitted by local linear kernel smoothing in the two parts, respectively. The jump-preserving surface estimator is defined by one of the three surface estimators provided by the three fitted local planes, depending on their weighted residual mean square values. The estimated surface in this stage preserves the jumps well, but its ability to remove noise is limited. Thus, to further remove noise, the estimated surface values at the design points are used as new data in the second stage of the procedure. The same procedure described above is applied to this new data, except that the surface estimator is defined by one of the three estimators provided by the three fitted local planes depending on their estimated variances. The overall surface estimator of the two-stage procedure preserves the jumps well and also removes the noise efficiently. To use this procedure, users only need to choose two bandwidth parameters, and the surface estimator has an explicit formula. So this procedure is convenient to use and compute. The adaptive weights smoothing procedure discussed in Section 5.5 estimates jump surfaces by an iterative algorithm. This procedure performs well when the true surface contains large homogeneous regions.

Many issues related to these procedures need further investigation. Some of them are common to all these procedures, including selection of the procedure parameters, function estimation in border regions of the design space, etc. Other issues are specific to each individual procedure. We discuss some of them below.

At the end of Subsection 5.2.3, we mentioned that the PC line fitted in a neighborhood of a given point provided a first-order approximation to the true JLC in the neighborhood. Second- or higher-order approximations to the JLC might be possible, although based on our numerical experience the first-order approximation often provides a better result. However, we have not tried more flexible curves (e.g., spline curves) for approximating the JLCs locally. Therefore, it may still be possible to improve the first-order approximation to the JLCs.

With regard to the local piecewisely linear kernel smoothing procedure, several remarks were made at the end of Subsection 5.3.1 for possible improvements. Here, we mention another potential improvement. The major idea of this procedure is to use a piecewisely linear function with possible jumps at the boundaries of the four quadrants of a neighborhood at a given point to approximate the true regression surface, which may have jumps in the neighborhood. Note that the use of the four quadrants separated by the two lines parallel to the $x-$ and $y-$axes is for convenience only. Alternatively, the neighborhood could be divided into m parts by half-lines starting from the given point along m directions, where $m > 1$ is an integer. With larger m, the procedure should become more flexible. However, it might be a challenging task to propose an appropriate data-driven algorithm for determining m and the related directions at any given point.

As pointed out at the beginning of Subsection 5.4.4, the signal-to-noise ratio should play a central role in determining the performance of procedures (5.16) and (5.17). At this time, however, little is known about the *specific* relationship among this ratio, the optimal window widths, and the performance of procedures (5.16) and (5.17).

Procedure (5.18) can be generalized in several directions. One possibility is that, after the first step, procedure (5.17) can be applied to the estimated surface a number of times in an iterative fashion. The second possibility is that procedure (5.16) alone can be applied to the data a number of times in an iterative way. We expect that such iterative procedures would improve the overall performance of the estimated surface, but the computation involved is much expensive. So there is a trade-off between performance improvement and computational complexity of such procedures.

For adaptive weights smoothing procedure, much research is needed on its theoretical properties and on selection of its procedure parameters.

Finding a general criterion for measuring the fit of jump-preserving surface estimation procedures is another topic of significant interest. A natural criterion for evaluating performance of these procedures is the mean squared error (MSE) of the estimated surface; however, although this criterion can measure the overall performance of the estimated surface, it may not be a reliable way of measuring how well the jumps are preserved. Because of the importance of jumps, a different criterion for measuring jump preservation might be needed, which is not available in the literature yet. The local MSE of the estimated surface defined in a neighborhood of the true JLCs might work for this purpose, but its theoretical and numerical properties are not yet clear. After such performance measures become available, different jump-preserving surface estimation procedures can be compared with regard to both their abilities of removing noise and of preserving jumps.

Problems

5.1 In Subsection 3.4.3, the sigma filter and the M smoother (cf. expressions (3.28) and (3.29)) are discussed for estimating regression curves with possible jumps preserved. Discuss how to generalize these two procedures for estimating regression surfaces with possible jumps preserved.

5.2 With regard to the three-stage procedure discussed in Section 5.2, consider the following problems.

5.2.1. Verify that the solution of the minimization problem (5.1) is the local PC line with expression (5.2).

5.2.2. In the neighborhood $N^*(x_i, y_j)$ of a given design point (x_i, y_j), suppose that a local quadratic curve going through the center of the detected jump points in $N^*(x_i, y_j)$ is fitted for approximating the JLCs. Derive an expression for the fitted local quadratic curve.

5.2.3. If the local linear kernel smoothing is used in the third stage of the procedure for defining the jump-preserving surface estimator $\widehat{f}$, give an expression for $\widehat{f}$.

5.3 The original global topographical elevation data can be downloaded from the ftp site of this book (see front cover for address). Using this data, solve the following problems.

5.3.1. Reproduce Figure 5.3.

5.3.2. It is mentioned in Subsection 5.2.1 that the detected jumps obtained in the first stage of the three-stage procedure do not need to be modified by any jump modification procedures. Reproduce Figure 5.3(b) based on the modified version of the detected jumps shown in Figure 5.3(f). Comment on the differences between the two fitted surfaces with and without modifications of the detected jumps.

5.4 Verify that the solution of the minimization problem (5.5) is given by the expressions in (5.6).

5.5 When expression (5.7) holds, there are several ways to define the surface estimator $\widehat{f}_{PLK}$. One is given in equation (5.8). Two other ways are discussed at the end of Subsection 5.3.1. Compare these three approaches, discussing their strengths and limitations.

5.6 Suppose that the procedure discussed in Section 5.3 is generalized in the following way. At a given point (x, y) in the design space, we consider a circular neighborhood $N_n(x, y) = \{(u, v) : (u, v) \in [0, 1] \times [0, 1], \sqrt{(u - x)^2 + (v - y)^2} \leq h_n\}$, where h_n is a bandwidth parameter. Then $N_n(x, y)$ is divided into m sectors along m given directions $\{\theta_j, j = 1, 2, \ldots, m\}$, where $m \geq 2$ is an integer. In each sector, a local linear kernel estimator of $f(x, y)$ is constructed. If the range of the m local linear kernel estimators, obtained in the m sectors respectively, is larger than a threshold value, then the surface estimator at (x, y) of the generalized procedure is defined by the one of the m local linear kernel estimators with smallest weighted residual sum of squares (cf. equation (5.9) for their definitions). Otherwise, the surface estimator is defined by the simple average of the m local linear kernel estimators.

5.6.1. Derive a formula for the threshold value used in the generalized procedure, in a similar way to that of formula (5.10).

5.6.2. Derive an expression for the surface estimator of the generalized procedure.

5.6.3. Apply the generalized procedure to the global topographical elevation data (cf. Problem 5.3 for downloading information) in the following three cases. (i) $m = 2$, $\theta_1 = 0$, and $\theta_2 = \pi$; (ii) $m = 3$, $\theta_1 = 0, \theta_2 = 2\pi/3$, and $\theta_3 = 4\pi/3$; (iii) $m = 4$, $\theta_1 = 0, \theta_2 = \pi/2, \theta_3 = \pi$, and $\theta_4 = 3\pi/2$. Use $\alpha_{n_1} = 0.001$ and the same kernel function as in Example 5.3. The bandwidth h_n can be determined by cross-validation. Comment on the differences among the three fitted surfaces.

5.7 With regard to estimation of surface gradient direction as discussed in Subsection 5.4.1, consider the following problems.

5.7.1. Show that the estimated gradient direction $G(x, y)$ at a given point (x, y) converges to the gradient direction of the true regression function f at (x, y) almost surely if f has continuous second-order derivatives in the neighborhood $N_n(x, y)$.

5.7.2. If the point (x, y) is on a single JLC and the JLC has a unique tangent line at (x, y), show that $G(x, y)$ is asymptotically and almost surely perpendicular to the tangent line, when f is assumed to have continuous second-order derivatives in $N_n(x, y)$ except on the JLC, at which the second-order derivatives of f are assumed to have one-sided limits.

5.8 Verify that

$$e(x, y) \geq \left(e^{(1)}(x, y) + e^{(2)}(x, y)\right) \Big/ 2$$

is true when some regularity conditions are satisfied. Therefore, the estimator $\widehat{f}_1(x, y; \underline{Z})$ can also be obtained by equation (5.17) after the quantity $e(x, y)/2$ is replaced by $e(x, y)$.

5.9 Prove that $e(x, y)$, $e^{(1)}(x, y)$, and $e^{(2)}(x, y)$, as defined by equations (5.13) and (5.15), are all L_2 consistent estimators of σ^2 when f has continuous second-order derivatives in $N_n(x, y)$ and $E(\varepsilon_{11}^4) < \infty$.

5.10 The original circle image used in Example 5.4 can be downloaded from the ftp site of this book (see front cover for address). Using this test image, do the following problems.

5.10.1. Reproduce Figure 5.11.

5.10.2. Apply the estimation procedure (5.16) to the noisy test image shown in Figure 5.11(a) twice. Namely, compute the following surface estimator:

$$\widehat{f}(x, y; \underline{Z}) = \widehat{f}_1(x, y; \widehat{\underline{f}}_1),$$

where $\widehat{\underline{f}}_1$ is defined in equation (5.18). Compare this estimator with the three estimators defined by equations (5.16)–(5.18).

6

Edge Detection In Image Processing

6.1 INTRODUCTION

As discussed in Chapter 1, an image can be regarded as a surface of the image intensity function. In order to process images using computers, from image storage to image analysis, images need to be digitized in both spatial location and brightness measure. A conventional digital image consists of pixels regularly arranged in rows and columns, as demonstrated in Section 1.1. Its gray level at each pixel takes integer values in the range $[0, L-1]$, where L is the number of gray levels, which is often selected to be an integer power of 2 (e.g., $L = 2^8 = 256$).

For better human or machine perception, images often need to be refined by various image processing techniques. For instance, in order to see some features or parts of an image better, these features or parts can be enhanced by an appropriate *image enhancement* procedure; to study a specific object in an image, its boundary can be highlighted by an *image segmentation* procedure. In the literature, many image processing techniques have been proposed for various purposes (see e.g., Gonzalez and Woods 1992, 2002, Pratt 1991, Rosenfeld and Kak 1982). In this and the next chapter, we focus mainly on two types of image processing techniques: those for *edge detection* (Chapter 6) and those for *image restoration* (Chapter 7), since these techniques are widely used in applications and are closely related to the jump regression analysis discussed in the previous three chapters.

In the literature, several different types of *edges* have been defined. One type of edges refers to places in an image at which the image intensity function has jumps. Such edges are often called *step edges* (see e.g., Haralick 1984). In many cases step edges represent object boundaries or changes in reflectance, illumination, or distance from the viewer. The 1-D profile of an ideal step edge is shown in Figure 6.1(a). In applications, step edges might look messier than the one in the plot because image gray levels may have noise involved and may also be blurred in the image acquisition process. See Section 7.1 in Chapter 7 for more explanation about noise and spatial blurring degradations in an observed image.

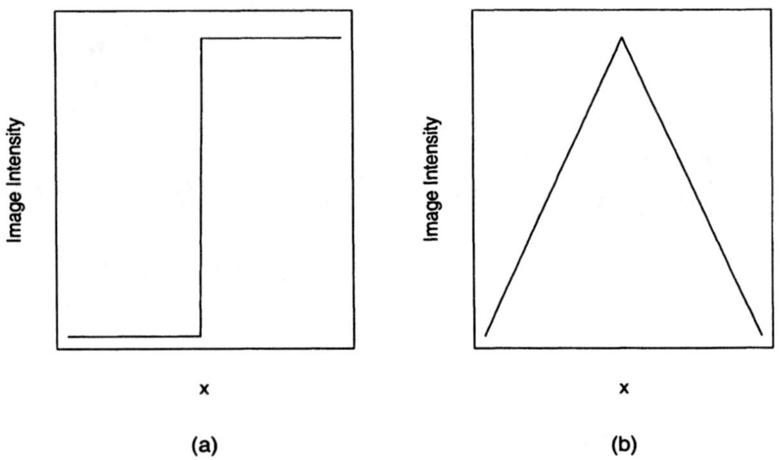

Fig. 6.1 (a): 1-D profile of an ideal step edge. (b): 1-D profile of an ideal roof edge.

For digital images, the gray levels at two neighboring pixels could be slightly different due to image digitization, especially when the number of possible gray levels L is large, even if the image intensity function is actually continuous around the two pixels in question. Obviously, step edges of interest should not include such slight jumps caused by image digitization. Jumps at step edges should be more dramatic than those tiny artificial jumps.

If an image is regarded as a surface of the image intensity function, then clearly step edges in the image correspond to jumps in the image intensity function; however, in some applications, we are interested in detecting jumps in the first-order derivatives of the image intensity function. Those jumps are called *roof edges* in the image processing literature. They correspond to places in the image at which the image intensity surface has sharp peaks, which look like roofs, or valleys. The 1-D profile of an ideal roof edge is displayed in Figure 6.1(b). Again, roof edges in applications might be messier for reasons similar to those given about step edges.

Besides step edges and roof edges, there are some other types of edges defined in the literature. For example, some authors distinguish step edges from the so-called *ramp edges* (e.g., Peli and Malah 1982). Around a ramp edge, image intensity changes from one level to another much higher or lower level through a ramp. Obviously, ramp edges have no essential difference from step edges if the image resolution is so low that the ramps are hardly visible. Other types of edges frequently discussed in the literature include spike edges (Davis 1975), line edges (e.g., Nalwa and Binford 1986), texture edges (e.g., Rosenfeld and Kak 1982), and so forth.

In most applications, the edges that we are interested in detecting are mainly step edges, because they often represent outlines of image objects or some sudden structural changes of the related process represented by the image. Thus, most edge detection techniques in the literature are designed specifically for detecting step edges. For this reason, in this chapter and in Chapter 7, edges refer to step edges if there is no further specification.

Most traditional edge detectors are based on estimation of the first- or second-order derivatives of the image intensity function. Some of their representations are introduced in Section 6.2. Canny (1986) suggested three criteria for optimal edge detection, which are discussed in Section 6.3. The idea of detecting edges based on multilevel pixels is discussed in Section 6.4. A cost minimization approach for edge detection is introduced in Section 6.5. Some postprocessing methods for linking detected edges are discussed in Section 6.6. Finally, several remarks conclude the chapter in Section 6.7.

6.2 EDGE DETECTION BASED ON DERIVATIVE ESTIMATION

The motivation for procedures introduced in this section is best illustrated by several plots regarding the step edge. Figure 6.2(a) displays a 1-D profile of a step edge which is slightly blurred during image acquisition. Its first- and second-order derivatives are displayed in Figures 6.2(b) and 6.2(c), respectively. It can be seen from the plots that (1) the first-order derivative peaks at the step edge, (2) the second-order derivative equals zero at the step edge, and (3) the second-order derivative changes its sign from positive to negative when the image intensity jumps from a lower level to a higher level. In the literature, the second and third properties are called *zero-crossing* properties of the second-order derivative. In the 2-D setup, the directional first- and second-order derivatives in the direction perpendicular to a given edge segment have similar properties. Therefore, derivatives of the image intensity function carry useful information about edges; so, many traditional edge detectors are constructed based on estimation of these derivatives. Some of their representations are introduced below, in three subsections: edge detectors based on first-order derivatives are introduced in Subsection 6.2.1, edge detectors based on second-order derivatives are discussed in Subsection 6.2.2, and edge detectors based on local surface estimation are discussed in Subsection 6.2.3.

6.2.1 Edge detection based on first-order derivatives

We first discuss how we expect the image intensity function $f(x, y)$ to behave at the edges of an image, where (x, y) denotes spatial location. Intuitively, if a pixel (x, y) is on an edge segment which is parallel to the y-axis, then the first-order partial derivative of f with respect to x at (x, y), denoted as $f_x(x, y)$, should be large in

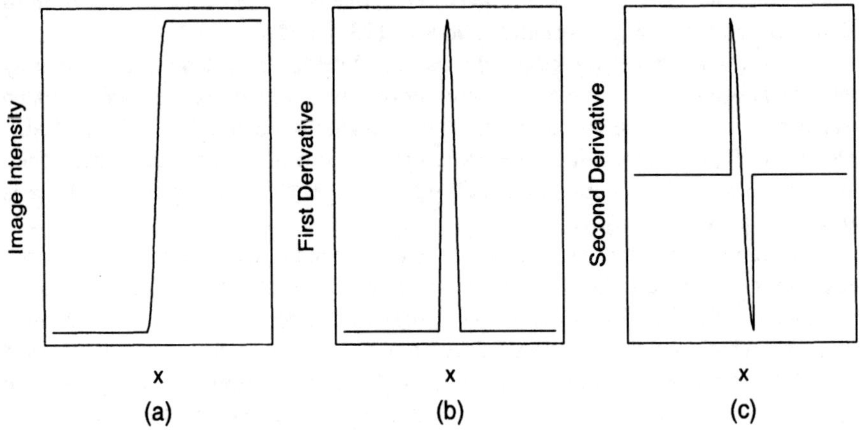

Fig. 6.2 (a): 1-D profile of a step edge that is slightly blurred. (b): First-order derivative of the 1-D profile. (c): Second-order derivative of the 1-D profile.

magnitude, as demonstrated in Figure 6.2(b). Similarly, the magnitude of $f_y(x, y)$ should be large if (x, y) is on an edge segment parallel to the x-axis. If (x, y) is on an edge segment with arbitrary orientation, then the magnitude of the gradient vector $G_f(x, y) = (f_x(x, y), f_y(x, y))'$, defined by

$$M_f(x, y) = \sqrt{f_x^2(x, y) + f_y^2(x, y)} \tag{6.1}$$

should be large. Thus, $M_f(x, y)$ can be used as an edge detection criterion, with large values indicating possible edge pixels. However, this criterion has some square and square-root operations involved, which are relatively difficult to implement. To simplify its computation, in the image processing literature, people often replace $M_f(x, y)$ with the following criterion:

$$A_f(x, y) = |f_x(x, y)| + |f_y(x, y)|. \tag{6.2}$$

To use either criterion (6.1) or (6.2) for edge detection, the first-order partial derivatives of f should be estimated first because f is unobservable and also does not have any mathematical expression in most applications. Toward this end, many different estimators of the first-order partial derivatives have been suggested in the image processing literature. Several such estimators are introduced below.

Suppose that the observed gray levels of the image are $\{Z_{ij}, i = 1, 2, \ldots, n_1; j = 1, 2, \ldots, n_2\}$. Then at the (i, j)-th pixel, the two partial derivatives $f_x(i, j)$ and $f_y(i, j)$ can be estimated simply by the differences

$$Z_{i+1,j} - Z_{ij} \quad \text{and} \quad Z_{i,j+1} - Z_{ij},$$

0	1
-1	0

-1	0
0	1

Fig. 6.3 Two Roberts masks.

respectively. Consequently, the two edge detection criteria $M_f(i,j)$ and $A_f(i,j)$ can be estimated by

$$\widehat{M}_f(i,j) = \sqrt{(Z_{i+1,j} - Z_{ij})^2 + (Z_{i,j+1} - Z_{ij})^2}$$

and

$$\widehat{A}_f(i,j) = |Z_{i+1,j} - Z_{ij}| + |Z_{i,j+1} - Z_{ij}|.$$

Clearly, the two differences mentioned above can also be replaced by $Z_{ij} - Z_{i-1,j}$ and $Z_{ij} - Z_{i,j-1}$, respectively, without losing any power for edge detection. To avoid this kind of ambiguity, Roberts (1965) suggested using the following two diagonal differences instead

$$Z_{i+1,j+1} - Z_{ij} \quad \text{and} \quad Z_{i+1,j} - Z_{i,j+1} \tag{6.3}$$

as estimators of the first-order directional derivatives of f at the center of the four related pixels, in the two diagonal directions. Then, either $\widehat{M}_f(i,j)$ or $\widehat{A}_f(i,j)$ based on these two diagonal differences can be used for edge detection. The differences in (6.3) are often called the *Roberts operators*, which can be denoted by the two masks shown in Figure 6.3. The convolution of the first mask with the observed gray levels in a 2×2 window, namely, the sum of the products between the numbers in the mask and the observed gray levels at the corresponding locations in the window, results in the first diagonal difference in (6.3); the convolution of the second mask with the same observations results in the second diagonal difference in (6.3).

Masks of even sizes, such as the Roberts masks (cf. Figure 6.3), are not convenient to implement in applications, because centers of the even-sized windows are not exactly at pixel locations. For this reason, various 3×3 masks have been suggested in the literature. Some commonly used 3×3 masks include the Prewitt masks (Prewitt 1970) shown in Figure 6.4, the Sobel masks (see Duda and Hart 1973, page 271) shown in Figure 4.15 in Subsection 4.4.3, and the Frei-Chen masks (Frei and Chen 1977). The Frei-Chen masks are the same as the Sobel masks except that the ± 2 values in the latter are replaced by $\pm\sqrt{2}$ values in the former. The major difference between the Prewitt masks and the Sobel or Frei-Chen masks is that the weights on the related pixels are the same in the Prewitt masks, but are different in the Sobel or Frei-Chen masks. In the Sobel or Frei-Chen masks, weights are larger for the four neighboring

1	1	1
0	0	0
-1	-1	-1

-1	0	1
-1	0	1
-1	0	1

Fig. 6.4 Two Prewitt masks.

pixels in the x- and y-axis directions than weights for the four diagonally neighboring pixels, which is reasonable because the former neighboring pixels are closer to the given pixel and are expected to provide more useful information for edge detection.

Masks of sizes 3×3 or smaller are simple to compute, but they have limited power for removing noise; thus, some larger masks have also been suggested in the literature. For example, Abdou (1973) suggested using two truncated pyramid operators that give linearly decreasing weights to pixels farther away from the given pixel. The 7×7 truncated pyramid mask in the x-direction is displayed in Figure 6.5, and the 7×7 truncated pyramid mask in the y-direction can be constructed similarly. Argyle (1971) and Macleod (1970, 1972) proposed their large masks based on the Gaussian weighting functions; these masks are further studied by Canny (1986) and are well-known as the *derivative of Gaussian* (DoG) operators (see Section 6.3 for related discussion). Let

$$g(x, s) = \frac{1}{\sqrt{2\pi}s} \exp\left(-\frac{x^2}{2s^2}\right), \tag{6.4}$$

be the probability density function (pdf) of the 1-D Gaussian (i.e., normal) distribution with mean 0 and standard deviation s (cf. Figure 2.1 in Subsection 2.2.1 for its plot when $s = 1$). Then the continuous version of the DoG operator in the horizontal direction is

$$H_{DoG}(x, y) = \frac{\partial[g(x, s)g(y, t)]}{\partial x} = \frac{-x[g(x, s)g(y, t)]}{s^2},$$

where $g(y, t)$ is the pdf of the 1-D Gaussian distribution with mean 0 and standard deviation t. Note that $g(x, s)g(y, t)$ is the pdf of a 2-D Gaussian distribution which is defined as a product of two 1-D Gaussian pdf's. The DoG operator in the vertical direction can be defined similarly. The discrete version of this operator can be obtained by sampling H_{DoG} over a window of given size. For example, if windows of size 3×3 are used in edge detection, then a discrete version of the DoG operator in the horizontal direction could be $\{H_{DoG}(i, j), i, j = -1, 0, 1\}$, for given scales s and t.

For each edge detection criterion (e.g., $\widehat{M}_f(i, j)$ or $\widehat{A}_f(i, j)$ based on one set of difference operators), large values of the criterion imply possible edge pixels.

-1	-1	-1	0	1	1	1
-1	-2	-2	0	2	2	1
-1	-2	-3	0	3	2	1
-1	-2	-3	0	3	2	1
-1	-2	-3	0	3	2	1
-1	-2	-2	0	2	2	1
-1	-1	-1	0	1	1	1

Fig. 6.5 The 7×7 truncated pyramid mask in the x direction.

However, we need to choose a threshold to specify how large is "large". In the literature, there are some theoretical discussions about threshold selection based on hypothesis testing theory, classical decision theory, or pattern recognition theory (e.g., Abdou 1973; Abdou and Pratt 1979; Fukunaga 1972). In applications, instead of using a single threshold, we can also try a sequence of threshold values and then choose the one which gives results with best visual impression.

Example 6.1 The image presented in Figure 1.1 is the famous Lena image, which is one of the most widely used test images in the image processing literature. It has 512×512 pixels. After adding independent and identically distributed (i.i.d.) noise with $N(0, 10^2)$ distribution to each pixel, the noisy Lena image is presented in Figure 6.6(a). Then the Roberts, Prewitt, Sobel, and 7×7 truncated pyramid operators are applied to the noisy Lena image, respectively. Their edge detection criteria based on the estimated image gradient magnitude (i.e., M_f in equation (6.1), with f_x and f_y estimated by the corresponding operators) are shown in Figures 6.6(b)–6.6(e), respectively, in which darker colors represent larger values. For the edge detection criterion based on the 7×7 truncated pyramid operators, several threshold values are tried, from which a threshold value of 1,250 is finally selected; the detected edges with this threshold value are shown in Figure 6.6(f) with dark pixels. It can be seen that the Roberts edge detection criterion is quite noisy, the Prewitt and Sobel edge detection criteria are slightly better in removing noise, and the 7×7 truncated pyramid operators remove the noise well.

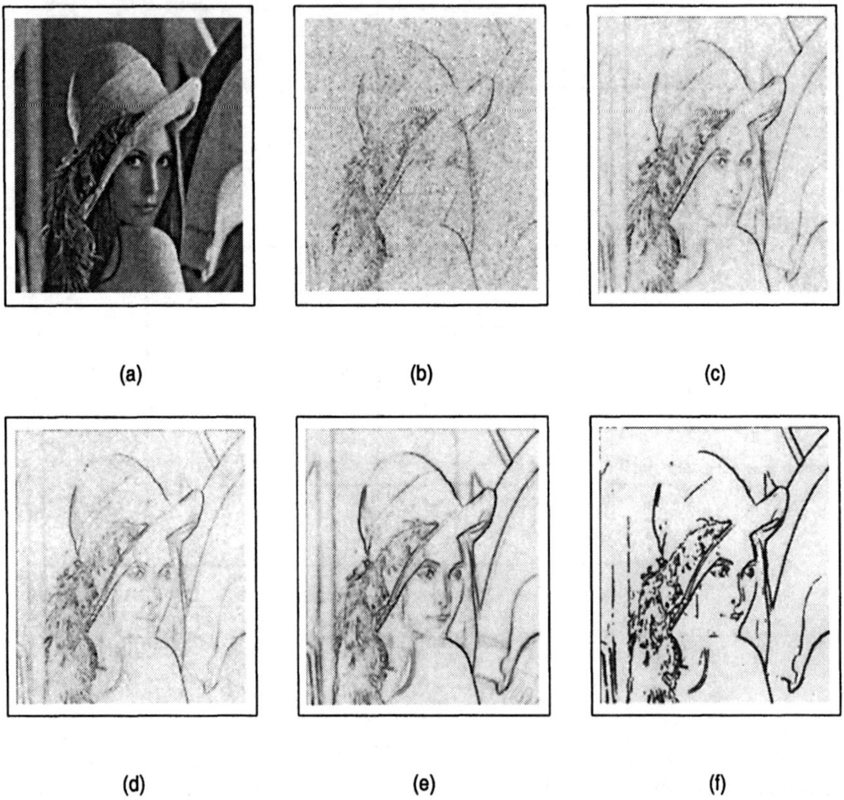

Fig. 6.6 (a): The noisy Lena image contaminated by i.i.d. noise with $N(0, 10^2)$ distribution. (b)–(e): Edge detection criteria based on the estimated image gradient magnitudes by the Roberts, Prewitt, Sobel, and 7×7 truncated pyramid operators, respectively. (f): Detected edges by the 7×7 truncated pyramid operators with a threshold value of 1,250.

6.2.2 Edge detection based on second-order derivatives

In 2-D setup, a commonly used operator based on second-order derivatives is the following Laplacian operator:

$$\nabla^2 = \frac{\partial^2}{\partial x^2} + \frac{\partial^2}{\partial y^2}.$$

For the image intensity function $f(x, y)$, if a given pixel (x_0, y_0) is on an edge segment, then $\nabla^2 f(x, y)$ has the zero-crossing properties around (x_0, y_0) in the following sense: $\nabla^2 f(x, y)$ would be positive on one side of the edge segment, negative on the other side, and zero at (x_0, y_0) or at some place(s) between (x_0, y_0) and its neighboring

0	-1	0
-1	4	-1
0	-1	0

(a)

-1	-1	-1
-1	8	-1
-1	-1	-1

(b)

Fig. 6.7 (a): A four-neighbor Laplacian mask. (b): An eight-neighbor Laplacian mask.

pixels. This is similar to the zero-crossing properties in the 1-D case, as illustrated in Figure 6.2(c).

There are many discrete versions of the Laplacian operator. The simplest Laplacian mask is the one shown in Figure 6.7(a), by which only the given pixel and its four closest neighboring pixels in the x- and y-axis directions are involved in computation. Prewitt (1970) suggested an eight-neighbor Laplacian mask, shown in Figure 6.7(b). Larger Laplacian masks can be defined similarly.

Compared to edge detectors based on the first-order derivatives of f, the Laplacian operator has an advantage in that it localizes the edges well. More specifically, its detected edges often look thinner than the detected edges of the edge detectors based on the first-order derivatives of f. However, it also has a major drawback for edge detection – it is sensitive to noise (Torre and Poggio 1986). To balance localization of the detected edges and ability of removing noise, Marr and Hildreth (1980) proposed the *Laplacian of Gaussian* (LoG) edge detector: the image is first smoothed by a weighted averaging procedure with the weights determined by a 2-D Gaussian density function, and then the edges are detected by applying the Laplacian operator to the smoothed image. From this definition, it can be seen that the LoG operator sacrifices some of its ability in localizing the detected edges compared to the Laplacian operator, since the image is smoothed in the first step, blurring some edges, and consequently the detected edges obtained in the second step look relatively thick. However, the LoG operator improves in noise removal at the same time.

We now specify the form of the LoG operator. Let $G(x, y)$ denote a 2-D Gaussian pdf. In applications, $G(x, y)$ is often selected to be a product of two 1-D Gaussian pdf's as follows:

$$G(x, y) = g(x, s)g(y, s),$$

where $g(x, s)$ is the 1-D Gaussian pdf defined in equation (6.4). Then the smoothed image can be obtained by

$$S\{f\}(x, y) = \int_{-\infty}^{\infty} \int_{-\infty}^{\infty} f(x - u, y - v)G(u, v) \, dudv,$$

which is the *convolution* of the image intensity function f and the weight function G, denoted as $S\{f\} = G \otimes f$. Thus, the LoG operator is defined by

$$\nabla^2(G \otimes f) = (\nabla^2 G) \otimes f, \tag{6.5}$$

where the equality holds due to the linearity property of the Laplacian operator. It can be checked that

$$\nabla^2 G(x, y) = \frac{1}{\pi s^4}\left(\frac{x^2+y^2}{2s^2} - 1\right)\exp\left(-\frac{x^2+y^2}{2s^2}\right).$$

In application, we can also use $-\nabla^2 G(x, y)$ for edge detection without losing any useful information. When $s = 1$, the cross-section of $-\nabla^2 G(x, y)$ at $y = 0$ is shown in Figure 6.8, where it looks like a Mexican hat. For this reason, the operator $-\nabla^2 G$ is often called the *Mexican hat* filter. For a given scale s, the discrete version of the LoG operator can be obtained by sampling the continuous version $\nabla^2 G$ over a $k \times k$ window. Huertas and Medioni (1986) pointed out that the size of the window should be at least $k = 6\sqrt{2}s$ in order to avoid deleterious truncation effects.

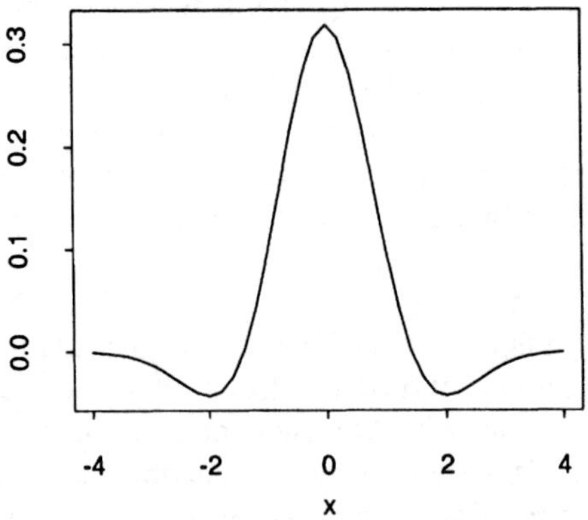

Fig. 6.8 Cross-section of $-\nabla^2 G(x, y)$ at $y = 0$ when $s = 1$, which looks like a Mexican hat.

The scale parameter s in the LoG operator controls the degree of smoothing in the first step of edge detection. When s is chosen large, edges with small jump magnitudes are blurred completely in this step, and consequently they are missed in the second step. In such a case, only those edges with large intensity jumps can be detected by the LoG operator. On the other hand, if s is chosen small, edges with small intensity jumps could be detected, but so could some false edges, due to noise. In this case, the detected edges would include many details.

The LoG edge detector often offers reliable edge location and is robust to noise as well, compared to the Laplacian operator. However, the prices paid for these advantages include (1) a little worse localization of the detected edges (as pointed out above), and (2) relatively expensive computation. To help reduce the impact of the latter of these, Marr and Hildreth (1980) showed that the LoG operator $\nabla^2 G$ can be well approximated by the following difference of two Gaussian operators:

$$g(x, s_1)g(y, s_1) - g(x, s_2)g(y, s_2),$$

where g is the 1-D Gaussian pdf defined in equation (6.4) and $s_1 < s_2$ are two scale parameters. It is demonstrated that this approximation is good when $s_2/s_1 = 1.6$. Because of its "parallel" structure, computation involved in this "difference of Gaussian" operator is simpler than that involved in the LoG operator.

Example 6.2 Figure 6.9 shows the detected edges of the noisy Lena image shown in Figure 6.6(a), by the eight-neighbor Laplacian mask (cf. Figure 6.7(b)), the LoG operator (6.5) with $s = 7$ and $k = 65$, and the "difference of Gaussian" operator with $s_2/s_1 = 1.6$, $s_1 = 7$ and $k = 65$. The panels in the first column show the three edge detection criteria: the darker the color, the larger the value. To find the zero-crossings of these criteria, it is often convenient to first transform them to their binary versions, in which black pixels denote those with positive values of the criteria and white pixels denote the remaining ones. The binary versions of the three edge detection criteria are shown by the panels in the middle column of Figure 6.9. Then the zero-crossings of a given criterion can be identified as those pixels whose eight immediately neighboring pixels in the vertical, horizontal, and two diagonal directions have different colors in the binary version of the criterion. The detected edges by the three criteria are shown in the panels in the last column of Figure 6.9. It can be seen from the plots that the eight-neighbor Laplacian mask cannot detect anything interesting because it is very sensitive to noise. The detected edges by the LoG and "difference of Gaussian" operators look much thinner than the detected edges by the operators based on the first-order derivatives (cf. Figure 6.6). However, it seems that some unnecessary details are also detected by the LoG operator, which might be related to selection of the scale parameter s, as mentioned above.

Huertas and Medioni (1986) showed that the Mexican hat filter $-\nabla^2 G(x, y)$ can be expressed in the following form:

$$-\nabla^2 G(x, y) = h_1(x)h_2(y) + h_2(x)h_1(y),$$

where

$$h_1(x) = \frac{1}{\sqrt{2\pi s^2}}\left(1 - \frac{x^2}{s^2}\right)\exp\left(-\frac{x^2}{2s^2}\right)$$

and

$$h_2(x) = \frac{1}{\sqrt{2\pi s^2}}\exp\left(-\frac{x^2}{2s^2}\right).$$

They pointed out that computation involved in the LoG edge detection could be substantially simplified by using this expression due to its "parallel" structure, and therefore the LoG operator does not need to be approximated by the "difference of Gaussian" operator. More discussion about this additive expression can be found in Chen *et al.* (1987).

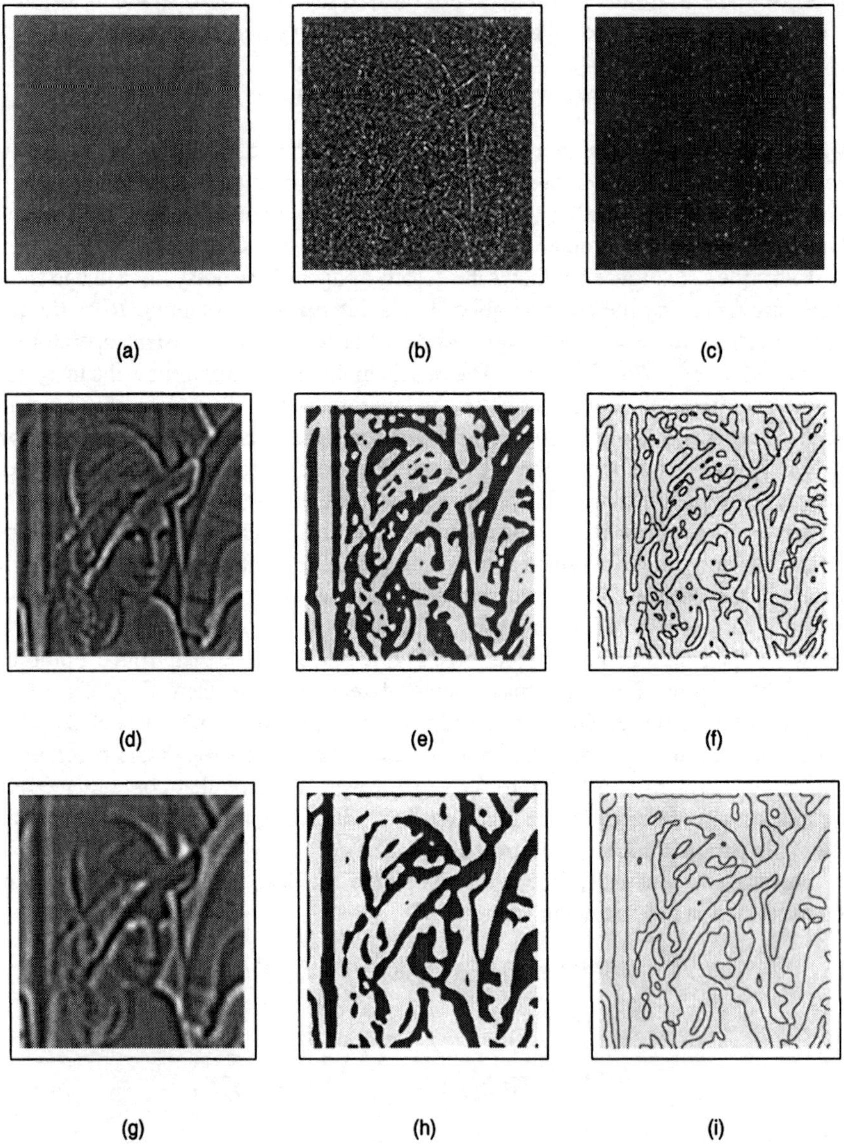

Fig. 6.9 (a): Edge detection criterion based on the eight-neighbor Laplacian mask. (b): Binary version of (a), with black pixels denoting those having positive values in (a) and white pixels denoting the remaining ones. (c): Zero-crossings of the edge detection criterion based on the eight-neighbor Laplacian mask. (d)-(f): Corresponding results of the LoG edge detector with $s = 7$ and $k = 65$. (g)-(i): Results of the "difference of Gaussian" operator with $s_2/s_1 = 1.6$, $s_1 = 7$, and $k = 65$.

The Laplacian, LoG, and "difference of Gaussian" operators all detect edges without using any information about the edge orientations. They are called *rotationally invariant* operators in the literature. On the other hand, at a given pixel on an edge segment, if we know the orientation of that edge segment, then the given edge pixel can also be detected by using the zero-crossing properties of the directional second-order derivative of the image intensity function in the direction perpendicular to the edge segment. Toward this end, the orientation of the edge segment can be estimated by using an edge detector based on the first-order derivatives (cf. some related discussions in Subsection 6.2.1), or by the local surface estimation methods introduced in Subsection 6.2.3 below. In comparing these two methods, Torre and Poggio (1986) showed that (1) the Gaussian filter G used in the LoG operator (6.5) is near-optimal in balancing localization of the detected edges and noise removal; (2) rotational invariant differential operators such as the LoG operator ensure closed edge contours; and (3) it depends on the specific application whether a rotationally invariant operator or a directional operator should be used.

6.2.3 Edge detection based on local surface estimation

Some edge detection masks introduced in the previous two subsections use differences among neighboring pixels as estimators of the corresponding first- or second-order derivatives of the image intensity function for edge detection. Since the individual pixels may be contaminated by noise and such estimators do not have much smoothing involved when sizes of the edge detection masks are small, they could be affected by noise quite dramatically.

To overcome this difficulty, Haralick (1984) suggested first fitting a local polynomial surface in a neighborhood of a given pixel to approximate the image intensity surface locally, and then the derivatives of the fitted polynomial surface can be used as estimators of the corresponding derivatives of the true image intensity surface. Noise is partially removed in the first step, and consequently the derivative estimators would be less sensitive to noise.

As an example, suppose that we consider a 3×3 neighborhood of a given pixel. In this neighborhood, the following quadratic polynomial surface is fitted:

$$q(r,c) = k_1 + k_2 r + k_3 c + k_4 r^2 + k_5 rc + k_6 c^2 + k_7 rc^2 + k_8 r^2 c + k_9 r^2 c^2, \quad (6.6)$$

where r is the row index of the neighborhood, c is the column index, and k's are coefficients. Without loss of generality, we assume that the origin of the coordinate system is at the center of the neighborhood, and thus $r, c \in \{-1, 0, 1\}$. For convenience of computation, Haralick considered using the following 1-D Chebyshev's orthogonal polynomials for surface fitting:

$$\{1, r, r^2 - 2/3\} \text{ and } \{1, c, c^2 - 2/3\}.$$

Their tensor products

$$\{1, r, c, r^2 - 2/3, rc, c^2 - 2/3, c(r^2 - 2/3), r(c^2 - 2/3), (r^2 - 2/3)(c^2 - 2/3)\}$$

are used as basis functions of all 2-D quadratic surfaces and are denoted as $P_j(r,c)$ for $j = 1, 2, \ldots, 9$. It can be easily checked that they are orthogonal to each other. Then, the quadratic function defined in equation (6.6) can be expressed by

$$q(r,c) = \sum_{j=1}^{9} a_j P_j(r,c), \tag{6.7}$$

where a_j denote coefficients which have the following relationship with the k_j coefficients in equation (6.6):

$$\begin{aligned}
k_1 &= a_1 - \frac{2}{3}a_4 - \frac{2}{3}a_6 + \frac{4}{9}a_9 \\
k_2 &= a_2 - \frac{2}{3}a_7 \\
k_3 &= a_3 - \frac{2}{3}a_8 \\
k_4 &= a_4 - \frac{2}{3}a_9 \\
k_6 &= a_6 - \frac{2}{3}a_9 \\
k_j &= a_j, \text{ for } j = 5, 7, 8, 9.
\end{aligned} \tag{6.8}$$

The a_j coefficients in equation (6.7) can be estimated by

$$\widehat{a}_j = \frac{\sum_{r,c\in\{-1,0,1\}} P_j(r,c) Z_{rc}}{\sum_{r,c\in\{-1,0,1\}} P_j^2(r,c)}, \text{ for } j = 1, 2, \ldots, 9,$$

where Z_{rc} denotes the observed image intensity at the (r,c)-th pixel in the neighborhood. Finally, the fitted quadratic surface is

$$\widehat{q}(r,c) = \sum_{j=1}^{9} \widehat{a}_j P_j(r,c).$$

After the local quadratic surface is fitted, the gradient direction of the true image intensity function f at the given pixel can be estimated by

$$\widehat{\theta} = \tan^{-1}\left(\widehat{k}_2/\widehat{k}_3\right),$$

where $\widehat{k}_j$ are estimators of k_j obtained from equation (6.8) after a_j are replaced by $\widehat{a}_j$. Haralick then suggested using the directional second-order derivative of $\widehat{q}$ in the direction of $\widehat{\theta}$ for edge detection. It can be checked that the directional second-order derivative of $\widehat{q}$ has the expression

$$\begin{aligned}
\widehat{q}''_{\widehat{\theta}}(r,c) &= 2\widehat{k}_4 \sin^2(\widehat{\theta}) + 2\widehat{k}_5 \sin(\widehat{\theta})\cos(\widehat{\theta}) + 2\widehat{k}_6 \cos^2(\widehat{\theta}) + \\
&\quad \left[4\widehat{k}_7 \sin(\widehat{\theta})\cos(\widehat{\theta}) + 2\widehat{k}_8 \cos^2(\widehat{\theta})\right] r +
\end{aligned}$$

$$\left[2\widehat{k}_7 \sin^2(\widehat{\theta}) + 4\widehat{k}_8 \sin(\widehat{\theta})\cos(\widehat{\theta})\right] c +$$
$$\left[2\widehat{k}_9 \cos^2(\widehat{\theta})\right] r^2 + \left[8\widehat{k}_9 \sin(\widehat{\theta})\cos(\widehat{\theta})\right] rc + \left[2\widehat{k}_9 \sin^2(\widehat{\theta})\right] c^2.$$

Therefore, at any point $(\rho\cos(\widehat{\theta}), \rho\sin(\widehat{\theta}))$ on the line passing the given point and in the direction of $\widehat{\theta}$, the above directional second-order derivative can be expressed by

$$\begin{aligned} \widehat{q}''_{\widehat{\theta}}(\rho\cos(\widehat{\theta}), \rho\sin(\widehat{\theta})) &= 2\left[\widehat{k}_4 \sin^2(\widehat{\theta}) + \widehat{k}_5 \sin(\widehat{\theta})\cos(\widehat{\theta}) + \widehat{k}_6 \cos^2(\widehat{\theta})\right] + \\ &\quad 6\sin(\widehat{\theta})\cos(\widehat{\theta})\left[\widehat{k}_7 \sin(\widehat{\theta}) + \widehat{k}_8 \cos(\widehat{\theta})\right]\rho + \\ &\quad 12\left[\widehat{k}_9 \sin^2(\widehat{\theta})\cos^2(\widehat{\theta})\right]\rho^2. \end{aligned}$$

Then, we can determine whether the given pixel is an edge pixel by checking if there is a zero-crossing of $\widehat{q}''_{\widehat{\theta}}(\rho\cos(\widehat{\theta}), \rho\sin(\widehat{\theta}))$ in the range of $-0.5 \leq \rho \leq 0.5$.

Nalwa and Binford (1986) pointed out several limitations of the edge detectors using the zero-crossing properties of the second-order derivatives of the image intensity function. In addition to second-order derivatives being sensitive to noise in general, they may have zero-crossings at places without any edges, which would result in false edge detection. If local surface fitting is used for estimating the derivatives, and the basis functions are chosen improperly, then zero-crossings of the estimated directional derivatives could result in extremely bad edge localization. As a remedy, they suggest using directional basis functions, which are constants in one direction and simple univariate functions in the orthogonal direction.

6.3 CANNY'S EDGE DETECTION CRITERIA

We have introduced some edge detectors in the previous section. How can we measure their performance? In general, what properties should a good edge detector possess? Canny (1986) studied these issues systematically and suggested three criteria for measuring performance of an edge detector. These three criteria are introduced in Subsection 6.3.1. Derivation of optimal edge detectors based on these criteria is discussed in Subsection 6.3.2, with some modifications discussed in Subsection 6.3.3.

6.3.1 Three criteria for measuring edge detection performance

Intuitively, a good edge detector should have the following three properties.

Good detection: There should be a low probability of missing real edge pixels and a low probability of detecting false edge pixels.

Good localization: The detected edge pixels should be close to the true edge pixels.

Single response: There should not be multiple responses to a true edge pixel.

The third property is more or less included in the first property implicitly. Canny made it explicit for convenience of mathematical processing.

Canny's 1986 paper made several important contributions, including (1) mathematical expressions for describing the above three criteria, (2) optimal edge detectors in some special cases based on the three criteria, and (3) a connection between Canny's optimal edge detectors and the derivative of Gaussian (DoG) operator (cf. Subsection 6.2.1). The optimal edge detection problem is first discussed in 1-D cases (cf. Figure 6.1(a) for the 1-D profile of an ideal edge), and then the 1-D methodologies are generalized to cases of 2-D or higher dimensions. This strategy is also adopted in this book.

We begin by assuming that the 1-D profile of a true image intensity function is $f(x) = AI(x \geq 0)$, which has a step edge at $x = 0$ with magnitude $A > 0$, and the observed image intensity at a given pixel x equals $f(x)$ plus an error term with $N(0, \sigma^2)$ distribution. The error terms at different pixels are assumed to be independent and identically distributed.

A method for evaluating the first criterion is derived as follows. For a given edge detection operator $u(x)$ having finite values in the window $[-W, W]$, its signal-to-noise ratio (SNR) is defined by

$$\mathrm{SNR}(u) = \frac{\left|\int_{-W}^{W} f(-x)u(x)\,dx\right|}{\sigma\sqrt{\int_{-W}^{W} u^2(x)\,dx}} = \frac{A\left|\int_{-W}^{0} u(x)\,dx\right|}{\sigma\sqrt{\int_{-W}^{W} u^2(x)\,dx}}. \tag{6.9}$$

Intuitively, the numerator of the last term of equation (6.9) is the absolute value of the mean of the convolution between the observed image intensity and the edge detection operator, and the denominator is the standard deviation of the convolution. If the SNR is larger, then both the probability of missing real edge pixels and the probability of detecting false edge pixels are smaller, and consequently the edge detection operator would perform better with respect to the first criterion mentioned previously. Therefore, SNR can be used as a measure of "good edge detection" for the edge detector $u(x)$.

Evaluation of the second criterion is based on the distribution of x_0, a local maximum of the convolution between the observed image intensity and the edge detection operator around the true edge position 0. By the Taylor's expansion and the fact that the first-order derivative of the convolution at x_0 is zero, Canny showed that the variance of x_0 is about

$$\frac{\sigma^2 \int_{-W}^{W}[u'(x)]^2\,dx}{\left[\int_{-W}^{W} f'(-x)u'(x)\,dx\right]^2}.$$

Since x_0 would be flagged as an edge pixel by the edge detector $u(x)$ and should be an unbiased estimator of the true edge position under some regularity conditions, localization of the detected edge can be measured by the reciprocal of the standard deviation of x_0. That is, localization is measured by

$$L(u) = \frac{\left|\int_{-W}^{W} f'(-x)u'(x)\,dx\right|}{\sigma\sqrt{\int_{-W}^{W}[u'(x)]^2\,dx}} = \frac{A\,|u'(0)|}{\sigma\sqrt{\int_{-W}^{W}[u'(x)]^2\,dx}}, \tag{6.10}$$

where the first-order derivative of $I(x \geq 0)$ at $x = 0$ has been defined as 1.

To avoid multiple responses to a single edge pixel (the third criterion), Canny suggested controlling the distance between peaks of the convolution between pure noise and the edge detection operator. He claimed that this distance is approximately twice the distance between two adjacent zero-crossings of the first-order derivative of the convolution, which is about

$$\pi\frac{\sqrt{\int_{-W}^{W}[u'(x)]^2\,dx}}{\sqrt{\int_{-W}^{W}[u''(x)]^2\,dx}}.$$

Thus, the distance between two peaks of the convolution is about

$$x_{max}(u) = 2\pi\frac{\sqrt{\int_{-W}^{W}[u'(x)]^2\,dx}}{\sqrt{\int_{-W}^{W}[u''(x)]^2\,dx}}. \tag{6.11}$$

If we set this distance to be some fraction k of the window width W, namely, we let

$$x_{max}(u) = kW, \tag{6.12}$$

then the number of maxima of the convolution, which could lead to false edge detections, can be controlled by fixing k. Therefore, $x_{max}(u)$ can be used for measuring multiple responses to a single edge pixel.

For a given edge detection operator u, it is easy to evaluate its performance in detecting a given edge by the above three criteria. If its SNR, localization and $x_{max}(u)$ measures are larger (i.e., their values defined in equations (6.9)-(6.11) are larger), then it performs better.

6.3.2 Optimal edge detectors by the three criteria

Canny pointed out that it is difficult (or impossible) to find an optimal edge detector by the above criteria when detecting a general edge (e.g., roof edge, line edge, etc.). However, in the special case that the edge to detect is an ideal step edge $f(x) = AI(x \geq 0)$, he derived an optimal edge detector in the following way.

First, it is easy to check that if we make a scale transformation on an edge detector $u(x)$, namely, we consider $u_w(x) = u(x/w)$ where $w > 0$ is a scale parameter, then

$$\text{SNR}(u_w) = \sqrt{w}\,\text{SNR}(u), \text{ and } L(u_w) = \frac{1}{\sqrt{w}}L(u).$$

Therefore, if we increase the scale parameter of the edge detection operator, then the SNR measure would be increased, but the localization measure would be decreased. Clearly, the two measures cannot be both improved by changing w.

Second, if u is the solution of the following problem:

$$\text{SNR}(u^*) = c_1, \text{ and } L(u^*) \text{ is maximized over all admissible functions } u^*,$$

then it can be checked that u_w is the solution of the following problem:

$$\text{SNR}(u^*) = c_2, \text{ and } L(u^*) \text{ is maximized over all admissible functions } u^*,$$

where $c_1, c_2 > 0$ are two constants and $w = c_2^2/c_1^2$.

For the above two reasons, Canny suggested considering the product of the criteria SNR(u) and $L(u)$ for finding optimal edge detectors, which is proportional to

$$\frac{\left|u'(0)\int_{-W}^{0} u(x)\,dx\right|}{\sqrt{\int_{-W}^{W} u^2(x)\,dx}\sqrt{\int_{-W}^{W}[u'(x)]^2\,dx}}. \tag{6.13}$$

Obviously, the quantity in expression (6.13) is invariant to scale transformations. Then the problem of finding an optimal edge detector becomes the problem of selecting an admissible function u such that the objective function (6.13) is maximized under the restriction of equation (6.12).

Under the assumption that the potential optimal edge detector u is an antisymmetric function on $[-W, W]$, which is intuitively reasonable, Canny derived the solution for the above maximization problem by the so-called *variational approach*. The resulting optimal edge detector is defined by

$$\begin{aligned} u(x) &= a_1 \exp(\alpha x)\sin(\omega x) + a_2 \exp(\alpha x)\cos(\omega x) + \\ &\quad a_3 \exp(-\alpha x)\sin(\omega x) + a_4 \exp(-\alpha x)\cos(\omega x) + c, \end{aligned}$$

where α, ω, c and a's are unknown parameters, and they can be chosen such that expression (6.13) is maximized under the restriction of equation (6.12).

Canny showed that, when x_{max} is chosen small, the optimal edge detection operator looks similar to a boxcar function, which equals a constant C on $[0, W]$, $-C$ on $[-W, 0)$, and 0 otherwise. When x_{max} is reasonably large, the optimal edge detection operator can be well approximated by the first-order derivative of a Gaussian pdf with mean 0 and variance s^2, that is, by

$$-\frac{x}{\sqrt{2\pi}s^3}\exp\left(-\frac{x^2}{2s^2}\right).$$

Because the local maxima of the convolution between the above first-order derivative of the Gaussian pdf and the observed image intensity correspond to zero-crossings of the convolution between the second-order derivative of the Gaussian pdf and the observed image intensity, the optimal edge detection operator by Canny's three criteria is also closely related to the Marr-Hildreth LoG operator (cf. Subsection 6.2.2).

In the 2-D setup, suppose that f is an image intensity function and $G(x, y)$ is the following 2-D Gaussian pdf (note that a constant multiplier has been ignored without sacrificing any power for edge detection):

$$G(x, y) = \exp\left(-\frac{x^2 + y^2}{2s^2}\right),$$

where s is a scale parameter. Then at a given edge pixel (x, y), the unit gradient direction $\vec{n}(x, y)$ of the convolution $G \otimes f$ is a good estimator of the direction perpendicular to the edge line on which (x, y) is located. So the directional first-order derivative of G in the direction of $\vec{n}(x, y)$ is a good edge detection operator. That is, a pixel (x, y) could be declared as an edge pixel if the directional second-order derivative of $G \otimes f$ in the direction of $\vec{n}(x, y)$ is zero, and the corresponding jump magnitude can be estimated by the Euclidean length of the gradient direction $\vec{n}(x, y)$. Obviously, this edge detection method is similar to those procedures based on zero-crossings of the directional second-order derivatives of f, which were briefly discussed at the end of Subsection 6.2.2. For selection of the scale parameter s in the above operator and other similar operators, please read Elder and Zucker (1998), Morrone *et al.* (1995), and the references cited there.

6.3.3 Some modifications

Canny's optimal edge detection operators are derived on a given window $[-W, W]$. Sarkar and Boyer (1991) pointed out that finite windows would cause some inconvenience in computation, and so they extended Canny's theory to the case when edge detection operators are defined on the entire real line $(-\infty, \infty)$. Canny's first two criteria (6.9) and (6.10) are easy to modify in such cases, by replacing W with ∞ in the integrations. However, the third criterion (6.12) is not well-defined if W is replaced by ∞. To overcome this difficulty, Sarkar and Boyer defined the *width* of a given edge detection operator u defined on $(-\infty, \infty)$ by

$$W^* = \frac{\sqrt{\int_{-\infty}^{\infty} x^2 u^2(x)\, dx}}{\sqrt{\int_{-\infty}^{\infty} u^2(x)\, dx}}.$$

Then W in equations (6.11) and (6.12) can be replaced by ∞ and W^*, respectively, and consequently equation (6.12) is changed to

$$2\pi \frac{\sqrt{\int_{-\infty}^{\infty} [u'(x)]^2\, dx}}{\sqrt{\int_{-\infty}^{\infty} [u''(x)]^2\, dx}} \cdot \frac{\sqrt{\int_{-\infty}^{\infty} u^2(x)\, dx}}{\sqrt{\int_{-\infty}^{\infty} x^2 u^2(x)\, dx}} = k.$$

After the three criteria are well-defined, they then derive optimal edge detection operators by the variational approach in a way similar to that of Canny's and show that their optimal edge detectors can also be approximated by the first-order derivative of a Gaussian pdf. However, they recommend using a recursive algorithm to implement their optimal edge detectors.

In the literature, there are some other modifications and generalizations of Canny's edge detection criteria: Tagare and deFigueiredo (1990) suggested an alternative localization measure; Petrou and Kittler (1991) tried to adapt Canny's criteria for detecting roof edges and ramp edges; Demigny and Kamlé (1997) proposed a discrete expression of Canny's criteria; and Cho *et al.* (1997) suggested using bootstrap to evaluate the performance of edge detection and other image-understanding tasks. More recent discussions about assessing the performance of edge detectors can be found in Heath *et al.* (1997, 1998), Shin *et al.* (2001), Yitzhaky and Peli (2003), and the references cited there.

6.4 EDGE DETECTION BY MULTILEVEL MASKS

Qiu and Bhandarkar (1996) introduced two edge detection procedures for detecting step edges and roof edges, respectively. A major feature of each of these two procedures is the use of multilevel masks in edge detection, which makes it possible to eliminate effects of local linear trends of the true image intensity function on edge detection. The two procedures are introduced below, each in a subsection.

6.4.1 Step edge detection

For convenience of presentation, sometimes both the image gray levels and pixel locations are called pixels in this section. Hopefully, this will not cause any confusion.

We begin by introducing some notation related to the set up for the procedure. At a given pixel Z_{ij}, let us consider a 9×9 mask centered at Z_{ij}. For pixels in the boundary regions, their masks are constructed assuming that the image is wrapped around at the boundaries. The 9×9 mask can be regarded as a union of nine 3×3 submasks (cf. Figure 6.10 below). We refer to these submasks as *second order* (SO) pixels. For convenience, the original 81 pixels are sometimes called the *first order* (FO) pixels, or simply pixels. The SO pixel that includes Z_{ij} is denoted as $Z_{ij}^{(2)}$. The nine SO pixels are denoted as $Z_{i+r,j+t}^{(2)}$, for $r, t = -1, 0, 1$, as demonstrated by Figure 6.10. $Z_{ij}^{(2)}$ has two pairs of diagonally neighboring SO pixels, comprising the set $N_D = \{Z_{i-1,j-1}^{(2)}, Z_{i-1,j+1}^{(2)}, Z_{i+1,j-1}^{(2)}, Z_{i+1,j+1}^{(2)}\}$, and two pairs of directly neighboring SO pixels, comprising the set $N_4 = \{Z_{i-1,j}^{(2)}, Z_{i+1,j}^{(2)}, Z_{i,j-1}^{(2)}, Z_{i,j+1}^{(2)}\}$. The subscript D in N_D denotes "diagonal" and the subscript 4 in N_4 denotes that the corresponding elements in N_4 are the 4 connected neighbors of $Z_{ij}^{(2)}$.

The basic idea is to compute the weighted average of the gray levels of the 3×3 FO pixels within a SO pixel, and to regard the weighted average as the gray level of the SO pixel. For each pair of SO pixels in either N_D or N_4, the absolute difference of their gray levels is then computed, and the maximum absolute difference of the four pairs of SO pixels can be used for edge detection.

Next, we illustrate the procedure for computing the weighted average by considering the SO pixel $Z^{(2)}_{i+1,j+1}$, which is shown in Figure 6.10 with the weights $\{a_1, a_2, a_3, a_4, a_5\}$ at the upper-right corner. The set of weights is also depicted in Figure 6.11(a). The FO pixels in $Z^{(2)}_{i+1,j+1}$ which are equidistant from Z_{ij} in the sense of the D_4 *distance* (also called *city-block distance* or *Manhattan distance* in the literature) are given the same weight. The D_4 distance between the (i, j)-th and (i', j')-th pixels is defined by $|i - i'| + |j - j'|$ (see e.g., Subsection 2.4.5, Gonzalez and Woods 1992). The set of weights are assumed to satisfy the following conditions:

(i) $a_1 + 2a_2 + 3a_3 + 2a_4 + a_5 = 1;\ a_i \geq 0,\ i = 1, 2, 3, 4, 5;$

(ii) $\{a_1, a_2, a_3, a_4, a_5\}$ is a linearly decreasing sequence and has a ratio relation as shown in Figure 6.11(a). In Figure 6.11(a), the numbers $\{4, 5, 6, 7, 8\}$ on the horizontal axis denote the fact that the pixels with weight a_1 are 4 D_4 units away from Z_{ij}, and so on.

Condition (i) ensures that the weights are nonnegative and normalized. Condition (ii) ensures that the weights decrease linearly with increasing D_4 distance from pixel Z_{ij}. These two conditions actually correspond to the triangular kernel density function used in statistical kernel smoothing techniques (cf. Chapters 2–5). We impose an additional constraint that the hypotenuse of the right-angled triangle shown in Figure 6.11(a) intersect with the horizontal axis at point 9. This implies that pixels at D_4 distance of 9 or greater from pixel Z_{ij} lie outside the neighborhood of interest. The unique weight sequence $\{a_1, a_2, a_3, a_4, a_5\}$ that satisfies all the above requirements can then be determined to be

$$\{a_1, a_2, a_3, a_4, a_5\} = \{5/27, 4/27, 3/27, 2/27, 1/27\}. \tag{6.14}$$

The same weights are used for the other SO pixels in N_D, namely, $Z^{(2)}_{i-1,j-1}$, $Z^{(2)}_{i-1,j+1}$, and $Z^{(2)}_{i+1,j-1}$.

Similarly, the weights $\{b_1, b_2, b_3\}$ for SO pixels in N_4 are assumed to satisfy the assumptions demonstrated by Figures 6.10 and 6.11(b), and the solution turns out to be

$$\{b_1, b_2, b_3\} = \{1/6, 1/9, 1/18\}. \tag{6.15}$$

Given the weights, the gray levels of the SO pixels $Z^{(2)}_{i+1,j+1}$ and $Z^{(2)}_{i,j+1}$ can be computed by the following formulas:

$$\begin{aligned} Z^{(2)}_{i+1,j+1} &= a_1 Z_{i+2,j+2} + a_2[Z_{i+2,j+3} + Z_{i+3,j+2}] + \\ &\quad a_3[Z_{i+2,j+4} + Z_{i+3,j+3} + Z_{i+4,j+2}] + \\ &\quad a_4[Z_{i+3,j+4} + Z_{i+4,j+3}] + a_5 Z_{i+4,j+4} \end{aligned}$$

and

$$\begin{aligned} Z^{(2)}_{i,j+1} &= b_1[Z_{i-1,j+2} + Z_{i,j+2} + Z_{i+1,j+2}] + \\ &\quad b_2[Z_{i-1,j+3} + Z_{i,j+3} + Z_{i+1,j+3}] + \\ &\quad b_3[Z_{i-1,j+4} + Z_{i,j+4} + Z_{i+1,j+4}]. \end{aligned}$$

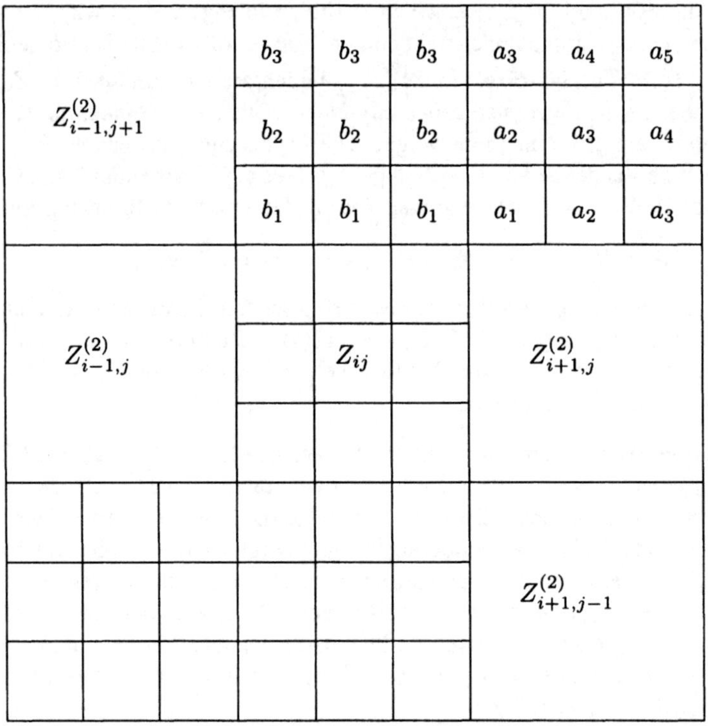

Fig. 6.10 The 9 × 9 mask centered at pixel Z_{ij} as a union of nine SO pixels. Weights $\{a_1, a_2, a_3, a_4, a_5\}$ and $\{b_1, b_2, b_3\}$ are used in computing the gray levels of the diagonally neighboring SO pixels and directly neighboring SO pixels, respectively, of the SO pixel $Z^{(2)}_{ij}$.

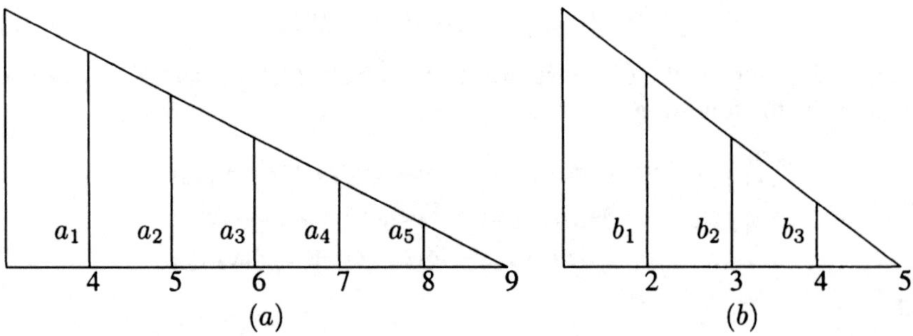

Fig. 6.11 Weights $\{a_1, a_2, a_3, a_4, a_5\}$ and $\{b_1, b_2, b_3\}$ satisfy the ratio relations as depicted in plots (a) and (b), respectively.

The gray levels of other N_D and N_4 SO pixels can be computed similarly.

Let us now consider the differences $[Z^{(2)}_{i+1,j+1} - Z^{(2)}_{i-1,j-1}]$, $[Z^{(2)}_{i,j+1} - Z^{(2)}_{i,j-1}]$, $[Z^{(2)}_{i-1,j+1} - Z^{(2)}_{i+1,j-1}]$ and $[Z^{(2)}_{i-1,j} - Z^{(2)}_{i+1,j}]$. Intuitively, if there is no step edge pixel in the 9×9 mask centered at pixel Z_{ij}, then all of the differences are expected to be small. On the other hand, if Z_{ij} is a step edge pixel, then under some regularity conditions we can find at least one pair of SO pixels such that the two SO pixels in the pair are located on two different sides of the edge segment on which Z_{ij} is located. Therefore, some of the differences are large in such cases.

However, two other factors should be taken into account in using these differences for detecting edges. One is the continuous variation of the underlying image intensity function in the 9×9 mask, and the other is the variation of the gray levels due to noise.

Intuitively, the variation of the image intensity function along a direction can be measured by its directional first-order derivative in that direction. As an example, for the direction from $Z_{i-3,j-3}$ to $Z_{i+3,j+3}$, the directional first-order derivative can be estimated by

$$D_1 = (d_1 + d_2)/2,$$

where

$$\begin{aligned} d_1 &= \frac{1}{3}[Z_{i-3,j-2} + Z_{i-2,j-2} + Z_{i-2,j-3}] - \\ &\quad \frac{1}{3}[Z_{i-4,j-3} + Z_{i-4,j-4} + Z_{i-3,j-4}] \\ d_2 &= \frac{1}{3}[Z_{i+4,j+3} + Z_{i+4,j+4} + Z_{i+3,j+4}] - \\ &\quad \frac{1}{3}[Z_{i+3,j+2} + Z_{i+2,j+2} + Z_{i+2,j+3}]. \end{aligned} \tag{6.16}$$

The above d_1 and d_2 can be regarded as the estimators of the directional first-order derivatives of the image intensity function at $(i-3, j-3)$ and $(i+3, j+3)$, respectively, and D_1 is their average. Similarly, D_2, D_3, and D_4 can be defined to measure the variation of the image intensity function along the directions from $Z_{i,j-3}$ to $Z_{i,j+3}$, from $Z_{i+3,j-3}$ to $Z_{i-3,j+3}$, and from $Z_{i+3,j}$ to $Z_{i-3,j}$, respectively. To eliminate the effects of the four directional first-order derivatives on edge detection, we define

$$\begin{aligned} \xi_1 &= Z^{(2)}_{i+1,j+1} - Z^{(2)}_{i-1,j-1} - \frac{25}{6} D_1 \\ \xi_2 &= Z^{(2)}_{i,j+1} - Z^{(2)}_{i,j-1} - \frac{8}{3} D_2 \\ \xi_3 &= Z^{(2)}_{i-1,j+1} - Z^{(2)}_{i+1,j-1} - \frac{25}{6} D_3 \\ \xi_4 &= Z^{(2)}_{i-1,j} - Z^{(2)}_{i+1,j} - \frac{8}{3} D_4. \end{aligned} \tag{6.17}$$

The last terms on the right hand sides of the above expressions are referred to as the *correction terms*. The multiplicative constants 25/6 and 8/3 are chosen such that the values of ξ's all equal zero when the image intensity function is linear, no edge pixels exist in the 9×9 neighborhood of pixel Z_{ij}, and no noise is involved in the observed gray levels. In such cases, the linear variation of the image intensity function has no effect on ξ's, and consequently it has no effect on edge detection using ξ's.

We now consider the effect of noise on the ξ's. Toward this end, we assume that the observed gray levels of the image follow the model

$$Z_{ij} = f(i,j) + \varepsilon_{ij},\ i = 1, 2, \ldots, n_1,\ j = 1, 2, \ldots, n_2,$$

where $\{f(i,j), i = 1, 2, \ldots, n_1,\ j = 1, 2, \ldots, n_2\}$ are the true gray levels and $\{\varepsilon_{ij}, i = 1, 2, \ldots, n_1,\ j = 1, 2, \ldots, n_2\}$ are i.i.d. random errors with mean zero and variance σ^2. By some simple algebraic manipulations, we can show that

$$\begin{aligned} \text{std}(\xi_1) &= \text{std}(\xi_3) = 2.6202\sigma, \\ \text{std}(\xi_2) &= \text{std}(\xi_4) = 1.7951\sigma, \end{aligned}$$

where $\text{std}(\xi_i)$ denotes the standard deviation of ξ_i. This leads to the definitions

$$\begin{aligned} \eta_i &= \xi_i/2.6202, \text{ for } i = 1, 3; \\ \eta_i &= \xi_i/1.7951, \text{ for } i = 2, 4. \end{aligned}$$

If there is no edge pixel in the 9×9 mask centered at Z_{ij}, then by the Central Limit Theorem (cf. Subsection 2.2.3), $\{\eta_1, \eta_2, \eta_3, \eta_4\}$ are all approximately normally distributed with mean zero and variance σ^2, since each of them is a linear combination of 18 observed gray levels.

We then define

$$M_{ij} = \max\{|\eta_1|, |\eta_2|, |\eta_3|, |\eta_4|\}$$

as the step edge detection criterion, with large values indicating possible jumps. We can derive threshold values for "large" values of M_{ij} as follows. Suppose that $C_{1-\alpha}$ is the $1 - \alpha$ quantile of the distribution of M_{ij}. Then

$$\begin{aligned} & P(M_{ij} > C_{1-\alpha}) = \alpha \\ \Leftrightarrow\quad & 1 - [P(|\eta_1| < C_{1-\alpha})]^4 = \alpha \\ \Leftrightarrow\quad & P(|\eta_1| < C_{1-\alpha}) = (1-\alpha)^{1/4} \\ \Leftrightarrow\quad & C_{1-\alpha} \approx Z_{[1+(1-\alpha)^{1/4}]/2} \cdot \sigma, \end{aligned}$$

where $\Leftrightarrow$ is a notation denoting equivalence and $Z_{[1+(1-\alpha)^{1/4}]/2}$ is the $[1 + (1 - \alpha)^{1/4}]/2$ quantile of the standard normal distribution. Note that in most applications, σ is unknown and it needs to be estimated from the image data. Statistically speaking, when there is no edge pixel in the 9×9 mask centered at pixel Z_{ij} and α is very small, M_{ij} has little chance of exceeding $C_{1-\alpha}$. In other words, if M_{ij} is larger than $C_{1-\alpha}$, then we have sufficient reason to conclude that there are edge pixels in the 9×9 mask. In such cases, we can flag Z_{ij} as a step edge pixel. Thus, an applicable threshold value for M_{ij} can be computed as

$$C_{1-\alpha} = Z_{[1+(1-\alpha)^{1/4}]/2} \cdot \widehat{\sigma},$$

Table 6.1 Several significance levels α and the corresponding $Z_{[1+(1-\alpha)^{1/4}]/2}$ values. The notation $Ae-B$ denotes $A \times 10^{-B}$.

α	$Z_{[1+(1-\alpha)^{1/4}]/2}$	α	$Z_{[1+(1-\alpha)^{1/4}]/2}$
5.0e-01	1.408093	1.0e-05	4.708129
4.5e-01	1.480175	5.0e-06	4.847543
4.0e-01	1.555243	1.0e-06	5.157701
3.5e-01	1.634761	5.0e-07	5.286029
3.0e-01	1.720681	1.0e-07	5.573271
2.5e-01	1.815839	5.0e-08	5.692763
2.0e-01	1.924768	1.0e-08	5.961456
1.5e-01	2.055659	5.0e-09	6.073695
1.0e-01	2.226268	1.0e-09	6.326984
5.0e-02	2.490915	1.0e-10	6.673367
1.0e-02	3.022202	1.0e-11	7.003303
5.0e-03	3.226681	1.0e-12	7.318952
1.0e-03	3.662164	1.0e-13	7.622118
5.0e-04	3.836061	1.0e-14	7.905250
1.0e-04	4.214791	1.0e-15	8.209103
5.0e-05	4.368675	1.0e-16	Infinite

where $\widehat{\sigma}$ is an estimator of σ. Table 6.1 lists several α values and their corresponding $Z_{[1+(1-\alpha)^{1/4}]/2}$ values.

By the definition of $C_{1-\alpha}$, the probability of detecting a false edge at each pixel is approximately α. On the other hand, those edge pixels whose jump magnitudes in image intensity are so small that $M_{ij} \leq C_{1-\alpha}$ are missed. Thus, $C_{1-\alpha}$ works as a tuning parameter, by which to balance missing-edge/false-edge possibilities. Although it is related to the confidence level α, this relationship is only *approximate* because it is based on the Central Limit Theorem, which provides an *approximate* normal distribution in a finite sample case (the sample size is 18 in the current case). In typical applications, the threshold value still needs to be heuristically adjusted. Qiu and Bhandarkar (1996) suggested choosing $\alpha \in [0.05, 0.0005]$ based on empirical experience.

An estimator of σ is now derived. A natural estimator of σ is the sample standard deviation of the gray levels in the 9×9 mask centered at Z_{ij}. However, if there are edge pixels in the mask, this estimator would perform poorly, because it would include variation due to jumps and thus overestimate σ. We therefore propose to estimate σ in the following manner. In each of the eight neighboring SO pixels, we first compute the sample variance of the gray levels of its 3×3 FO pixels. These variances are denoted as $s^2_{i-r,j-t}$, for $r, t = -1, 0, 1$ and $(r, t) \neq (0, 0)$. We then define

$$\widehat{\sigma}_{ij} = \text{median}\{\tau_{i-1,j}, \tau_{i-1,j+1}, \tau_{i,j+1}, \tau_{i+1,j+1}\},$$

where

$$\begin{aligned}
\tau_{i-1,j} &= \sqrt{[s^2_{i-1,j} + s^2_{i+1,j}]/2},\\
\tau_{i-1,j+1} &= \sqrt{[s^2_{i-1,j+1} + s^2_{i+1,j-1}]/2},\\
\tau_{i,j+1} &= \sqrt{[s^2_{i,j+1} + s^2_{i,j-1}]/2},\\
\tau_{i+1,j+1} &= \sqrt{[s^2_{i+1,j+1} + s^2_{i-1,j-1}]/2}.
\end{aligned}$$

Obviously, $\tau_{i-1,j}, \tau_{i-1,j+1}, \tau_{i,j+1}$, and $\tau_{i+1,j+1}$, which are derived, respectively, from the SO pixel pairs $[Z^{(2)}_{i-1,j}, Z^{(2)}_{i+1,j}]$, $[Z^{(2)}_{i-1,j+1}, Z^{(2)}_{i+1,j-1}]$, $[Z^{(2)}_{i,j+1}, Z^{(2)}_{i,j-1}]$, and $[Z^{(2)}_{i+1,j+1}, Z^{(2)}_{i-1,j-1}]$, are natural estimators of σ, when no edge pixels exist in the 9×9 mask centered at Z_{ij}. So $\widehat{\sigma}_{ij}$ should be a good estimator of σ in such cases. On the other hand, when Z_{ij} is an edge pixel and there is only one edge segment in its 9×9 neighborhood, then some of the SO pairs cited above are such that one of the SO pixels in the pair is on one side of the edge segment and the other SO pixel is on the other side. The τ quantities based on these SO pixel pairs are not much affected by the presence of the edge segment and thus would provide good estimators of σ. Other τ quantities, however, are not good estimators because they include variation due to jumps. We address this difficulty by taking the median of all four τ values, so $\widehat{\sigma}_{ij}$ can still provide a good estimator of σ, regardless of whether there are edge pixels in the 9×9 mask centered at Z_{ij}. We term $\widehat{\sigma}_{ij}$ as the *zeroth-order estimator of* σ because the sample variance $s^2_{i-r,j-t}$ can be regarded as the residual mean squares after the true FO pixel gray levels in the SO pixel $Z^{(2)}_{i-r,j-t}$ are estimated by their sample mean, which is a zeroth-order estimator in the sense that it is a zeroth-order polynomial.

Clearly, each of $s^2_{i-r,j-t}$, for $r, t = -1, 0, 1$ and $(r, t) \neq (0, 0)$, is a rough estimator of σ^2, because the sample mean is a rough estimator of the true FO pixel gray levels in the corresponding SO pixel. A more accurate estimator of σ can be constructed as follows. In each of the eight neighboring SO pixels $Z^{(2)}_{i-r,j-t}$, we perform a least-squares planar fit and then obtain the residual mean squares $\tilde{s}^2_{i-r,j-t}$. After some algebraic manipulations, we have the following expression for $\tilde{s}^2_{i-r,j-t}$:

$$\tilde{s}^2_{i-r,j-t} = \frac{1}{6}(\mathbf{Z}'\mathbf{Z} - \mathbf{Z}'P_X\mathbf{Z}), \tag{6.18}$$

where $\mathbf{Z}$ is a 9×1 vector whose elements are the gray levels of the 3×3 FO pixels in $Z^{(2)}_{i-r,j-t}$ scanned in a row-major fashion (i.e. left-to-right then top-to-bottom). The

9×9 matrix P_X is given by

$$P_X = \frac{1}{18} \times \begin{bmatrix} 8 & 5 & 2 & 5 & 2 & -1 & 2 & -1 & -4 \\ 5 & 5 & 5 & 2 & 2 & 2 & -1 & -1 & -1 \\ 2 & 5 & 8 & -1 & 2 & 5 & -4 & -1 & 2 \\ 5 & 2 & -1 & 5 & 2 & -1 & 5 & 2 & -1 \\ 2 & 2 & 2 & 2 & 2 & 2 & 2 & 2 & 2 \\ -1 & 2 & 5 & -1 & 2 & 5 & -1 & 2 & 5 \\ 2 & -1 & -4 & 5 & 2 & -1 & 8 & 5 & 2 \\ -1 & -1 & -1 & 2 & 2 & 2 & 5 & 5 & 5 \\ -4 & -1 & 2 & -1 & 2 & 5 & 2 & 5 & 8 \end{bmatrix}.$$

We then define

$$\widehat{\sigma}^*_{ij} = \text{median}\{\tilde{\tau}_{i-1,j}, \tilde{\tau}_{i-1,j+1}, \tilde{\tau}_{i,j+1}, \tilde{\tau}_{i+1,j+1}\}$$

where $\tilde{\tau}$ values are defined in the same way as τ values except that the s^2 sample variances need to be replaced by the $\tilde{s}^2$ residual mean squares. We term $\widehat{\sigma}^*_{ij}$ as the *first-order estimator of* σ.

Example 6.3 Figure 6.12(a) presents the gray-scale peppers image of size 512×512. Its gray-scale resolution is 8 bits per pixel (i.e., the gray levels are in the range $[0, 255]$). The above step edge detection procedure is applied to this image with the first-order estimator of σ. The significance level α is chosen to be 0.05, 0.005, or 0.0005. The detected edges with these three α levels are presented in Figures 6.12(b)-6.12(d), respectively. These results show that the value of α controls the degree of details in the detected edges. If it is too large, or equivalently the threshold $C_{1-\alpha}$ is too small, then many false edges would be detected. On the other hand, if it is too small, then many true edges would be missed. Qiu and Bhandarkar (1996) suggested choosing $\alpha \in [0.05, 0.0005]$ based on our empirical experience.

6.4.2 Roof edge detection

As already mentioned in Section 6.1, a roof edge signifies a jump in the first-order derivative but continuity in the image intensity function itself. Figure 6.1(b) shows the 1-D profile of the image intensity surface in the vicinity of an ideal roof edge. Suppose that s is the position of the roof edge in that plot, s_1 is a nearby position on its left, and s_2 is a nearby position on its right. We make the following observations:

(1) The quantity $|f'(s_2) - f'(s_1)|$ is relatively large because f' has a step discontinuity at s. Here, f' denotes the first-order derivative of the 1-D profile of the image intensity function.

(2) There is a relation $f'(s_1) \geq f'(s) \geq f'(s_2)$. Furthermore, $f'(s_1) + f'(s_2) - 2f'(s) = 0$. Here, $f'(s)$ is defined by $f'(s) = \lim_{\Delta s \to 0}[f(s + \Delta s) - f(s - \Delta s)]/2\Delta s$ and $\Delta s > 0$ is a small increment of s.

The above two observations are the basis of our roof edge detection procedure, developed as follows. At a given pixel Z_{ij}, we consider the 9×9 mask centered at

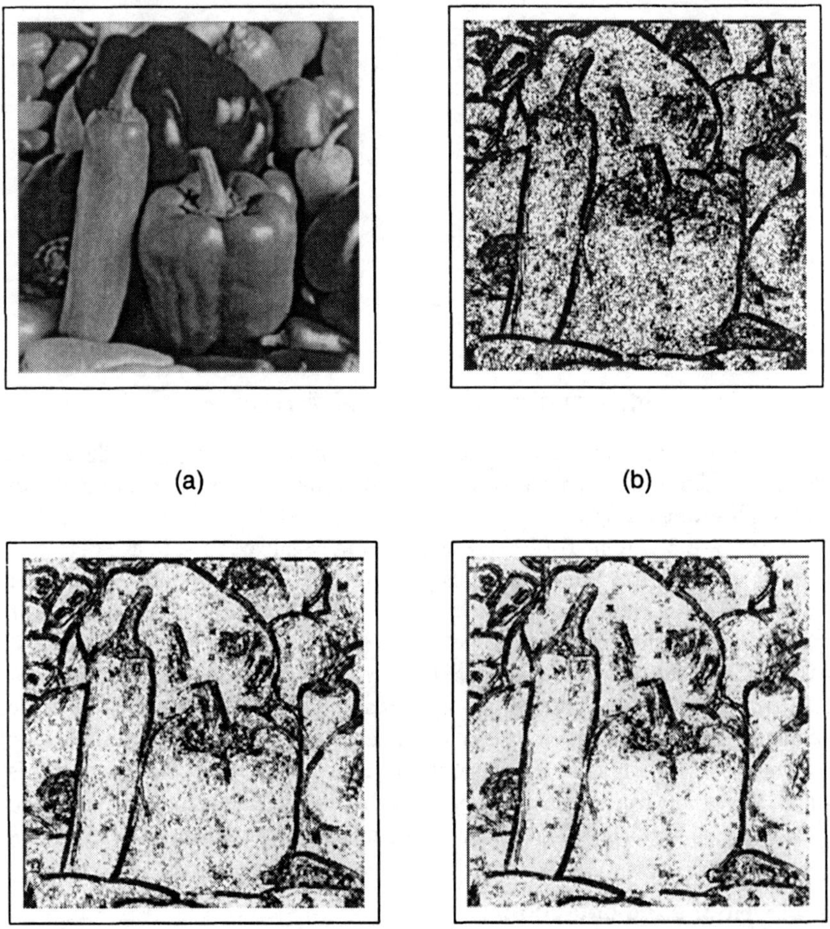

(a) (b)

(c) (d)

Fig. 6.12 (a): Gray-scale pepper image. (b): Detected edges with $\alpha = 0.05$. (c): Detected edges with $\alpha = 0.005$. (d): Detected edges with $\alpha = 0.0005$. In the step edge detection procedure, the first-order estimator of σ has been used.

that pixel, as before (cf. Figure 6.10), which consists of nine SO pixels $Z^{(2)}_{i+s,j+t}$, for $s, t = -1, 0, 1$. As in the case of step edge detection, we consider four directions: the two diagonal directions and the two directions along the image rows and columns. For the direction from $Z_{i-3,j-3}$ to $Z_{i+3,j+3}$, we use d_1 and d_2 as defined in equation (6.16) for estimating the directional first-order derivatives at pixels $Z_{i-3,j-3}$ and

$Z_{i+3,j+3}$, respectively. Similarly, we compute

$$\begin{aligned} d_3 &= \frac{1}{3}[Z_{i,j+1} + Z_{i+1,j+1} + Z_{i+1,j}] - \\ &\quad \frac{1}{3}[Z_{i-1,j} + Z_{i-1,j-1} + Z_{i,j-1}], \end{aligned} \tag{6.19}$$

which is used as an estimator of the directional first-order derivative in the same direction at pixel Z_{ij}. An additional correction term is needed to eliminate the effect of variation of the directional first-order derivative of the image intensity function along that direction on roof edge detection. Let

$$\begin{aligned} c_1 &= \frac{1}{3}[Z_{i+3,j+4} + Z_{i+4,j+4} + Z_{i+4,j+3} + \\ &\quad Z_{i+2,j+3} + Z_{i+2,j+2} + Z_{i+3,j+2}] - \\ &\quad \frac{2}{3}[Z_{i+2,j+4} + Z_{i+3,j+3} + Z_{i+4,j+2}], \\ c_2 &= \frac{1}{3}[Z_{i-3,j-2} + Z_{i-2,j-2} + Z_{i-2,j-3} + \\ &\quad Z_{i-4,j-3} + Z_{i-4,j-4} + Z_{i-3,j-4}] - \\ &\quad \frac{2}{3}[Z_{i-4,j-2} + Z_{i-3,j-3} + Z_{i-2,j-4}]. \end{aligned} \tag{6.20}$$

Then c_1 and c_2 can be regarded as estimators of the directional second-order derivatives of the image intensity function at $Z_{i+3,j+3}$ and $Z_{i-3,j-3}$, respectively. Their average

$$c = \frac{c_1 + c_2}{2} \tag{6.21}$$

can be used as an estimator of the directional second-order derivative at Z_{ij}. We could choose the multiplicative constant in the correction term as we did in equation (6.17), to delete the effect of quadratic variation of the image intensity function from the process of roof edge detection. Given the complexity of this computation, we suggest using 5 as the multiplicative constant, which is the Manhattan distance from $M_{i+3,j+3}$ to $M_{i-3,j-3}$. Then we define

$$\delta_1 = d_1 - d_2 - 5c, \tag{6.22}$$

$$\gamma_1 = d_1 + d_2 - 2d_3. \tag{6.23}$$

The corresponding quantities $\delta_2, \delta_3, \delta_4$ and $\gamma_2, \gamma_3, \gamma_4$ along the other three directions can be defined similarly. Clearly, the quantities $\{\delta_1, \delta_2, \delta_3, \delta_4\}$ correspond to observation (1) made at the beginning of this subsection, and the quantities $\{\gamma_1, \gamma_2, \gamma_3, \gamma_4\}$ correspond to observation (2).

If there are no step edge pixels and roof edge pixels in the 9×9 mask centered at pixel Z_{ij}, then $\delta_1, \delta_2, \delta_3$, and δ_4 are all approximately normally distributed with mean 0 and standard deviation σ_δ. From equations (6.16) and (6.20)–(6.22), we can easily compute the value of σ_δ to be $\sigma_\delta = 5.1316\sigma$. By the same reasoning as in the case of step edge detection, if there is some δ_i such that

$$|\delta_i| > Z_{1-\alpha/2} \cdot 5.1316\widehat{\sigma}, \tag{6.24}$$

then we have sufficient reason to believe that Z_{ij} is a roof edge pixel. However, if there are some step edge pixels in the 9×9 mask, they too could result in large values for some of the δ_i. We therefore should check an additional condition, derived from the quantities $\{\gamma_1, \gamma_2, \gamma_3, \gamma_4\}$, before making a final decision on whether Z_{ij} is a roof edge pixel. As pointed out earlier, if Z_{ij} is a roof edge pixel, then some of $\{\gamma_1, \gamma_2, \gamma_3, \gamma_4\}$ should be small. From equations (6.16), (6.19), and (6.23), we can easily show that $\text{std}(\gamma_i) = 2\sigma$, for $i = 1, 2, 3, 4$. Therefore, γ_i can be regarded as small if

$$|\gamma_i| \leq Z_{1-\alpha/2} \cdot 2\widehat{\sigma}. \tag{6.25}$$

Thus, the overall test proceeds as follows. At the pixel Z_{ij}, let us assume that $\{\delta_{i_1}, \delta_{i_2}, \ldots, \delta_{i_r}\} \subseteq \{\delta_1, \delta_2, \delta_3, \delta_4\}$ contains all δ values satisfying condition (6.24), where $r \leq 4$. Then we check the corresponding γ values $\{\gamma_{i_1}, \gamma_{i_2}, \ldots, \gamma_{i_r}\}$ using condition (6.25). If at least one of these γ values satisfies condition (6.25), we decide that Z_{ij} is a roof edge pixel.

6.5 EDGE DETECTION BASED ON COST MINIMIZATION

Tan *et al.* (1992) formulated the edge detection problem as a problem of cost minimization. They first provided a mathematical description about edges. Then they defined a cost function, which is a linear combination of five cost factors. Finally, they detected edges by minimizing the cost function using a simulated annealing algorithm. These steps are discussed below in three subsections.

6.5.1 A mathematical description of edges

Figure 6.1(a) demonstrates the 1-D profile of an ideal step edge, but we have not yet provided any precise description about edges in the 2-D setup. Intuitively, edges should be thin and continuous curves with sufficient length which separate regions in an image having significantly dissimilar characteristics. Tan *et al.* (1992) formulated this intuitive concept of edges in mathematical terms.

First, mathematical notation for edge pixel representation is introduced. For simplicity, let us assume that the numbers of pixel rows and columns are the same, that is, $n_1 = n_2 = n$. A true image F is a 2-D array of pixels

$$F = \{f(i,j) : i,j = 1,2,\ldots,n\},$$

where $f(i,j)$ denotes the gray level of the (i,j)-th pixel in the range $[0, 255]$. An *edge configuration* S is defined as a 2-D array

$$S = \{s(i,j) : i,j = 1,2,\ldots,n\},$$

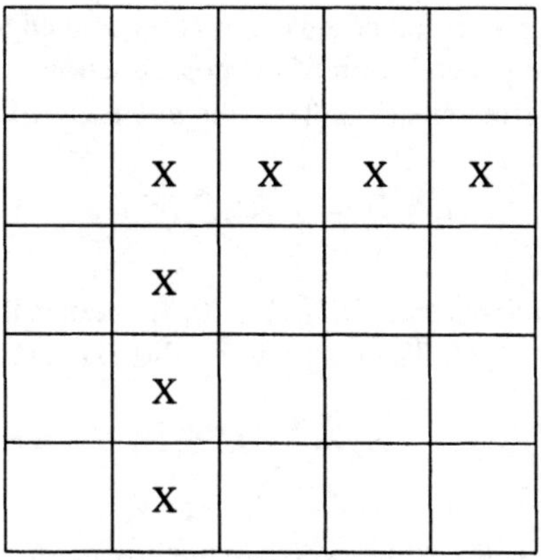

Fig. 6.13 An edge includes a circle of length 3 at the upper-left corner of the edge pixels marked by "X".

where $s(i, j)$ is a binary value 0 or 1. When $s(i, j) = 1$, the (i, j)-th pixel is called an *edge pixel*. Otherwise, it is a *nonedge pixel*. Let **S** denote the set of all possible edge configurations.

Next, several definitions about some special edge structures are made. A *walk* is a nonempty sequence of edge pixels $e_1, e_2, \ldots, e_K$ such that e_k is one of the eight neighboring pixels of e_{k+1}, for $1 \leq k < K - 1$. A *cycle* is a walk such that the origin and terminus are the same, the origin and internal pixels are distinct, and there is at least one internal pixel. A *path* is a walk that does not include any cycle in it. A collection of edge pixels is called *connected* if, for any two different edge pixels in the collection, there exists a path in the collection connecting them. Let B be any collection of edge pixels. Then it can be partitioned into nonempty connected subsets $B_1, B_2, \ldots, B_M$ such that $B_j \bigcup B_k$ is not connected for any $1 \leq j \neq k \leq M$. These subsets $B_1, B_2, \ldots, B_M$ are called *components* of B. An *edge* is defined as a component of the set of all edge pixels in an edge configuration S. A *segment* of an edge E is a subset of E that is connected. For a given edge or edge segment, it can include circles of length 3 (cf. Figure 6.13), at which the edge or edge segment looks thick. An edge pixel that is not contained in any circle of length 3 is called a *thin edge pixel*. Otherwise, it is called a *thick edge pixel*. An edge that contains only thin edge pixels is called a *thin edge*. Otherwise, it is called a *thick edge*.

6.5.2 Five cost factors and the cost function

By the edge detection scheme of Tan *et al.* (1992), the best edge configuration is the one that minimizes a well-defined cost function among all possible edge configurations in **S**. Their proposed cost function depends on the following five cost factors:

edge curvature (C_c), dissimilarity of the regions separated by the edges (C_d), number of edge pixels (C_e), fragmentation of the edges (C_f), and edge thickness (C_t).

We first introduce the general form of the cost function. Let L denote the set of all pixel sites and l be a given site in L. Then the *point cost* of the edge configuration $S \in \mathbf{S}$ is defined by

$$G(S,l) = \sum_j w_j C_j(S,l), \tag{6.26}$$

where $w_j > 0$ are coefficients and $C_j(S,l) \in [0,1]$ denotes the jth point cost factor of S at l, for $j \in \{c,d,e,f,t\}$. The *total cost* of $S \in \mathbf{S}$ is then

$$G(S) = \sum_{l \in L} G(S,l),$$

which is the objective function to minimize in the procedure by Tan *et al.*

Next, we introduce the definitions of $C_j(S,l)$ for $j \in \{c,d,e,f,t\}$. In the discussion below, an endpoint of an edge refers to an edge pixel that has at most one neighboring edge pixel, a nonisolated endpoint has exactly one neighboring edge pixel, and an isolated endpoint has no neighboring edge pixels. The cost for edge curvature assigns a cost to any edge pixel that is not an endpoint edge pixel. Let $s(l)$ be such an edge pixel. Then it must be the connection point of several (at least two) straight edge segments (e.g., the edge pixel located at the upper-left corner of the edge segment shown in Figure 6.13 is a connection of a horizontal edge segment and a vertical edge segment). Suppose that there are m pairs of such straight edge segments connected at $s(l)$. For the jth pair, let $\phi_j(l)$ be the smaller angle (in degrees) between the two edge segments of the pair, and $\theta_j(l) = 180 - \phi_j(l)$. Then the *curvature* of S at $s(l)$ is defined by

$$\theta(l) = \begin{cases} \max_{1 \le j \le m} \theta_j(l), & \text{if } s(l) \text{ is an edge pixel, but not an endpoint} \\ 0, & \text{otherwise.} \end{cases}$$

For an image with equally spaced pixel sites, the curvature at any pixel site can only take four possible values: 0, 45, 90, and 135°. The local point cost factor $C_c(S,l)$ is then defined by

$$C_c(S,l) = \begin{cases} 0, & \text{if } \theta(l) = 0^o \\ 0.5, & \text{if } \theta(l) = 45^o \\ 1.0, & \text{if } \theta(l) > 45^o. \end{cases}$$

This cost factor tends to remove curvy edges.

To define the cost factor C_d, Tan *et al.* first proposed an iterative algorithm to assign a region dissimilarity measure $0 \le d(l) \le 1$ to any pixel site $l \in L$. If the value of $d(l)$ is larger, then the pixel at l has a larger chance to be an edge pixel because the regions separated by a potential edge passing l are more dissimilar. The iterative algorithm initially assigns $d(l) = 0$ for all l. At a given pixel site l, there are 12 potential thin edge segments, each of which has the length of 3 and is centered at l. For each potential thin edge segment, two regions R_1 and R_2 of the same size are defined on its two sides. The dissimilarity between R_1 and R_2 is measured by

$a(R_1, R_2)$, which could be a simple difference between gray level averages in R_1 and R_2 or some other well-defined measure. Then, the one of the 12 potential thin edge segments which results in the maximum value of $a(R_1, R_2)$ is selected. The three pixel sites on this potential thin edge segment are denoted as l, l_1, and l_2. To delete some false edge pixels, the selected thin edge segment needs to be shifted in a direction determined by the edge structure. If the selected thin edge segment is vertical, horizontal, or diagonal, then it is shifted by one pixel in the two directions perpendicular to the segment. Otherwise, it is shifted by one pixel in each of the four directions: up, down, left, and right. For each shifted edge segment, two new one-sided regions R_1 and R_2 should be determined and the dissimilarity measure $a(R_1, R_2)$ should be computed again. If there is a larger value of $a(R_1, R_2)$ from one of the shifted edge segments, then we do not change any $d(l)$ values. Otherwise, we increase the values of $d(l), d(l_1)$, and $d(l_2)$ by $a(R_1, R_2)/3$, where $a(R_1, R_2)$ is the dissimilarity measure at l of the selected thin edge segment before shifting. The current iteration continues until all pixel sites have been considered. The iterative algorithm stops when the current iteration does not change any $d(l)$ value. Finally, the cost factor C_d is defined by

$$C_d(S, l) = \begin{cases} 0, & \text{if } s(l) \text{ is an edge pixel} \\ d(l), & \text{if } s(l) \text{ is not an edge pixel.} \end{cases}$$

This cost factor tends to place edge pixels at sites of high region dissimilarity.

The cost factor C_e, which accounts for the number of edge pixels, is simply defined by

$$C_e(S, l) = \begin{cases} 1, & \text{if } s(l) \text{ is an edge pixel} \\ 0, & \text{if } s(l) \text{ is not an edge pixel.} \end{cases}$$

The purpose of introducing this factor is to avoid too many edge pixels being detected.

The cost factor C_f for measuring edge fragmentation is defined by

$$C_f(S, l) = \begin{cases} 1.0, & \text{if } s(l) \text{ is an isolated endpoint edge pixel} \\ 0.5, & \text{if } s(l) \text{ is a nonisolated endpoint edge pixel} \\ 0, & \text{otherwise.} \end{cases}$$

The purpose of including this factor as a component of the cost function is to remove fragmented edges or to locally link them.

The cost factor C_t, which accounts for edge thickness, is defined by

$$C_t(S, l) = \begin{cases} 1, & \text{if } s(l) \text{ is a thick edge pixel} \\ 0, & \text{if } s(l) \text{ is a thin edge pixel.} \end{cases}$$

This cost factor aims to produce thin edges.

6.5.3 Minimization using simulated annealing

The detected edges by the procedure of Tan *et al.* (1992) are defined by one of the minimizers of the total cost function $G(S)$ over $S \in \mathbf{S}$. To search for the minimizers, Tan *et al.* suggested using the following simulated annealing procedure.

Let $\{T_k, k = 1, 2, \ldots, \infty\}$ be a sequence of strictly positive numbers such that $T_1 \geq T_2 \geq \ldots$ and $\lim_{k\to\infty} T_k = 0$, Q be a probability mass function defined on $\mathbf{S}$ (i.e., $Q(S) \geq 0$ for any $S \in \mathbf{S}$, and $\sum_{S\in\mathbf{S}} Q(S) = 1$), and R be a bivariate probability functional on $\mathbf{S}$ such that for any $S, S' \in \mathbf{S}$, $R(S, S') \geq 0$ and $\sum_{S'\in\mathbf{S}} R(S, S') = 1$. In the literature, the cost function G is often called the *energy function*, the sequence $\{T_k, k = 1, 2, \ldots, \infty\}$ is called the *temperature schedule*, and the bivariate probability function R is called the *transition probability function*. The *simulated annealing algorithm* is iterative and works as follows.

(1) Initially, set $k = 0$ and generate an initial state X_0 with probability $P(X_0 = S) = Q(S)$, for any $S \in \mathbf{S}$.

(2) Generate a candidate state Y_k with conditional probability $P(Y_k = S'|X_k = S) = R(S, S')$.

(3) Define

$$X_{k+1} = \begin{cases} Y_k, & \text{with probability } \exp(-[G(Y_k) - G(X_k)]_+/T_k) \\ X_k, & \text{with probability } 1 - \exp(-[G(Y_k) - G(X_k)]_+/T_k) \end{cases}$$

where $[x]_+ = \max(x, 0)$ and $G(\cdot)$ is the total cost.

(4) Set $k = k + 1$ and go to step (2). The algorithm stops when the difference between X_k and X_{k+1} is smaller than a prespecified level.

In applications, Tan *et al.* suggested choosing the components of the simulated annealing algorithm as follows. Q is chosen to be the uniform probability mass function on $\mathbf{S}$, and

$$T_k = \frac{d_0}{\log(n(k)k_s + 2)}, \quad \text{for } k \geq 1 \tag{6.27}$$

where

$$d_0 = \max\{w_f, 2w_c, 2w_f + w_d - w_e\},$$

$$n(k) = \left[\frac{k}{n^2}\right] n^2,$$

$[x]$ denotes the integer part of x, and $k_s \geq 0$ is a scaling constant. The transition probability function R is determined as follows. First, a pixel site l is randomly selected from L. Then, the current state S is changed to a candidate state S' in one of the following five ways with probabilities $\{p_j, j = 1, 2, 3, 4, 5\}$, respectively: (1) S' is the same as S except that it takes the opposite edge labeling at the site l; (2) the two states are the same except that S' takes the opposite edge labelings at l and another randomly selected site in the 3×3 neighborhood of l; (3) the two states are the same except that when the edge structure in the 3×3 neighborhood of l is one of 14 given patterns (cf. Tan *et al.* (1992) for details), then it is changed to a different pattern by shifting one edge pixel by a unit; (4) this change is similar to (3) except that when the edge structure in the 3×3 neighborhood of l is one of 10 given patterns, then it needs to be changed to a pattern more dramatically different than the change mentioned in (3); and (5) S' and S are the same except that the edge labelings of S' in the 3×3 neighborhood of l are chosen randomly. Tan *et al.* pointed out that the sequence $\{X_k, k = 0, 1, \ldots, \infty\}$ will converge to a global minimizer of the cost function if the related parameters and probability functions are chosen in the way as described above, based on Hajek's (1988) theorem about simulated annealing.

However, the probabilities $\{p_j, j = 1, 2, 3, 4, 5\}$ mentioned in the previous paragraph, the parameters $\{w_j, j = c, d, e, f, t\}$ in equation (6.26), and the scaling constant k_s in equation (6.27) still need to be selected. Tan *et al.* derived some conditions on $\{w_j, j = c, d, e, f, t\}$, under which it is proved that the detected edge has some good properties. For example, they proved that the detected edge does not include any thick edge pixels if

$$w_t > 2w_f - w_c + w_d - w_e.$$

By taking these conditions into account, in their numerical examples, Tan *et al.* selected $w_e = 1, w_d = 2, w_c \in \{0.25, 0.5, 0.75\}, w_f \in \{2, 3, 4\}$, and $w_t = 2w_f - w_c + w_d - w_e + 0.01$. The five probabilities are fixed at $p_1 = 200/1024, p_2 = 300/1024, p_3 = 200/1024, p_4 = 200/1024$, and $p_5 = 124/1024$. The constant k_s is chosen empirically.

To simplify computation, Tan *et al.* suggested using a decision tree for computing the point cost $G(S, l)$ in equation (6.26). They also pointed out that if S' and S are the same except at a single site l, then $\Delta G(S, S') = G(S') - G(S)$ only depends on the point costs of S and S' in the 3×3 neighborhood of l. If S and S' differ at m sites, then it is easy to find a sequence of edge configurations $S = S_1, S_2, \ldots, S_m = S'$ such that S_k and S_{k+1} differ at only a single site, and

$$\Delta G(S, S') = \sum_{k=1}^{m-1} \Delta G(S_k, S_{k+1}).$$

Note that most pairs of edge configurations to be compared in Tan *et al.*'s procedure differ only at a few sites. The above strategies can indeed substantially reduce the computing burden.

In the literature, there are several different versions of the edge detection procedure described above. For example, instead of using the simulated annealing procedure for finding an optimal edge configuration, Tan *et al.* (1989) suggested using a *hill climbing* or *greedy search* technique, which compares the current edge configuration with all edge configurations that differ from the current configuration only at a single pixel and then selects the best one. Bhandarkar *et al.* (1994) suggested using a genetic algorithm-based optimization procedure for finding a best edge configuration of the cost minimization problem.

6.6 EDGE LINKING TECHNIQUES

Detected edges from some edge detection procedures may have breaks at some places due to noise involved in the observed image intensities or other factors, including the use of threshold values in edge detection, digitization of image intensities in the image acquisition process, and so forth. In this section, we introduce several edge linking techniques, which are a type of *post-processing* techniques, so-called because they are often used after the edges are detected.

As mentioned in Subsection 4.3.3, the detected edges may also include some isolated deceptive edge pixels due to noise. There may also be some deceptive edge pixels close to true edge pixels, making the detected edges appear thick, due to the nature of local smoothing involved in most edge detection procedures. Thus, deletion of scattered deceptive edge pixels and edge thinning are also relevant post-processing problems. To handle such problems, two edge modification procedures were discussed in Subsection 4.3.3. Although we will not discuss these problems again in this section, this does not mean that they are less important. Interested readers can also read some standard text books on image processing, which usually include some discussions about image thinning (e.g., Section 11.2, Rosenfeld and Kak 1982), to which edge thinning is related.

6.6.1 Edge linking by curve estimation

In some image applications, true edges correspond to boundaries of a single or a few image objects. In such cases, the edge structure is relatively simple because the true edges consist of one or a few smooth curves, and the detected edges can be linked by fitting smooth curves.

Example 6.4 Figure 6.14(a) presents a squared unit of a gene microarray image, which consists of 48×48 pixels. It is known that the squared unit has a foreground region (also called a spot) in the middle and a background region toward the borders; these regions are separated by a closed boundary curve. Pixels in the foreground region represent abundance of cDNA sequences of a given gene and are expected to be brighter than those in the background region. Hence, the boundary curve is a step edge curve. In order to prepare the gene expression data for further statistical analysis, the boundary curve needs to be detected, which is the *image segmentation* problem for analyzing microarray images. The dark pixels in Figure 6.14(b) denote the detected boundary curve by the Sobel edge detector with thresholding value 300 (cf. Subsections 4.4.3 and 6.2.1 for related discussions). It can be seen that the detected boundary curve has some breaks. Figure 6.14(c) presents the detected boundary curve linked by the edge linking procedure described below.

A general microarray image consists of thousands of squared units arranged regularly in rows and columns. Centers of the squared units can be roughly known from the *image addressing* step. It is also known that the centers are inside the boundary curves. Thus, it is convenient to analyze individual squared units separately. For this reason, we focus on a single squared unit here without losing any generality.

It is known that the true boundary curve in the squared unit is a closed curve with the following expression:

$$\begin{cases} x(\theta) &= r(\theta)\cos(\theta) \\ y(\theta) &= r(\theta)\sin(\theta), \end{cases}$$

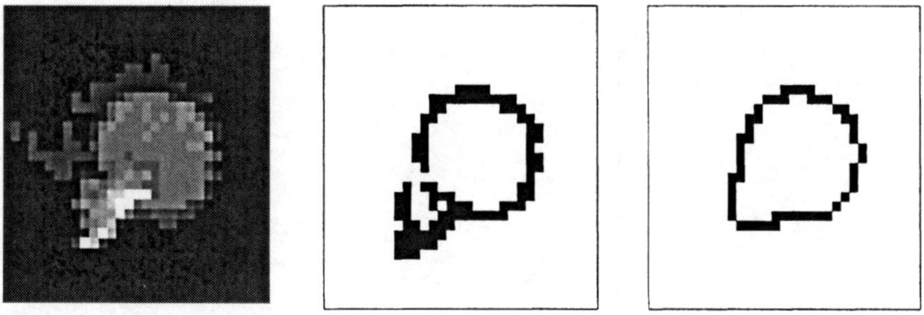

(a) (b) (c)

Fig. 6.14 (a): A squared unit of a microarray image consisting of 48 × 48 pixels. (b): Detected edges by the Sobel edge detector with thresholding value 300. (c): Detected edges linked by the local linear kernel smoothing procedure.

where $\theta \in [0, 2\pi)$ denotes the angle formed by the positive part of the x-axis and the line segment between the unit center and the point $(x(\theta), y(\theta))$ on the curve, and $r(\theta)$ is the Euclidean length of that line segment. Estimation of the true boundary curve is then equivalent to estimation of $r(\theta)$ for any $\theta \in [0, 2\pi)$.

Suppose that the detected edge pixels are $\{(r_i \cos(\theta_i), r_i \sin(\theta_i)), i = 1, 2, \ldots, m\}$ with $0 \le \theta_1 \le \theta_2 \le \ldots \le \theta_m < 2\pi$. Then $r(\theta)$ can be estimated by the following local linear kernel smoothing procedure:

$$\min_{a,b} \sum_{i=1}^{m} \{r_i - [a + b(\theta_i - \theta)]\}^2 K\left(\frac{\theta_i - \theta}{h_\theta}\right),$$

where K is a kernel function and h_θ is a bandwidth. Note that there is no "boundary problem" in this example, because the left part of a neighborhood of $\theta = 0$ and the right part of a neighborhood of $\theta = 2\pi$ can both be properly defined. In Figure 6.14(c), the dark curve denotes the fitted boundary curve from the detected edges shown in Figure 6.14(b), using the Epanechnikov kernel function $K(u) = \frac{3}{4}(1 - u^2)I(-1 \le u \le 1)$ and $h_\theta = 0.5$.

6.6.2 Local edge linking based on image gradient estimation

Here, we describe a simple local edge linking procedure based on estimation of image gradient. Suppose that $\widehat{G}_f(x, y) = (\widehat{f}_x(x, y), \widehat{f}_y(x, y))'$ is an estimator of the image gradient at the pixel (x, y) (cf. Subsection 6.2.1 for introduction). If (x, y) is an edge pixel, then any other edge pixel (x', y') in a predefined neighborhood of (x, y) can be linked with (x, y) by a line if

$$|\widehat{M}_f(x, y) - \widehat{M}_f(x', y')| < M_0$$

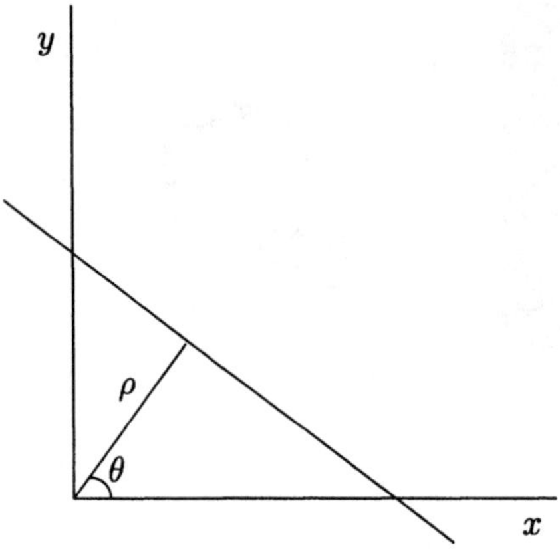

Fig. 6.15 Any line in the (x, y) Cartesian coordinate system corresponds to a point (ρ, θ) in the polar coordinate system by the Hough transform.

and

$$|\widehat{D}_f(x,y) - \widehat{D}_f(x',y')| < D_0$$

where $\widehat{M}_f$ denotes the magnitude of the estimated image gradient $\widehat{G}_f$ (cf. equation (6.1)), $\widehat{D}_f$ is its direction, and M_0 and D_0 are two positive thresholding constants. When (x, y) moves through all the detected edge pixels, most small breaks in the detected edges can be linked. To avoid duplicate linkings, a simple bookkeeping procedure is needed to keep a record of linked edge pixels.

6.6.3 Global edge linking by the Hough transform

In the (x, y) Cartesian coordinate system, any straight line can be expressed by

$$\rho = x\cos(\theta) + y\sin(\theta), \tag{6.28}$$

where $\rho \geq 0$ denotes the Euclidean distance of the line from the origin and $\theta \in [-\pi, \pi)$ is the angle formed by the positive part of the x-axis and the line passing the origin and perpendicular to the line expressed in equation (6.28), as demonstrated by Figure 6.15. Therefore, equation (6.28) defines a transformation of a line in the (x, y) Cartesian coordinate system to a point (θ, ρ) in the polar coordinate system, which is called the *Hough transform* (Hough 1962) in the literature.

For a given point (x^*, y^*) in the (x, y) Cartesian coordinate system, all lines passing through this point correspond to a curve in the polar system, which has the expression

$$\rho = x^* \cos(\theta) + y^* \sin(\theta).$$

Therefore, the Hough transform also transforms a point in the (x, y) Cartesian coordinate system to a curve in the polar system. To illustrate this idea, in Figure 6.16(a), there are five points labeled A, B, C, D, and E. Each point corresponds to a curve in the polar system, as shown in Figure 6.16(b). The curves in Figure 6.16(b) cross each other at several places. For example, the curves corresponding to A, C, and E cross each other at the boundary point $(3\pi/4, 0)$ in the polar system, which implies that the points A, C, and E in Figure 6.16(a) are located on a single line with coordinate $(3\pi/4, 0)$ in the polar system. This is clearly true from Figure 6.16(a). Other crossings in Figure 6.16(b) can be explained similarly. Thus, from Figure 6.16(b), it is possible to figure out which points in Figure 6.16(a) are located on a common line; this idea is the basis of the edge linking technique using the Hough transform.

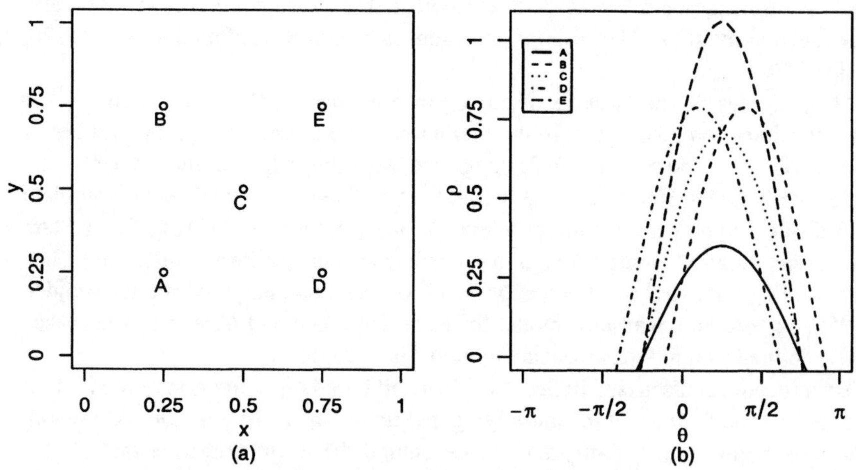

Fig. 6.16 (a): Five points in the (x, y) Cartesian coordinate system are shown, labeled A, B, C, D, and E. (b): Five curves in the polar system correspond to the five points in plot (a).

The edge linking proceeds in the following manner. Suppose that there are N detected edge pixels; these correspond to N curves in the polar system. Let ρ^* be the maximum ρ value of these curves. We then divide the region $[0, \rho^*] \times [-\pi, \pi)$ into K^2 equally spaced subregions. For a given subregion, suppose there are N_0 curves going through it. Then the N_0 detected edge pixels corresponding to these curves are approximately on a straight line with its polar coordinate equal to the coordinate of the subregion center. Therefore, these edge pixels can be linked.

In applications, we usually link edge pixels in the following way. The subregion in the polar system with most curves going through it is found first, and the edge pixels corresponding to the curves in that subregion are linked by a line with polar coordinate equal to the central coordinate of the subregion. Then the subregion in the polar system with second most curves going through it can be found, the corresponding edge pixels are linked, and so on. This process continues until most detected edge pixels are linked and further linking would not much improve the appearance of the detected edges.

In the literature, there are some different versions of the Hough transform, such as those suggested by Duda and Hart (1972), Illingworth and Kittler (1988), Kiryati *et al.* (1991), and O'Gorman and Clowes (1976). For a survey on the Hough transform, please read Palmer *et al.* (1994).

6.7 SOME DISCUSSIONS

We have discussed some edge detection procedures in the image processing literature, primarily those for detecting step edges. In the literature, there are some existing procedures for detecting other types of edges, including ramp edges, roof edges, line edges, texture edges, etc. For more discussion on this topic, please read standard text books such as Pratt (1991, Sections 16.7 and 18.5) and Rosenfeld and Kak (1976, Section 8.2).

Also, besides the procedures discussed in this chapter, there are various other step edge detection procedures in the literature. Several important examples are as follows: Hansen and Elliot (1982) suggested detecting edges based on a Markov random field model; Huang and Tseng (1988) formulated the edge detection problem as a statistical hypothesis testing problem; Perona and Malik (1990) and Saint-Marc *et al.* (1991) suggested edge detection procedures using anisotropic diffusion; Chen *et al.* (1991) discussed edge detection based on residual analysis; van der Heijden (1995) suggested a covariance model for edge detection; and Meer and Georgescu (2001) defined a confidence measure to help detect edges.

The procedures discussed in Section 6.2 are all based on estimation of the first- or second-order derivatives of the underlying true image intensity function. As demonstrated by Figure 6.2, the major rationale behind these procedures is that (1) the absolute difference between two one-sided averages of the image gray levels of the pixels located on two different sides of a given edge segment is large, and (2) the second-order derivatives of the image intensity function have the zero-crossing properties around edge pixels.

Most edge detectors based on first-order derivatives are constructed from two masks of fixed size. The two masks are usually defined along two orthogonal directions (in many cases, the x-axis and y-axis directions) for estimating the two directional first-order derivatives along these directions. The Euclidean length of the vector consisting of the two directional derivative estimators, which is also a good estimator of the image gradient, is often used as a criterion for edge detection. Many edge detectors are based on masks of size 3×3, although masks of larger sizes are also used sometimes (cf. Figure 6.5). In Subsection 4.4.3, we have explained that the Sobel edge detector is a special case of the kernel-type jump detectors discussed in Section 4.4. It is easy to see that most other edge detectors based on first-order derivatives are also special cases of kernel-type jump detectors, although kernel functions are not defined explicitly in these edge detectors.

However, there are some differences between edge detectors and kernel-type jump detectors. The first difference is that many edge detectors in the image processing literature (e.g., the Prewitt, Sobel, and Frei-Chen edge detectors) assign weights based on intuition, while the kernel-type jump detectors in the statistical literature assign weights by kernel functions. In the statistical literature, there exist many discussions about selection of optimal kernel functions (e.g., Müller 1988). Second, most edge detectors use square-shaped masks for edge detection, which are convenient to use and simple to compute. However, to successfully detect edges with different curvatures by edge detectors based on two masks parallel to two orthogonal directions, as explained in Section 4.4 (cf. assumption (A 4.2)), the two masks should be narrow along the two orthogonal directions. Third, it is pointed out in Subsection 4.4.3 that the Euclidean length of the estimated image gradient may not be a good criterion for edge detection; instead, we recommend using the maximum value of the absolute components of the estimated image gradient vector for jump detection.

Edge detection based on zero-crossings of the second-order derivatives is quite an active research area in the edge detection literature mainly because the detected edges by such detectors have good localization properties. In the statistical literature, however, there is little discussion so far about jump detection based on estimation of second-order derivatives of the true regression function. One major reason is that estimators of the second-order derivatives are often much noisier than estimators of the first-order derivatives (cf. Figures 6.9(a)-(c)). To overcome this difficulty, the LoG edge detector discussed in Subsection 6.2.2 first presmooths the image by a Gaussian smoother and then computes the convolution between the smoothed image and the Laplacian operator for edge detection. By this procedure, some edges would be blurred in the presmoothing stage, which is especially true for those edges with large curvatures or small jump magnitudes. In my opinion, presmoothing should be performed by some edge-preserving smoothing procedures, such as those discussed in Chapter 5. Much research is needed on this topic.

Section 6.3 discusses Canny's (1986) three edge detection criteria, based on which an optimal edge detector is derived for detecting ideal step edges under some regularity conditions. This optimal edge detector can be well approximated by the directional first-order derivative of the 2-D Gaussian pdf in the image gradient direction. Obviously, such an edge detector is similar to the kernel-type jump detector discussed in Section 4.3 (cf. equation (4.18)), which is based on differences of image gray level averages along the estimated gradient direction. It is worth mentioning that Chapter 4 also discusses several other related kernel-type jump detectors, based on the maximum difference of two one-sided averages along all possible directions (cf. the RDKE in Subsection 4.2.1), along two specific directions (cf. the jump detector discussed in Subsection 4.4.1), and along more than two specific directions (cf. the jump detector discussed in Subsection 4.4.4).

Although it is difficult to derive optimal edge detectors in general cases based on Canny's three edge detection criteria (cf. Section III, Canny 1986), the three criteria are still important in the sense that they can be used for evaluating the performance of a given edge detector quite easily.

The edge detection procedure discussed in Section 6.4 has several interesting features. First, it suggests a concept of multilevel masks. By defining the second-order pixels, which are averages of some neighboring first-order pixels, noise can be removed to some degree, and consequently the edge detection procedure is robust to noise. Second, it considers four different directions for edge detection, instead of just two directions as most other edge detectors do. Therefore, it is flexible enough to detect edges with different shapes and orientations. Third, it eliminates most effects of the linear variation of the true image intensity function on edge detection by using the correction terms in its edge detection criterion. However, this procedure still has room for improvement. For instance, we have used a 9×9 mask in the procedure. In certain applications, masks with variable sizes may be more appropriate. In some planar regions of the image intensity surface, we could use relatively large masks, whereas in highly textured regions, the masks could be small. Also, when detecting a potential edge at a given location, this procedure uses only the pixels in the 9×9 neighborhood of the given location for making decisions. No other information has been used in its decision-making process. In many cases, some additional information, typically of a contextual nature, is available and helpful for edge detection. If we compare this procedure with the jump detection procedures discussed in Chapter 4, we can find some connections between them. For example, this procedure may be regarded as a modification of the procedure discussed in Subsection 4.4.4, which detects jumps along more than two specific directions. In the latter procedure, the linear variation of the true image intensity function may have some effect on edge detection, whereas such effect is mostly eliminated in the former procedure, as described in Section 6.4.

The procedure discussed in Section 6.5 formulates the edge detection problem as a cost minimization problem, with five cost factors being considered and quantified. The minimization problem is solved by using the hill climbing (Tan *et al.* 1989), simulated annealing (Tan *et al.* 1992), and genetic algorithm-based (Bhandarkar *et al.* 1994) procedures. At this moment, it is still unclear how to choose some of the procedure parameters properly (e.g., the weights in equation (6.26)). More practical guidelines are also needed for implementing the related searching algorithms for finding global maxima of the objective function.

Section 6.6 discusses several edge linking procedures. Similarly, back in Subsection 4.3.3, we introduced two modification procedures for deleting some isolated edge pixels and for edge thinning. It should be pointed out that most of these post-processing procedures are *ad hoc* in nature. Much research is needed to propose some post-processing procedures that are theoretically sound and practically convenient.

In comparing edge detection procedures in the image processing literature with jump detectors in the statistical literature, we would say that each of the two groups of methods has its own strengths and limitations. Most edge detection procedures are simple to use because (1) they treat edges in an image as a point set and their detected edges consist of individual pixels which have the flexibility that they need not form curves; (2) at a given pixel, small windows (e.g., the two 3×3 Sobel masks) are used for edge detection; and (3) only a few directions (in most cases the x-axis and y-axis directions) are searched at each pixel for a possible edge pixel. However, a major limitation of these edge detectors lies behind the fact that some of them are constructed based on intuition and they do not have enough theory to support them. For instance, with a given edge detection procedure, it is difficult to tell users where in an image the procedure would detect edges well and where it may fail to detect edges properly. As an example, the conventional Sobel edge detector (cf. Figure 4.15

in Subsection 4.4.3) uses two 3×3 Sobel masks at each pixel to obtain estimators of the first-order partial derivatives of the image intensity function in the x- and y-axis directions. If the window size increases to $k_1 \times k_2$, with $k_1, k_2 > 3$, then we should determine how the Sobel operator can be modified accordingly such that the detected edges are statistically consistent, in the sense that the detected edges would converge to the true edges when k_1, k_2, and the image resolution all tend to infinity. Although edge detectors can be evaluated by numerical experiments based on visual impression, we believe that theoretical justifications can help us to understand their strengths and limitations so that they can be further improved.

On the other hand, most jump detection procedures in the statistical literature have theories to support them, which is important for this area to further develop. Some recent jump detection procedures are already flexible enough to handle most applications, although there are still many research problems that need to be solved, as discussed in Section 4.5. However, some of the procedures require extensive computation because they are based on a number of maximizations/minimizations in searching for jumps. For example, the maximization procedures used in some kernel-type methods (e.g., those discussed in Section 4.2) require a great amount of computation. These procedures also need some restrictive model assumptions, such as knowing the number of jump location curves beforehand, which is hard to satisfy in most applications.

Problems

6.1 Figure 6.1 presents the 1-D profiles of an ideal step edge and an ideal roof edge. Give a mathematical expression for each of these two 1-D profiles. In 2-D setup, suppose that an ideal step edge segment and an ideal roof edge segment are both parallel to the y-axis. Give their mathematical expressions in such cases.

6.2 Figure 6.2(a) presents the 1-D profile of a step edge which is slightly blurred. If there is no blurring, the first- and second-order derivatives of the 1-D profile function, which is a step function, are not well defined at the edge location. In applications, we often use differences to replace derivatives. Suppose that f is a 1-D step function with the expression

$$f(x) = I(x > 0.5), \quad \text{for } x \in [0, 1].$$

Then we can use $[f(x+0.01) - f(x-0.01)]/0.02$ as an approximation to $f'(x)$, and $[f(x+0.01) + f(x-0.01) - 2f(x)]/0.01^2$ as an approximation to $f''(x)$. Check that these two differences have the appearances shown in Figures 6.2(b) and 6.2(c), respectively.

6.3 The Prewitt, Sobel, and Frei-Chen masks discussed in Subsection 6.2.1 are all of size 3×3. Comment on their differences in detecting edges.

6.4 The original Lena image can be downloaded from the ftp site of this book (see front cover for address). Using this image, do the following problems.

6.4.1. Reproduce Figure 6.6.

6.4.2. In making Figure 6.6(f), try two other threshold values, 500 and 2,000, and comment on the results.

6.4.3. Use the DoG operator discussed in Subsection 6.2.1 to detect edges. Its edge detection criterion is based on estimated image gradient magnitude, as in the previous two parts. This procedure has a number of parameters to choose: (i) the scale parameters s and t in the 2-D Gaussian pdf, (ii) the size of the window over which the discrete version of the DoG operator is obtained from its continuous version, and (iii) a threshold parameter. Try several different values for each parameter and comment on the results. Compared to the edge detectors considered in the previous two parts, does the DoG operator perform better?

6.5 Suppose that $G(x, y) = g(x, s)g(y, s)$, where g has the expression (6.4). Check that the following two expressions hold.

(i)

$$\nabla^2 G(x, y) = \frac{1}{\pi s^4}\left(\frac{x^2+y^2}{2s^2} - 1\right)\exp\left(-\frac{x^2+y^2}{2s^2}\right)$$

(ii)

$$-\nabla^2 G(x, y) = h_1(x)h_2(y) + h_2(x)h_1(y),$$

where the two 1-D functions h_1 and h_2 are defined in Subsection 6.2.2.

6.6 Do the following problems related to the LoG edge detector discussed in Subsection 6.2.2.

6.6.1. Reproduce Figure 6.9.

6.6.2. In making Figures 6.9(d)-(f), try the following parameter values: (1) $(s, k) = (3, 27)$, (2) $(s, k) = (5, 45)$, (3) $(s, k) = (7, 27)$, and (4) $(s, k) = (7, 85)$. In cases (1) and (2), the scale parameter s is quite small, and $k \geq 6\sqrt{2}s$ as suggested by Huertas and Medioni (1986). In case (3), $k < 6\sqrt{2}s$. In case (4), k is much larger than $6\sqrt{2}s$. Comment on the results.

6.6.3. Figure 6.9 did not consider the four-neighbor Laplacian mask. Without doing any computational work, what would you expect about its detected edges?

6.7 Check that the equations in (6.8) all hold.

6.8 Suppose that the 1-D profile of an image intensity function is $f(x) = 2I(x \geq 0)$. For the following three edge detection operators with the window $[-1, 1]$, compute the values of Canny's three edge detection criteria and comment on their performances in detecting that step edge.

(i) $u_1(x) = x$, for $x \in [-1, 1]$;

(ii) $u_2(x) = x^3$, for $x \in [-1, 1]$;

(iii) $u_3(x) = \sin(\pi x)$, for $x \in [-1, 1]$.

6.9 This problem is related to the edge detector discussed in Section 6.4.

6.9.1. Check that the weights in equations (6.14) and (6.15) satisfy the conditions specified in Figure 6.11.

6.9.2. Check that $\{\xi_1, \xi_2, \xi_3, \xi_4\}$ as defined in equation (6.17) are all zero if there is no noise involved in the observed image intensities, the true image intensity function is linear, and no edge pixels exist in the 9×9 neighborhood of the (i, j)-th pixel.

6.9.3. Check that equation (6.18) holds.

6.10 The original pepper image shown in Figure 6.12(a) can be downloaded from the ftp site of this book (see front cover for address). Using this image, do the following problems.

6.10.1. Reproduce Figure 6.12.

6.10.2. Apply the two modification procedures discussed in Subsection 4.4.4 to results shown in Figure 6.12(c), deleting some isolated edge pixels and making the detected edges thinner.

6.11 Use the edge detector discussed in Section 6.5, which is based on cost minimization and simulated annealing, to detect edges in the pepper image (cf. Problem 6.10 for downloading information). This edge detector has some procedure parameters involved. First, use the same values as those mentioned in Subsection 6.5.3, and then try some other choices. Comment on the results.

6.12 Apply the global edge linking technique based on the Hough transform to the detected edges of the pepper image obtained in Problem 6.10.2.

7

Edge-Preserving Image Restoration

7.1 INTRODUCTION

Observed images generated by image acquisition systems are usually not exactly the same as true images, but are instead degraded versions of their true images. Degradations can occur in the entire process of image acquisition, and there are many different sources of degradation. For example, in aerial reconnaissance, astronomy, and remote sensing, images are often degraded by atmospheric turbulence, aberrations of the optical system, or relative motion between the camera and the object. Other sources, such as the scattering of light between the target object and image acquisition system, limited spatial extent, and so forth, can introduce additional degradations.

Image degradations can be classified into several categories. Some of them affect image gray levels of individual pixels only, without introducing spatial blur. Degradations of this type are called *point degradations* in the literature. The degradations involving spatial blur are called *spatial degradations*. Other types of degradations involve chromatic or temporal effects. Since this book focuses on monochrome digital images, we only discuss image processing techniques dealing with point and spatial degradations in this chapter. For a detailed discussion about formation and description of various degradations, please read books on image restoration, such as Andrews and Hunt (1977) and Bates and McDonnell (1986).

Example 7.1 Figure 7.1(a) displays the original image of a mountain. A contaminated version of this image by point degradations is shown in Figure 7.1(b), and a spatially blurred version of this image is shown in Figure 7.1(c). In applications, an observed image can include both point and spatial degradations. A degraded version of the mountain image contaminated by both point and spatial degradations is displayed in Figure 7.1(d).

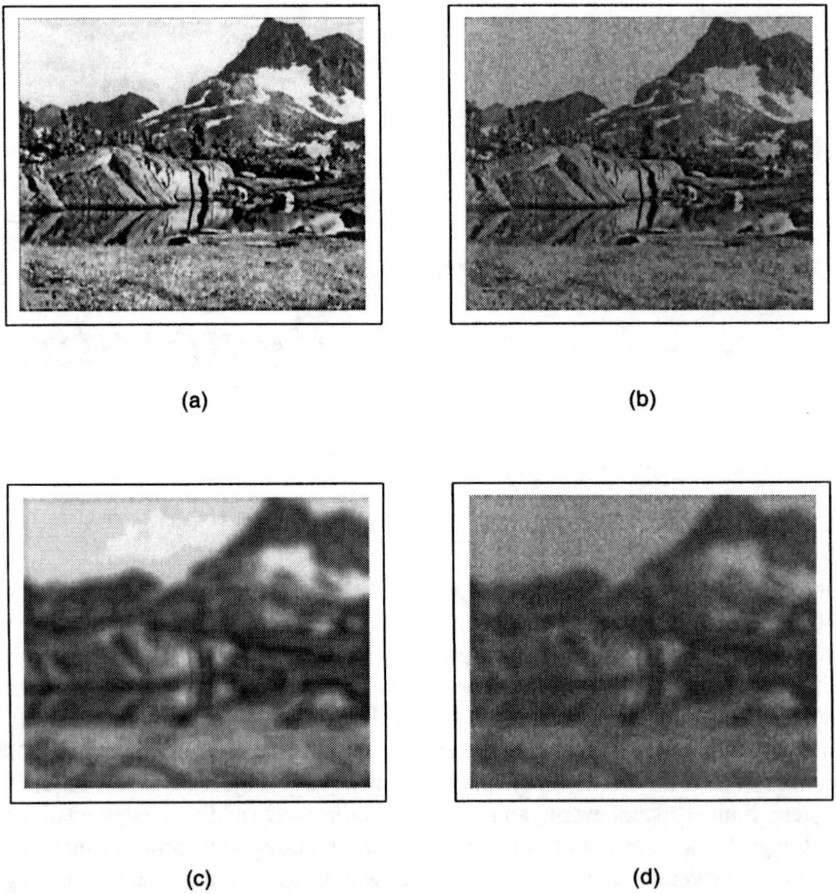

Fig. 7.1 (a): The original mountain image. (b): The mountain image contaminated by point degradations. (c): The mountain image contaminated by spatial degradations. (d): The mountain image contaminated by both point and spatial degradations.

In the literature, image processing techniques handling degradations are often referred to as *image restoration* or *image reconstruction* techniques, although some authors make a distinction between these two terminologies. For example, Bates and McDonnell (1986) pointed out that image restoration involves estimating the parameters of unwanted image distortion, the exact nature of which is unknown *a priori*, and using the estimated parameters to refurbish the original information; while image reconstruction is concerned with recovering detail in severely blurred images, the causes of whose imperfections are known *a priori*. Alternatively, like many other authors, Rosenfeld and Kak (1982) defined these techniques based on the major goal of the image processing; if the goal is to recover the original image from its degraded version, the technique is termed image restoration, and if the goal is to recover the original image from its projections along some specific directions, which is not covered by this book, the technique is termed image reconstruction. Because we only discuss image processing techniques handling point and spatial degradations in this chapter, no distinction between the two terminologies is made here or in the rest of the book.

Point degradations occurring naturally in the image acquisition process usually alter individual gray levels in a random pattern. As an example, refer to the synthetic aperture radar image shown in Figure 1.2 of Section 1.1. For this reason, point degradations are also called *noise* in the literature. In a noisy image, some image features, such as step and roof edges, are not as visible as they are in the original image. When the noise level is high, namely, the degree of pointwise gray level alterations is high, sometimes it is even visually or mechanically difficult to identify the objects in the noisy image. Many of us have this kind of experience when we watch a TV program without a cable connection or a powerful antenna.

In a spatially degraded image, major image features are often blurred by some smoothing processes. In this case, alterations of individual gray levels are different from those caused by point degradations, since they are usually spatially correlated instead of spatially independent. The mechanism behind this kind of spatial correlation is often unknown in applications, making spatial degradations relatively difficult to handle.

In the literature, several image restoration techniques have been suggested: some of them are designed for removing noise, or *denoising*, whereas others are mainly for *deblurring*. There are also some image restoration techniques that can be used for both denoising and deblurring. A major statistical tool for removing noise involves smoothing. However, conventional smoothing procedures result in spatial blurring at the same time that noise is smoothed away, as described in Chapters 3–5. Major image features such as edges are blurred by these smoothing procedures. Thus, edge-preserving image restoration techniques are needed for this purpose.

In the remaining parts of this chapter, we introduce several different types of image restoration techniques. Image restoration based on Fourier transformations is discussed in Section 7.2. Markov random field (MRF) modeling is a popular tool in this area, and several representative image restoration procedures based on MRF modeling are introduced in Section 7.3. In Section 7.4, we introduce some image restoration methods by robust local smoothing. Image restoration by nonlinear anisotropic diffusion filtering is discussed in Section 7.5. Finally, some remarks conclude the chapter in Section 7.6.

7.2 IMAGE RESTORATION BY FOURIER TRANSFORMATIONS

In this section, some traditional image restoration procedures based on Fourier transformations and basic algebra are introduced. The image restoration problem is first formulated in Subsection 7.2.1. Then, 2-D Fourier transformations are defined in Subsection 7.2.2. Image restoration based on Fourier transformations and inverse filtering is discussed in Subsection 7.2.3. Finally, image restoration by an algebraic approach is briefly discussed in Subsection 7.2.4.

7.2.1 An image restoration model

We begin by describing f and Z, the true and observed images of the same scene, respectively. Then, one reasonable model to describe the fact that Z is a degraded version of f is

$$Z(x,y) = H\{f\}(x,y) + \varepsilon(x,y), \text{ for } (x,y) \in \Omega, \tag{7.1}$$

where $\varepsilon(x,y)$ denotes pointwise degradation at the position (x,y), Ω is the design space of the image (i.e., space of all possible pixel locations), $H\{f\}$ is a functional transformation of f defined by

$$H\{f\}(x,y) = \int\int_{\Omega} h^*(x,y,x',y')f(x',y')\, dx'dy',$$

and h^* is a function related to spatial degradation. By its definition, $H\{f\}(x,y)$ is clearly a weighted average of gray levels of the true image f, with weights determined by the function h^*. Therefore, $H\{f\}(x,y)$ in equation (7.1) denotes the spatially degraded image. In equation (7.1), the observed image is assumed to be a summation of the spatially degraded image and the pointwise noise; that is, the two types of degradations are assumed to be *additive*.

The image restoration problem is to recover the original image f from the observed but degraded image Z. In applications, both the noise and h^* are unobservable, making this problem challenging.

It should be pointed out that the spatial degradation is assumed to be *linear* in equation (7.1) in the sense that

$$H\{a_1 f_1 + a_2 f_2\} = a_1 H\{f_1\} + a_2 H\{f_2\},$$

where a_1 and a_2 are two arbitrary constants, and f_1 and f_2 are any two images. However, in some applications, spatial degradations are nonlinear. For example, photocopies are often nonlinear transformations of their original documents. In such cases, linear spatial degradation models such as equation (7.1) can only provide an approximation for the true spatial degradation process, and this approximation is reasonably good only when the model is used locally in the design space. The additive noise assumption made in equation (7.1) is also questionable in some applications, especially when nonlinear spatial degradations are involved in the observed images. For more discussion about these issues, read Rosenfeld and Kak (1982, Chapters 6 and 7).

In engineering sciences, the *delta* (or *impulse*) function is frequently used. A 2-D delta function $\delta(x,y)$ is defined as a function on R^2 having an area of unity in an infinitesimal neighborhood of the origin and taking the value 0 everywhere else. That is,

$$\delta(x,y) = 0, \text{ if } (x,y) \neq (0,0), \ \ \delta(0,0) = \infty$$

and

$$\int\int_{R^2} \delta(u,v)\, dudv = \int\int_{N(0,0)} \delta(u,v)\, dudv = 1,$$

where $N(0,0)$ is any open neighborhood of the origin $(0,0)$. If an image has an impulse signal at a given point (x_0, y_0) and equals 0 elsewhere, this can be described by $f(x,y) = \delta(x - x_0, y - y_0)$ for $(x,y) \in \Omega$. Then the spatially degraded version of this *point source* image is $h^*(x, y, x_0, y_0)$; that is, the function h^* actually determines how a point source image is degraded. For this reason, h^* is called the *point spread function* (PSF).

If the PSF h^* has the form

$$h^*(x,y,x',y') = h(x - x', y - y'),$$

where h is a bivariate function, then the spatial degradation is *position* (or *space*) *invariant*, a property of considerable interest. If the spatial degradation is position invariant, then the degraded image has the expression

$$H\{f\}(x,y) = \int\int_{\Omega} h(x - x', y - y') f(x', y')\, dx'dy', \tag{7.2}$$

which is obviously a convolution between h and f, denoted by $H\{f\} = h \otimes f$. In such cases, $H\{f\}(x,y)$ is a weighted average of gray levels $f(x',y')$ of the original image, and the weights depend only on the position of (x',y') relative to (x,y).

A discrete version of equation (7.2) can be formulated as follows. Suppose that the true image f has $A \times B$ pixels and the PSF h is sampled at $C \times D$ positions. For simplicity, their discrete values are denoted by $\{f(i,j), 1 \leq i \leq A, 1 \leq j \leq B\}$ and $\{h(i,j), 1 \leq i \leq C, 1 \leq j \leq D\}$, respectively. First, we extend their values to dimensions $M \times N$ as follows, where $M \geq A + C - 1$ and $N \geq B + D - 1$ are two integers:

$$f_e(i,j) = \begin{cases} f(i,j), & \text{if } 1 \leq i \leq A \text{ and } 1 \leq j \leq B \\ 0, & \text{if } A + 1 \leq i \leq M \text{ or } B + 1 \leq j \leq N, \end{cases}$$

$$h_e(i,j) = \begin{cases} h(i,j), & \text{if } 1 \leq i \leq C \text{ and } 1 \leq j \leq D \\ 0, & \text{if } C + 1 \leq i \leq M \text{ or } D + 1 \leq j \leq N. \end{cases}$$

Then, the discrete version of equation (7.2) can be written as

$$H\{f\}_e(i,j) = \sum_{i'=1}^{M} \sum_{j'=1}^{N} h_e(i - i', j - j') f_e(i', j'),$$

for $1 \leq i \leq M$ and $1 \leq j \leq N$.

Thus, the discrete version of the degradation model (7.1) can be written as

$$Z_e(i,j) = \sum_{i'=1}^{M}\sum_{j'=1}^{N} h_e(i-i',j-j')f_e(i',j') + \varepsilon_e(i,j),$$

where $1 \le i \le M$, $1 \le j \le N$, and $\{\varepsilon_e(i,j), 1 \le i \le M, 1 \le j \le N\}$ is the extended version of the error terms, extended in the same way as $\{f_e(i,j), 1 \le i \le M, 1 \le j \le N\}$. Let $\mathbf{Z}, \mathbf{f}$, and ε denote MN-dimensional vectors obtained by stacking the rows of the $M \times N$ matrices $\{Z_e(i,j), 1 \le i \le M, 1 \le j \le N\}$, $\{f_e(i,j), 1 \le i \le M, 1 \le j \le N\}$, and $\{\varepsilon_e(i,j), 1 \le i \le M, 1 \le j \le N\}$, respectively. Then, the above equation can be expressed in matrix notation as

$$\mathbf{Z} = \mathbf{H}\mathbf{f} + \varepsilon, \tag{7.3}$$

where $\mathbf{H}$ is a matrix of dimension $MN \times MN$, defined by

$$\mathbf{H} = \begin{bmatrix} \mathbf{H}_1 & \mathbf{H}_M & \mathbf{H}_{M-1} & \cdots & \mathbf{H}_2 \\ \mathbf{H}_2 & \mathbf{H}_1 & \mathbf{H}_M & \cdots & \mathbf{H}_3 \\ \mathbf{H}_3 & \mathbf{H}_2 & \mathbf{H}_1 & \cdots & \mathbf{H}_4 \\ \vdots & \vdots & \vdots & \ddots & \vdots \\ \mathbf{H}_M & \mathbf{H}_{M-1} & \mathbf{H}_{M-2} & \cdots & \mathbf{H}_1 \end{bmatrix}$$

and

$$\mathbf{H}_j = \begin{bmatrix} h_e(j,1) & h_e(j,N) & h_e(j,N-1) & \cdots & h_e(j,2) \\ h_e(j,2) & h_e(j,1) & h_e(j,N) & \cdots & h_e(j,3) \\ h_e(j,3) & h_e(j,2) & h_e(j,1) & \cdots & h_e(j,4) \\ \vdots & \vdots & \vdots & \ddots & \vdots \\ h_e(j,N) & h_e(j,N-1) & h_e(j,N-2) & \cdots & h_e(j,1) \end{bmatrix}$$

for $1 \le j \le M$. Obviously, the matrix $\mathbf{H}_j$ is circulant, and the blocks of $\mathbf{H}$ are also subscripted in a circular manner. Such a matrix $\mathbf{H}$ is called *block-circulant matrix*, and it has some good algebraic properties. See Gonzalez and Woods (1992, Chapter 5) for related discussions.

7.2.2 2-D Fourier transformations

We now want to define and discuss a Fourier transformation in the 2-D setting. In Subsection 2.6.1, 1-D Fourier transformations are defined for functions in $L^2[0, 2\pi]$. In the 2-D setup, let $L^2(R^2)$ denote the collection of all measurable complex functions f on R^2 satisfying the condition that

$$\int\int_{R^2} f(x,y)\overline{f(x,y)}\, dxdy < \infty,$$

where $\overline{f(x,y)}$ denotes the complex conjugate of $f(x,y)$. Then the Fourier transformation of f is defined by

$$\mathcal{F}\{f\}(u,v) = \int\int_{R^2} f(x,y) \exp[-i2\pi(xu+yv)]\, dxdy.$$

The inverse Fourier transformation of a 2-D measurable complex function g in $L^2(R^2)$ is defined by:

$$\mathcal{F}^{-1}\{g\}(x,y) = \int\int_{R^2} g(u,v) \exp[i2\pi(xu + yv)]\, dudv.$$

The Fourier and inverse Fourier transformations have many good properties, including the following three.

(i) For any 2-D measurable complex function f in $L^2(R^2)$,

$$f = \mathcal{F}^{-1}\left\{\mathcal{F}\{f\}\right\}.$$

(ii) For any two 2-D measurable complex functions f and g in $L^2(R^2)$,

$$\mathcal{F}\{f \otimes g\} = \mathcal{F}\{f\}\mathcal{F}\{g\},$$

where $f \otimes g$ denotes the convolution of f and g.

(iii) For any two 2-D measurable complex functions f and g in $L^2(R^2)$ and any two complex constants a and b,

$$\mathcal{F}\{af + bg\} = a\mathcal{F}\{f\} + b\mathcal{F}\{g\}.$$

The first property says that the function f can be obtained from its Fourier transformation by taking an inverse Fourier transformation. Therefore, the true image can be estimated easily if its Fourier transformation has been estimated. The third property simply shows that the Fourier transformation is a linear operator. However, we are most interested in the second property, which is especially important for image processing. Recall that many image processing operators (cf. edge detectors introduced in Chapter 6) have the form of a convolution between an image processing mask and the observed image. Thus, the tasks of such image processing operators can also be accomplished in the following three steps: (i) the Fourier transformations of the image processing mask and the observed image are computed, respectively, (ii) the product of the two Fourier transformations is computed, and (iii) the inverse Fourier transformation of the product is computed to finish the related image processing task. In fact, many image processing techniques make use of these properties of the Fourier and inverse Fourier transformations, especially after the *fast Fourier transformation* (FFT) and *inverse FFT* techniques, which are numerical algorithms to compute the Fourier and inverse Fourier transformations in a significantly faster way, were proposed in the literature (see e.g., Gonzalez and Woods 1992, Chapter 3).

In discrete cases, suppose that $\{f(x,y), x = 1, 2, \ldots, A, y = 1, 2, \ldots, B\}$ are image gray levels at A rows and B columns of pixels. Then the *discrete Fourier transformation* of f is defined by

$$\mathcal{F}\{f\}(u,v) = \frac{1}{\sqrt{AB}} \sum_{x=1}^{A} \sum_{y=1}^{B} f(x,y) \exp\left[-i2\pi(xu/A + yv/B)\right],$$

for $u = 1, 2, \ldots, A$ and $v = 1, 2, \ldots, B$; the *discrete inverse Fourier transformation* of a function g is defined by

$$\mathcal{F}^{-1}\{g\}(x,y) = \frac{1}{\sqrt{AB}} \sum_{u=1}^{A} \sum_{v=1}^{B} g(u,v) \exp\left[i2\pi(xu/A + yv/B)\right],$$

for $x = 1, 2, \ldots, A$ and $y = 1, 2, \ldots, B$.

It can be checked that the discrete Fourier transformation $\mathcal{F}\{f\}(u,v)$ and the discrete inverse Fourier transformation $\mathcal{F}^{-1}\{g\}(x,y)$ are both *periodic* with periods A and B, respectively, for u (or x) and v (or y). That is,

$$\mathcal{F}\{f\}(u,v) = \mathcal{F}\{f\}(u+A,v) = \mathcal{F}\{f\}(u,v+B) = \mathcal{F}\{f\}(u+A,v+B);$$

$$\mathcal{F}^{-1}\{g\}(x,y) = \mathcal{F}^{-1}\{g\}(x+A,y) = \mathcal{F}^{-1}\{g\}(x,y+B) = \mathcal{F}^{-1}\{g\}(x+A,y+B),$$

for any $u, x = 1, 2, \ldots, A$ and $v, y = 1, 2, \ldots, B$. In addition, if f is real-valued, then

$$\mathcal{F}\{f\}(u,v) = \overline{\mathcal{F}\{f\}(-u,-v)},$$

which implies that

$$|\mathcal{F}\{f\}(u,v)| = |\mathcal{F}\{f\}(-u,-v)|,$$

where $|\mathcal{F}\{f\}(u,v)|$ denotes the magnitude of $\mathcal{F}\{f\}(u,v)$, namely, $|\mathcal{F}\{f\}(u,v)| = (\mathcal{F}\{f\}(u,v)\overline{\mathcal{F}\{f\}(u,v)})^{1/2}$. So the magnitude of the discrete Fourier transformation is *symmetric* about the origin in the (u,v) domain, which is often called the frequency domain for reasons explained in Subsection 2.6.1.

Example 7.2 In Figure 7.2, plot (a) presents an original image with a dot in the middle. It has $A = 64$ rows and $B = 64$ columns of pixels. A central dot has a gray level of 1, and the remaining pixels all have gray levels of 0. A blurred version of this image is shown in plot (b). The magnitude of the discrete Fourier transformation of the blurred image is shown in plot (c). It can be seen that it is nonzero only around the four corners of the plot. Plot (d) shows the magnitude of the shifted discrete Fourier transformation $\mathcal{F}\{f\}(u + A/2, v + B/2)$ (i.e., the origin of the (u,v) coordinate system is shifted to $(A/2, B/2)$ in plot (d)). Since the magnitude of the discrete Fourier transformation is periodic and symmetric in this case, plot (d) also shows the magnitude of the discrete Fourier transformation in one full period, with nonzero values in the middle area.

7.2.3 Image restoration by Fourier transformation

We now discuss how the properties for the 2-D Fourier transformation are applied. By the second and third properties of the Fourier transformation mentioned previously and equations (7.1) and (7.2), when the spatial degradation is position invariant,

$$\mathcal{F}\{Z\}(u,v) = \mathcal{F}\{h\}(u,v)\mathcal{F}\{f\}(u,v) + \mathcal{F}\{\varepsilon\}(u,v), \tag{7.4}$$

for any $(u,v) \in R^2$. In defining the Fourier transformations in equation (7.4), Z, h, f, and ε first need to be extended to R^2 by setting their values equal to zero beyond their original domains of definition.

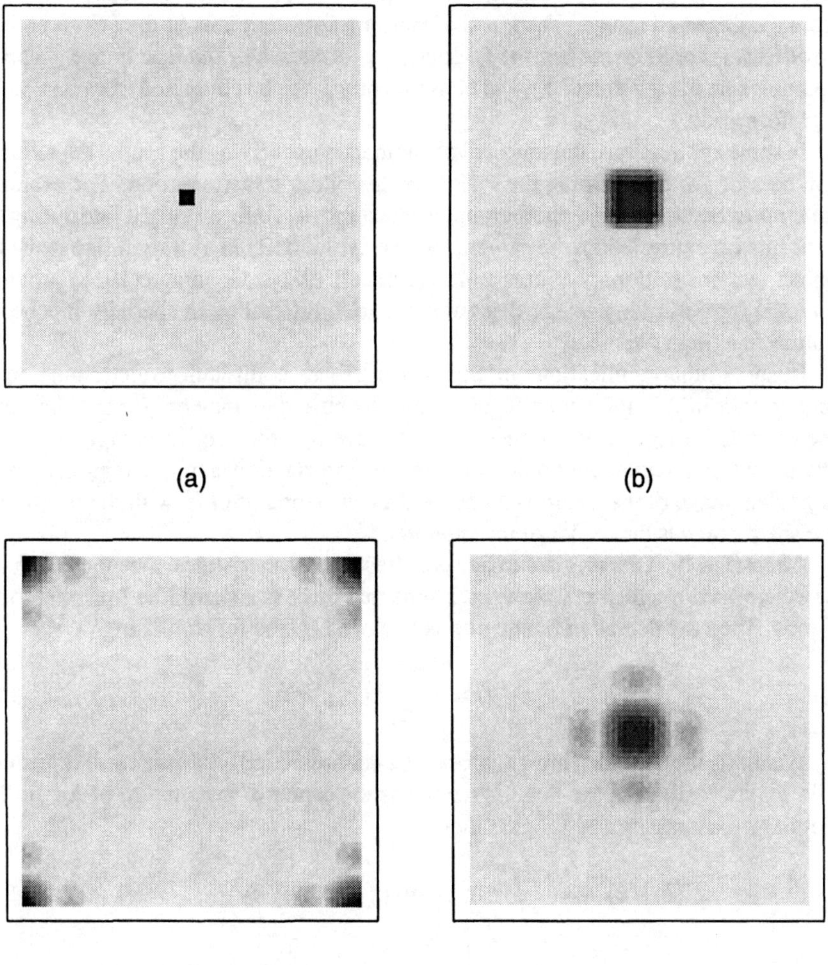

Fig. 7.2 (a): The original image of a dot. (b): Blurred version of the original image. (c): Magnitude of the discrete Fourier transformation of the blurred image in plot (b). (d): Magnitude of the shifted discrete Fourier transformation, with the origin of the (u, v) coordinate system shifted to $(A/2, B/2)$.

Equation (7.4) can also be written as:

$$\mathcal{F}\{f\}(u,v) = \frac{\mathcal{F}\{Z\}(u,v)}{\mathcal{F}\{h\}(u,v)} - \frac{\mathcal{F}\{\varepsilon\}(u,v)}{\mathcal{F}\{h\}(u,v)}, \tag{7.5}$$

as long as $\mathcal{F}\{h\}(u, v) \neq 0$. In equation (7.5), $\mathcal{F}\{Z\}(u, v)$ can be easily computed from the observed image. Thus, if the Fourier transformations of the PSF h and the error term ε are known, then the Fourier transformation of the true image f can be computed from equation (7.5), and consequently f can be computed from its Fourier transformation.

In some applications, the physical phenomenon underlying the spatial degradation can be used for determining the PSF h or its Fourier transformation. For example, *linear blur* occurs when a photographed object moves along a straight line during the exposure or, equivalently, the camera moves involuntarily in a straight line while the object remains stationary (Bates and McDonnell 1986). Goodman (1968) provided a model for describing image degradation due to diffraction in spatially incoherent optical imaging systems.

In some other applications in which h or $\mathcal{F}\{h\}$ is difficult to determine based on our prior information about the imaging systems, they may be obtained based on the degraded image. For instance, if we believe that the original image includes a sharp point in a region but no other interesting information is in that region, then the degraded image of that region can be used as an estimator of h, with the position of the sharp point as the center of the support of h.

Alternatively, if the original image contains several sharp lines, then the PSF h can be determined roughly as follows. Assume that there is a sharp line l parallel to the x-axis. Then the degraded image of this line can be used for estimating

$$h_l(y) = \int_{-\infty}^{\infty} h(x, y)\, dx$$

by averaging the degraded image across the x-axis direction for a given y-axis position. The y-axis position of the line l is treated as the center of the support of $h_l(y)$. The Fourier transformation of $h_l(y)$ is then

$$\begin{aligned} \mathcal{F}\{h_l\}(v) &= \int_{-\infty}^{\infty} h_l(y) \exp(-i2\pi v y)\, dy \\ &= \int_{-\infty}^{\infty} \left[\int_{-\infty}^{\infty} h(x, y)\, dx \right] \exp(-i2\pi v y)\, dy \\ &= \mathcal{F}\{h\}(0, v). \end{aligned}$$

Therefore, if the image and the PSF are assumed to be defined in the (x, y) coordinate system and their Fourier transformations are defined in the (u, v) coordinate system, then the image of a sharp line parallel to the x-axis can be used for estimating $\mathcal{F}\{h\}$ along the v-axis in the (u, v) coordinate system. Similarly, the image of a sharp line forming an angle θ with the positive x-axis can be used for estimating $\mathcal{F}\{h\}$ along a line in the (u, v) coordinate system forming an angle $\theta + \pi/2$ with the positive u-axis. More generally, suppose that the original image includes m sharp lines in the directions $\theta_1, \theta_2, \ldots, \theta_m$. Then the image of these lines can be used for estimating $\mathcal{F}\{h\}$ in the (u, v) coordinate system along m lines passing through the origin and in the directions $\theta_1 + \pi/2, \theta_2 + \pi/2, \ldots, \theta_m + \pi/2$. If $\mathcal{F}\{h\}$ can be assumed circularly symmetric, then its final estimator can be obtained by averaging the estimated values of $\mathcal{F}\{h\}$ along the m directions whose (u, v) coordinates are located on a same circle centered at the origin. Otherwise, the value of $\mathcal{F}\{h\}$ whose (u, v) coordinate is located on a line passing through the origin and in an arbitrary direction θ can be estimated by interpolation (when m is small) or local smoothing (when m is reasonably large) of its estimated values on some of the m lines in nearby directions.

After $\mathcal{F}\{h\}$ is determined, the Fourier transformation of f can be estimated by

$$\widehat{\mathcal{F}\{f\}}(u, v) = \frac{\mathcal{F}\{Z\}(u, v)}{\mathcal{F}\{h\}(u, v)}, \tag{7.6}$$

ignoring the noise in equation (7.5). Then the estimator of the true image can be obtained by inverse Fourier transformation; thus, image restoration procedures based on this idea are called *inverse filtering* procedures.

Comparing equations (7.6) with (7.5), we can find two obvious limitations in using inverse filtering procedures. One limitation is that noise is ignored completely in equation (7.6), which makes the restored image noisy. To examine the effect of this more carefully, equation (7.5) can be written as

$$\frac{\mathcal{F}\{Z\}(u, v)}{\mathcal{F}\{h\}(u, v)} = \mathcal{F}\{f\}(u, v) + \frac{\mathcal{F}\{\varepsilon\}(u, v)}{\mathcal{F}\{h\}(u, v)}.$$

Generally speaking, $\mathcal{F}\{h\}(u, v)$ drops rapidly when (u, v) moves away from the origin, and $\mathcal{F}\{\varepsilon\}(u, v)$ drops off much less rapidly. Thus, the second term would dominate the first term when (u, v) is a certain distance away from the origin. Therefore, much noise is involved in $\widehat{\mathcal{F}\{f\}}(u, v)$ as defined at equation (7.6). One way to overcome this problem is to define

$$M(u, v) = \begin{cases} 1/\mathcal{F}\{h\}(u, v), & \text{if } \max(u, v) \leq w_0 \\ 0, & \text{otherwise}, \end{cases} \tag{7.7}$$

where $w_0 > 0$ is an appropriately chosen constant. Then the Fourier transformation of f can be estimated by

$$\widehat{\mathcal{F}\{f\}}^*(u, v) = \mathcal{F}\{Z\}(u, v) M(u, v).$$

The second limitation in using equation (7.6) is that $\mathcal{F}\{h\}(u, v)$ could take the value of 0 at some places in the (u, v) coordinate system, in which case the recovered image would be affected by noise dramatically even when $\mathcal{F}\{\varepsilon\}(u, v)$ is small. Theoretically speaking, if $\mathcal{F}\{h\}(u, v) = 0$ only at countably many points, then it would not cause much problem because $\widehat{\mathcal{F}\{f\}}$ in equation (7.6) is not well-defined only at countable points. Thus, $\widehat{f}$, which is defined by the inverse Fourier transformation of $\widehat{\mathcal{F}\{f\}}$, is still well-defined, since the inverse Fourier transformation is an integration and would not be affected by countably many values of the integrand. However, in applications, if $\mathcal{F}\{h\}(u_0, v_0) = 0$ for a given point (u_0, v_0) in the (u, v) coordinate system, then $\mathcal{F}\{h\}(u, v)$ is usually small in a neighborhood of (u_0, v_0). The recovered image would be affected much by noise in such cases. One approach to overcome this difficulty is to choose w_0 in equation (7.7) such that $\mathcal{F}\{h\}(u,v) \neq 0$ when $\max(u,v) \leq w_0$. An alternative approach is to define $M(u,v)$ by

$$M(u,v) = \begin{cases} 1/\mathcal{F}\{h\}(u,v), & \text{if } \mathcal{F}\{h\}(u,v) \geq w_1 \\ 0, & \text{otherwise}, \end{cases} \tag{7.8}$$

where $w_1 > 0$ is a constant. Obviously, equations (7.7) and (7.8) can be used together. Namely, $M(u,v)$ can also be defined as follows:

$$M(u,v) = \begin{cases} 1/\mathcal{F}\{h\}(u,v), & \text{if } \max(u,v) \leq w_0 \text{ and } \mathcal{F}\{h\}(u,v) \geq w_1 \\ 0, & \text{otherwise.} \end{cases}$$

Example 7.3 Figure 7.3(a) presents an original image with 64×64 pixels. Its dark pixels all have gray levels of 1, and its white pixels all have gray levels of 0. A blurred version of this image is shown in Figure 7.3(b), where the smoothing procedure used for generating this blurred version is exactly the same as that used for generating the blurred image shown in Figure 7.2(b). Then the image shown in Figure 7.2(b), which is a blurred version of a dot, can be used as an estimator of the PSF h; the magnitude of its discrete Fourier transformation is shown in Figure 7.2(c). A restored image of the one shown in Figure 7.3(a) by procedure (7.7) with $w_0 = 6$ is shown in Figure 7.3(c), and a restored image by procedure (7.8) with $w_1 = 0.03$ is shown in Figure 7.3(d). It can be seen that both the procedures (7.7) and (7.8) have limited power for image restoration in this example.

7.2.4 Image restoration by algebraic approach

Equation (7.3) gives a matrix expression for the observed image intensities and has the form of a linear regression model (cf. Subsection 2.2.4 for a description). Thus, using the idea of least squares estimation in linear regression analysis, if the PSF h is determined, then the vector $\mathbf{f}$ of the true image intensities can be estimated by the solution of the following minimization problem:

$$\min_{\mathbf{f} \in R^{MN}} \|\mathbf{Z} - \mathbf{H}\mathbf{f}\|^2,$$

where $\|\cdot\|$ is the Euclidean norm. If the matrix $\mathbf{H}'\mathbf{H}$ has an inverse, then the solution of this problem is

$$\widehat{\mathbf{f}} = (\mathbf{H}'\mathbf{H})^{-1}\mathbf{H}'\mathbf{Z}.$$

Otherwise, the inverse matrix in the above expression should be replaced by a generalized inverse matrix (cf., e.g., Ben-Israel and Greville 1977). Furthermore, if $\mathbf{H}$ has an inverse, then the above expression can be simplified to

$$\widehat{\mathbf{f}} = \mathbf{H}^{-1}\mathbf{Z}.$$

The above expressions look neat, but they are often difficult to compute directly. For example, suppose that the image to recover has 256×256 pixels and the PSF h is sampled at 64×64 positions. Then the minimum values of M and N (cf.

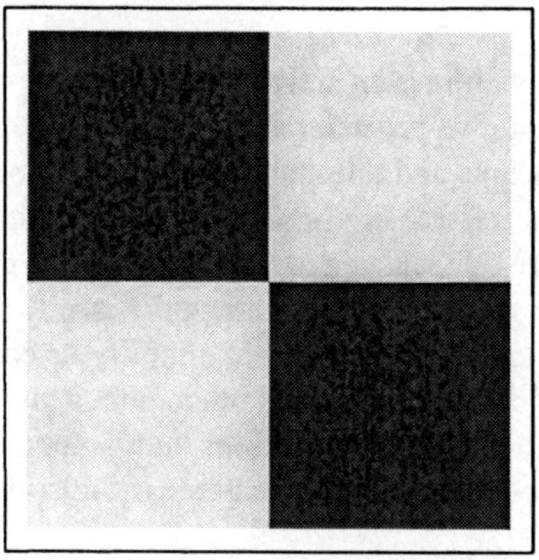

(a)

(b)

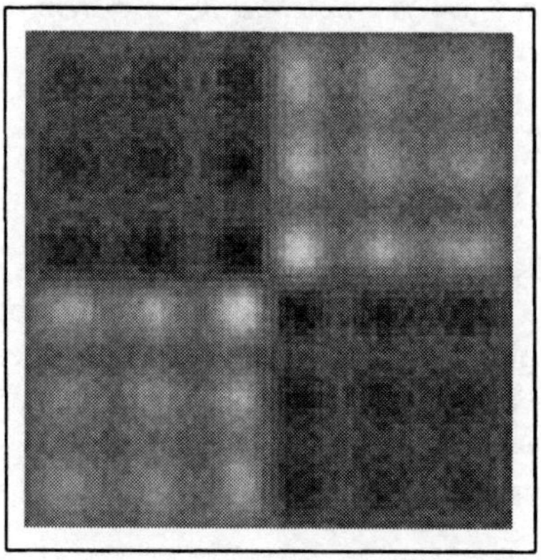

(c)

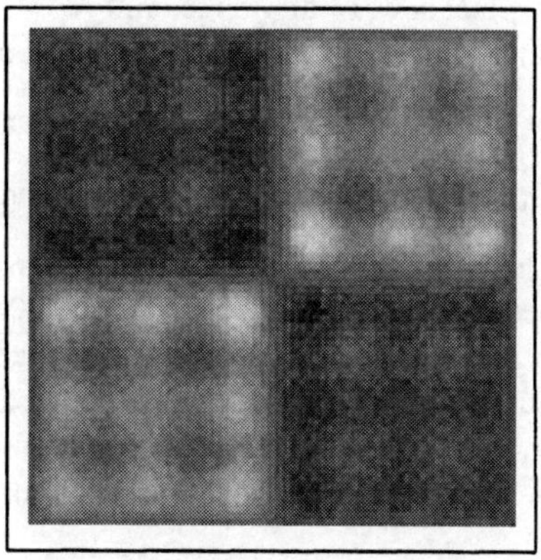

(d)

Fig. 7.3 (a): The original image. (b): Blurred version of the original image. (c): Restored image by procedure (7.7) with $w_0 = 6$. (d): Restored image by procedure (7.8) with $w_1 = 0.03$.

Subsection 7.2.1 for their definition) are both $256 + 64 - 1 = 319$, and the minimum dimension of the matrix $\mathbf{H}$ is $319^2 \times 319^2$, which is about $10^5 \times 10^5$. The computation involved in computing $\mathbf{H}^{-1}$ for a matrix this large is quite challenging. However, in Subsection 7.2.1, it has been pointed out that $\mathbf{H}$ is a block-circulant matrix, some of whose algebraic properties can be used for simplifying such computation, as shown by Gonzalez and Woods (1992, Chapter 5).

7.3 IMAGE RESTORATION BY MARKOV RANDOM FIELD MODELING

Image restoration by Markov random field modeling is an active research area in the image processing literature. Several representative procedures in this area are discussed in this section. After some basic concepts and terminologies about Markov random field (MRF) models and Bayesian estimation are introduced in Subsection 7.3.1, the popular image restoration procedure by Geman and Geman (1984) is discussed in Subsection 7.3.2. This procedure provides a general framework for restoring images by MRF modeling, but it is computationally expensive. To simplify its computation, Besag (1986) suggested the iterated conditional modes procedure, which is discussed in Subsection 7.3.3 along with some other modifications and generalizations. Image restoration based on regularization framework is discussed in Subsection 7.3.4.

7.3.1 Markov random field modeling and Bayesian estimation

We begin by defining some basic terms. Suppose that $\mathcal{S} = \{s_1, s_2, \ldots, s_n\}$ is the set of pixel sites of an image, and $\mathcal{N} = \{N_s : s \in \mathcal{S}\}$ is a *neighborhood system* of $\mathcal{S}$ satisfying the following two conditions:

(i) for any site $s \in \mathcal{S}$, $s \notin N_s$; and

(ii) for any two different sites s and t in $\mathcal{S}$, if $s \in N_t$, then $t \in N_s$.

Then all sites in N_s are called *neighbors* of s for any $s \in \mathcal{S}$, and the pair $(\mathcal{S}, \mathcal{N})$ becomes a *graph* (cf., e.g., Diestel 2000, for a general introduction of graph theory). A subset $C \subseteq \mathcal{S}$ is called a *clique* if any pair of pixel sites in C are neighbors of each other. The set of all cliques is denoted by $\mathcal{C}$. For example, suppose that an image has n_1 rows and n_1 columns of pixels so that $\mathcal{S} = \{(i,j) : 1 \leq i, j \leq n_1\}$. For a given positive integer $k < n_1$, one neighborhood system can be defined by

$$\mathcal{N}_k = \{N_{ij} : 1 \leq i, j \leq n_1\}, \tag{7.9}$$

where

$$N_{ij} = \{(i', j') : 0 < (i - i')^2 + (j - j')^2 \leq k\}.$$

When $k = 1$ or 2, the neighbors of a given site s are displayed in Figure 7.4, and the possible cliques are displayed in Figure 7.5.

We also need to use *conditional probability* notation, which has not been introduced before. For two random events A and B with $P(B) > 0$, the conditional probability of A conditional on B, denoted as $P(A|B)$, is defined by

$$P(A|B) = \frac{P(A \text{ and } B)}{P(B)}.$$

Intuitively, $P(A|B)$ tells us the likelihood that the event A will happen given that the event B has happened. If

$$P(A|B) = P(A),$$

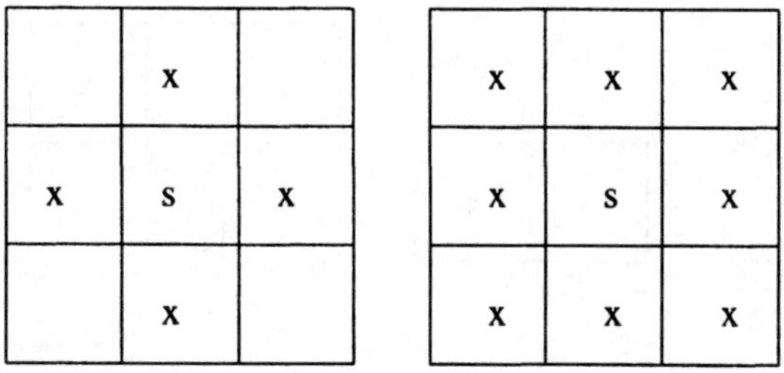

(a) (b)

Fig. 7.4 Sites labeled "x" are neighbors of the sites labeled "s" in the neighborhood system $\mathcal{N}_k$. Plot (a): $k = 1$. Plot (b): $k = 2$.

then the two events A and B are *independent*, in which case the chance that A will happen does not depend on whether B happens or not.

We are now able to define a Markov random field. Suppose that $\mathbf{X} = \{X_s, s \in \mathcal{S}\}$ is a family of random variables indexed in $\mathcal{S}$. Sometimes such a random variable family is also called a *random field* in $\mathcal{S}$. If each component of $\mathbf{X}$ takes values in a common discrete set Λ, then all possible *configurations* of $\mathbf{X}$ are in the set

$$\Psi = \{\mathbf{x} = (x_{s_1}, x_{s_2}, \ldots, x_{s_n}) : x_{s_j} \in \Lambda \text{ for } 1 \leq j \leq n\}.$$

The random variable family $\mathbf{X}$ is called a *Markov random field* (MRF) with respect to the neighborhood system $\mathcal{N}$ if it has the following two properties:

(i) for any $\mathbf{x} \in \Psi$,

$$P(\mathbf{X} = \mathbf{x}) > 0;$$

(ii) for any $s \in \mathcal{S}$,

$$P(X_s = x_s | X_t = x_t, t \neq s) = P(X_s = x_s | X_t = x_t, t \in N_s). \tag{7.10}$$

The first property about a MRF $\mathbf{X}$ says that $\mathbf{X}$ has a positive chance to take any possible configuration in Ψ. The second property describes the following *local* relationship among the components of $\mathbf{X}$: for any $s \in \mathcal{S}$, X_s depends only on its neighboring components. This property is often referred to as the *local characteristics* of the MRF in the literature. An important result derived by Besag (1974) and several other authors is that the joint distribution of $\mathbf{X}$ is uniquely determined by the set of all the conditional distributions defined by equation (7.10).

In applications, it is not convenient to describe a MRF by specifying its joint distribution or all of the conditional distributions described in (7.10). Fortunately, several authors, including Besag (1974), Hammersley and Clifford (1971), and Moussouris (1974), have proven that the random field $\mathbf{X}$ is a MRF with respect to the graph $(\mathcal{S}, \mathcal{N})$

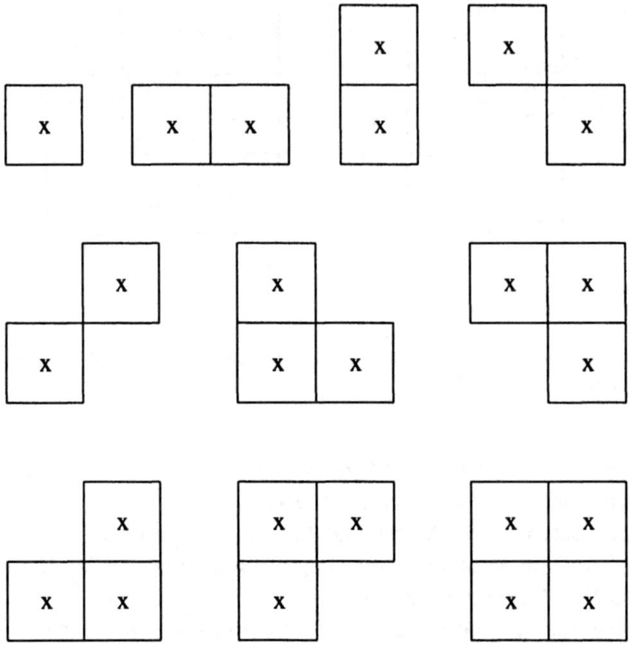

Fig. 7.5 When the neighborhood system is defined by equation (7.9) and $k = 2$, the 10 possible cliques are as shown here. When $k = 1$, there are three possible cliques, shown by the first three plots in the top row.

if and only if it has a *Gibbs distribution* relative to $(\mathcal{S}, \mathcal{N})$. The Gibbs distribution relative to $(\mathcal{S}, \mathcal{N})$ is defined by

$$P(\mathbf{X} = \mathbf{x}) = \frac{1}{Z} \exp\left(-\frac{U(\mathbf{x})}{T}\right), \tag{7.11}$$

where $U(\mathbf{x})$ is the *energy function* of the form

$$U(\mathbf{x}) = \sum_{C \in \mathcal{C}} V_C(\mathbf{x}), \tag{7.12}$$

which is a sum of all *clique potentials* $V_C(\mathbf{x})$ for $C \in \mathcal{C}$; Z is a normalizing constant determined by

$$Z = \sum_{\mathbf{x} \in \Psi} \exp\left(-\frac{U(\mathbf{x})}{T}\right);$$

and T is a *temperature* parameter.

From equation (7.11), it can be seen that a Gibbs distribution has larger chance to take a configuration with smaller energy. The temperature T also has a big impact on the shape of the distribution. If T is large, then the distribution tends to be flat, and thus all configurations tend to have similar chances to occur. On the other hand, the distribution would have sharp peaks at the minimizers of the energy function when T is small.

By using the equivalence between MRF models and Gibbs distributions, a MRF model can be defined by specifying the energy function $U(\mathbf{x})$ and the temperature parameter T in equation (7.11). For simplicity, T can sometimes be treated as a part of $U(\mathbf{x})$, or, equivalently, it can be assumed to be 1. In image processing applications, commonly used energy functions often take the form

$$U(\mathbf{x}) = \sum_{i=1}^{n} V_1(x_{s_i}) + \sum_{i=1}^{n} \sum_{s_j \in N_{s_i}} V_2(x_{s_i}, x_{s_j}) + \ldots,$$

where $V_1, V_2, \ldots$ are, respectively, clique potentials of cliques with sizes one, two, and so forth. Because the clique potentials in the above expression depend only on clique sizes instead of clique locations, MRF models with such energy functions are homogeneous in the sense that specified interactions among neighboring pixels do not depend on pixel locations in the image. One special case of the above energy function is the one with only two terms on the right-hand side in which pixels are assumed to have pairwise relationships only.

Once the energy function $U(\mathbf{x})$ and the temperature parameter T are specified, the conditional distributions of the MRF model can be computed by the expressions

$$P(X_s = x_s | X_t = x_t, t \neq s) = \frac{\exp\left(-\frac{U(\mathbf{x})}{T}\right)}{\sum_{\mathbf{x}'_s} \exp\left(-\frac{U(\mathbf{x}'_s)}{T}\right)},$$

where $\mathbf{x}'_s$ is any configuration that agrees with $\mathbf{x}$ at all sites except the site s.

In the remaining part of this subsection, let us briefly discuss Bayesian estimation of population parameters because image restoration by MRF modeling is based on this statistical estimation technique. This is another general method for estimating population parameters, besides several other methods discussed in Section 2.2. Suppose that a random variable X has a pdf $f_X(x; \theta)$, where θ is an unknown parameter or parameter vector and the form of f_X is assumed known. One example of such a pdf is the pdf of $N(\mu, \sigma^2)$ distribution, given in equation (2.1) in Section 2.2, in which $\theta = (\mu, \sigma^2)'$. In such cases, estimation of the distribution of X is equivalent to estimation of θ.

Although θ is unknown in this setup, we can sometimes specify a distribution for all its possible values based on our past experience about similar distributions. In that sense, θ can be treated as a random variable, denoted as Θ. Its distribution specified beforehand is called a *prior distribution*. Without loss of generality, we assume that the pdf of this prior distribution exists and is denoted as $p_\Theta(\theta)$. Then the pdf of X, $f_X(x; \theta)$, can be regarded as a *conditional* pdf of X, conditional on $(\Theta = \theta)$. For this reason, let us rewrite $f_X(x; \theta)$ as $f_{X|\Theta}(x|\theta)$ to express explicitly the conditional nature of this distribution.

The joint pdf of (X, Θ) (see the definition of the joint pdf of a random vector in Problem 2.2 at the end of Chapter 2) is then given by

$$f_{X|\Theta}(x|\theta)p_{\Theta}(\theta).$$

By the *Bayes Theorem* (cf., e.g., Lindgren 1993, Sections 2.8 and 8.7), the conditional distribution of Θ conditional on $(X = x)$ is

$$f_{\Theta|X}(\theta|x) = \frac{f_{X|\Theta}(x|\theta)p_{\Theta}(\theta)}{p_X(x)},$$

where $p_X(x)$ is the *marginal* pdf of X, namely, the pdf of X ignoring the existence of Θ. The distribution specified by the pdf $f_{\Theta|X}(\theta|x)$ is called the *posterior distribution* of Θ; it is the distribution of Θ knowing the observed value x of X.

Suppose that $\delta(X)$ is an estimator of Θ, and $L(\Theta, \delta(X))$ is a *loss function* measuring the potential "loss" in estimating Θ by $\delta(X)$. The smaller the value of $L(\Theta, \delta(X))$, the better. Commonly used loss functions include

(i) the square loss

$$L(\Theta, \delta(X)) = (\Theta - \delta(X))^2;$$

(ii) the absolute loss

$$L(\Theta, \delta(X)) = |\Theta - \delta(X)|;$$

(iii) the "0-1" loss

$$L(\Theta, \delta(X)) = \begin{cases} 0, & \text{if } \Theta = \delta(X) \\ 1, & \text{otherwise.} \end{cases}$$

For almost all x, suppose that $\delta^*(x)$ minimizes

$$\mathrm{E}(L(\Theta, \delta(X))|X = x),$$

among all possible estimators δ, where the expectation E is taken with respect to the conditional distribution of Θ conditional on $X = x$ (i.e., the posterior distribution of Θ). Then the estimator $\delta^*(X)$ is called the *Bayes estimator*. For instance, when the square loss function is used, the Bayes estimator is just the mean of the posterior distribution of Θ. Similarly, the Bayes estimator is the median of the posterior distribution of Θ when the absolute loss function is used, and it is one of the maximizers of the posterior distribution function (the pdf when the posterior distribution is absolutely continuous, or the probability mass function when the posterior distribution is discrete) if the "0-1" loss function is used.

7.3.2 Geman and Geman's MAP procedure

Geman and Geman (1984) suggested an image restoration procedure based on MRF modeling. One major feature of this procedure is that it introduces a *line process* for representing edges in an image, as described below. Suppose that a given image

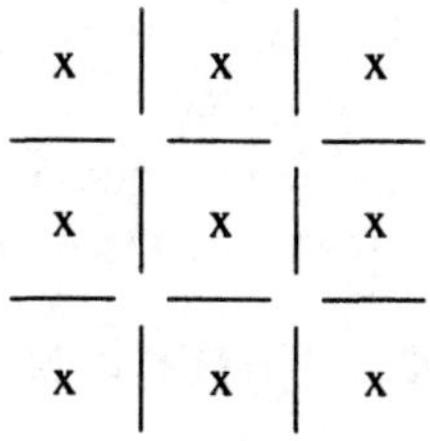

Fig. 7.6 The "x" symbols denote pixel sites. The horizontal or vertical lines between neighboring pixels denote unobservable edge elements. For the central pixel site, the four line sites immediately surrounding it are considered to be its neighboring line sites. For a horizontal line site between two neighboring pixel sites in the middle column, the six line sites surrounding it (three such line sites are on its left and the other three are on its right) are considered its neighboring line sites.

has n_1 rows and n_1 columns of pixels, $\mathbf{X} = \{X_{ij}; i,j = 1,2,\ldots,n_1\}$ is the matrix of observed image intensities at the pixel sites $\mathcal{S} = \{(i,j); i,j = 1,2,\ldots,n_1\}$, and $\mathbf{F} = \{F_{ij}; i,j = 1,2,\ldots,n_1\}$ is the matrix of true image intensities. Between any two neighboring pixels in either the horizontal or vertical direction, Geman and Geman assumed that there is an unobservable edge element, which is denoted by a horizontal or vertical line, as demonstrated in Figure 7.6. The edge element is assumed to be binary, with 1 denoting an existing edge between the two pixels and 0 denoting no edge. The line process refers to the matrix of all such edge elements, denoted by $\mathbf{L}$, and the set of all possible line sites is denoted by $\mathcal{D}$. We can think of any line site in $\mathcal{D}$ as placed midway between two vertically or horizontally neighboring pixels. To distinguish it from the line process $\mathbf{L}$, Geman and Geman also called $\mathbf{F}$ the *intensity process*.

The neighborhood system of $\mathcal{D}$ can be defined similarly to the neighborhood system of $\mathcal{S}$, as previously demonstrated by Figure 7.4. If $\mathcal{S}$ and $\mathcal{D}$ are taken into account simultaneously as a union, then one neighborhood system of the union can be defined as follows. For each pixel site $(i,j) \in \mathcal{S}$, it has four neighboring pixel sites, as displayed in Figure 7.4(a), and four neighboring line sites immediately surrounding it, as displayed in Figure 7.6. For each line site in $\mathcal{D}$, it has six neighboring line sites, as displayed in Figure 7.6, and two neighboring pixel sites on either site.

Next, the MRF part of the procedure is defined. Let $\mathbf{Y} = (\mathbf{F},\mathbf{L})$ and $\mathcal{G} = \mathcal{S} \bigcup \mathcal{D}$. Then $\mathbf{Y}$ can be regarded as a random field in $\mathcal{G}$. An appropriate neighborhood system can be defined for $\mathcal{G}$, such as the one mentioned above, and is denoted by $\mathcal{N}^* = \{N_g^*, g \in \mathcal{G}\}$. Geman and Geman assumed that $\mathbf{Y}$ is a MRF with respect to the graph $(\mathcal{G},\mathcal{N}^*)$. Thus, by the equivalence between a MRF model and a Gibbs distribution (cf. equations (7.11) and (7.12)), the *prior* distribution of $\mathbf{Y}$ has the expression

$$P(\mathbf{F} = \mathbf{f}, \mathbf{L} = \mathbf{l}) = \frac{1}{Z} \exp\left(-\frac{U(\mathbf{f},\mathbf{l})}{T}\right),$$

where

$$U(\mathbf{f},\mathbf{l}) = \sum_{C \in \mathcal{C}} V_C(\mathbf{f},\mathbf{l})$$

is an energy function and $\mathcal{C}$ is the set of all cliques in $\mathcal{G}$.

We assume that the matrices of observed and true image intensities, $\mathbf{X}$ and $\mathbf{F}$, have the following relationship:

$$\mathbf{X} = \phi(H(\mathbf{F})) \odot \mathbf{N},$$

where $\mathbf{N}$ denotes the matrix of noise involved in the observed image intensities, H is a blurring operator, ϕ is a possible nonlinear transformation, and $\odot$ is some suitably invertible operation such as addition or multiplication. Note that the above expression is quite flexible; also, when $\odot$ is the addition operator and ϕ is absent, it is similar to equation (7.1). From the above expression, $\mathbf{N}$ can be expressed as a function of $\mathbf{X}$ and $\phi(H(\mathbf{F}))$, denoted by

$$\mathbf{N} = \psi(\mathbf{X}, \phi(H(\mathbf{F}))),$$

where ψ is some function. For instance, when $\odot$ is the addition operator, $\psi(\mathbf{X}, \phi(H(\mathbf{F}))) = X - \phi(H(\mathbf{F}))$.

Geman and Geman did not discuss how to determine H, ϕ, and $\odot$ in applications. Instead, they assumed that these components had been all specified beforehand. In their numerical studies, they assumed the following: blurring is position invariant, it involves a simple convolution over a 3×3 window, and the convolution mask is specified by

$$H(k_1, k_2) = \begin{cases} 1/2, & \text{if } k_1 = k_2 = 0 \\ 1/16, & \text{if } |k_1| \leq 1, |k_2| \leq 1 \text{ and } (k_1, k_2) \neq (0, 0). \end{cases}$$

That is, the image intensity at a given pixel in a blurred image is assumed to be a weighted average of the original image intensities at the given pixel and its eight neighboring pixels in the 3×3 window. The transformation ϕ is assumed to be absent or $\phi(x) = \sqrt{x}$ and $\odot$ is either additive or multiplicative.

Under the assumption that the noise components in $\mathbf{N}$ are i.i.d. with $N(\mu, \sigma^2)$ distribution, Geman and Geman proved that the *posterior* distribution of $\mathbf{Y}$ given $\mathbf{X} = \mathbf{x}$ is also a Gibbs distribution with energy function

$$U^P(\mathbf{f},\mathbf{l}) = U(\mathbf{f},\mathbf{l}) + \|\mu_{\mathbf{N}} - \psi(\mathbf{x}, \phi(H(\mathbf{F})))\|^2/(2\sigma^2), \qquad (7.13)$$

where the superscript "P" denotes "posterior", $\mu_{\mathbf{N}}$ denotes a $n_1 \times n_1$ matrix with all entries equal to μ, and $\|\cdot\|$ is the Euclidean norm. The restored image is defined by one of the maximizers of the posterior distribution, which is called the maximum *a posteriori* (MAP) estimator. Obviously, the MAP estimator is also a minimizer of the energy function defined at equation (7.13).

If the MAP estimator is obtained by searching all configurations of $\mathbf{Y}$, the computation involved is enormous. As an example, suppose that we have a moderately sized image with 128×128 pixels. The image intensity at each pixel has 256 levels. Then, the total number of configurations of $\mathbf{Y}$ is about $(256 * 2)^{128*128}$, which is a number that is out of the capacity of most powerful computers that we have today.

To overcome this computational difficulty, Geman and Geman suggested using the *Gibbs sampler* scheme and an *annealing* algorithm for obtaining the MAP estimator, as follows.

Let π denote the posterior distribution of $\mathbf{Y}$. By the Gibbs sampler scheme, a sequence of samples of $\mathbf{Y}$, denoted by $\{\mathbf{Y}(t), t = 0, 1, 2, \ldots\}$ which is indexed by time t, are generated as follows. Let $\mathbf{Y}(0)$ be an arbitrary initial sample in the sequence. At a given time t, $\mathbf{Y}(t)$ is generated based on its predecessor $\mathbf{Y}(t-1)$ in the following way. First, $\mathbf{Y}(t)$ is assumed to be different from $\mathbf{Y}(t-1)$ in at most one component at the site s_t. Namely, $\mathbf{Y}(t)$ and $\mathbf{Y}(t-1)$ are the same at all sites except the site s_t. Second, at the site s_t we generate a new intensity level for $\mathbf{Y}(t)$ to replace the old intensity level of $\mathbf{Y}(t-1)$ at this site. The new intensity level is generated from the conditional distribution of Y_{s_t} given the observed intensities at the neighboring sites, namely, given $\{Y_g, g \in N^*_{s_t}\}$. In applications, the sequence of sites $\{s_t, t = 1, 2, \ldots\}$ can simply be chosen as the pixel and line sites in $\mathcal{G}$ gone through by a repeated raster scan, that is, the pixel and line sites from left to right on each row and from top to bottom of the image. These sites can be used periodically for any number of times.

For the Gibbs sampler scheme described above, Geman and Geman proved the following result.

Theorem 7.1 (Geman and Geman 1984) *Assume that for each $g \in \mathcal{G}$, the sequence $\{s_t, t = 1, 2, \ldots\}$ contains g infinitely often. Then for any given starting configuration $\mathbf{Y}(0) = \mathbf{y}(0)$ and any configuration $\mathbf{y}$ of $\mathbf{Y}$,*

$$\lim_{t \to \infty} P(\mathbf{Y}(t) = \mathbf{y} | \mathbf{Y}(0) = \mathbf{y}(0)) = \pi(\mathbf{y}).$$

This result guarantees that the distribution of $\mathbf{Y}(t)$, for any given starting configuration $\mathbf{y}(0)$, converges to the posterior distribution π. Geman and Geman called this property the *statistical relaxation* property of the Gibbs sampler scheme.

Suppose that T^* is the temperature of the posterior distribution π of $\mathbf{Y}$ given the observed image $\mathbf{X} = \mathbf{x}$. If T^* is replaced by $T(t)$ where $T(t) > 0$ decreases to zero when t tends to infinity, then it is easy to see from the form of the Gibbs distribution (cf. equation (7.11)) that the resulting posterior distribution would converge to a distribution with mass 1 on the minimizers of the energy function $U^P(\mathbf{f}, \mathbf{l})$ as defined in equation (7.13). Any one of these minimizers is just the MAP estimator we are trying to find.

Based on this intuition, Geman and Geman suggested using the following annealing schedule to obtain the MAP estimator. Let $\pi_{T(t)}$ denote the posterior Gibbs distribution with temperature $T(t)$ and energy function $U^P(\mathbf{f}, \mathbf{l})$, and let $\mathbf{Y}(t)$ be the Gibbs sample generated from $\pi_{T(t)}$, as described above. Namely, $\mathbf{Y}(t)$ is different from $\mathbf{Y}(t-1)$ only at one site s_t, and the value of $Y_{s_t}(t)$ is generated from the conditional distribution of $Y_{s_t}(t)$ conditional on $\{Y_g(t) = Y_g(t-1), g \in N^*_{s_t}\}$. If Ψ_0 denotes the set of all minimizers of $U^P(\mathbf{f}, \mathbf{l})$, π_0 is a uniform distribution on Ψ_0, and Δ is the range of all possible values of $U^P(\mathbf{f}, \mathbf{l})$, then we have the following result.

Theorem 7.2 (Geman and Geman 1984) *Assume that there exists an integer $\tau \geq n^*$ such that, for every $t = 0, 1, 2, \ldots$, we have*

$$\mathcal{G} \subseteq \{s_{t+1}, s_{t+2}, \ldots, s_{t+\tau}\},$$

where n^* is the total number of sites in $\mathcal{G}$. Let $\{T(t), t = 1, 2, \ldots\}$ be any decreasing sequence of temperatures such that

(a) $\lim_{t\to\infty} T(t) = 0$; *and*

(b) there exists an integer $t_0 \geq 2$ such that for all $t \geq t_0$, $T(t) \geq n^*\Delta/\log(t)$.

Then for any starting configuration $\mathbf{Y}(0) = \mathbf{y}(0)$ and any configuration $\mathbf{y}$ of $\mathbf{Y}$,

$$\lim_{t\to\infty} P(\mathbf{Y}(t) = \mathbf{y}|\mathbf{Y}(0) = \mathbf{y}(0)) = \pi_0(\mathbf{y}).$$

In the above theorem, the condition on $\{s(t), t = 1, 2, \ldots\}$ requires that there exists an integer τ such that any segment of $\{s(t), t = 1, 2, \ldots\}$ of the length τ covers all sites in $\mathcal{G}$. This condition is obviously satisfied if $\{s(t), t = 1, 2, \ldots\}$ are determined using the repeated raster scan, as described above. The conditions on the sequence of temperatures require that $T(t)$ converges to 0 when t tends to infinity, but the convergence rate should not be faster than $1/\log(t)$, and, more specifically, $T(t)$ should not be smaller than $n^*\Delta/\log(t)$. According to this theorem, as long as these conditions are all satisfied, the sequence of samples $\{\mathbf{Y}(t), t = 1, 2, \ldots\}$ converges in distribution to a MAP estimator.

In applications, Geman and Geman suggested using the following temperature scheme

$$T(k) = \frac{C}{\log(1+k)}, \text{ for } 1 \leq k \leq K,$$

where $T(k)$ is the temperature used during the kth full sweep of $\mathcal{G}$ (i.e., the kth time that all sites in $\mathcal{G}$ are gone through), K is the total number of full sweeps, and C is a constant selected empirically.

7.3.3 Besag's ICM procedure and some modifications

The MAP procedure provides a general approach for restoring images based on Bayesian estimation. However, this procedure makes enormous computational demands, although the Gibbs sampler scheme and an annealing schedule have been used to make the computing tasks possible. To simplify the computation involved in the MAP procedure, Besag (1986) suggested the *iterated conditional modes* (ICM) algorithm for image restoration, which is introduced below, along with some modifications and generalizations.

As in the previous subsection, let $\mathbf{F} = \{F_{ij}; i, j = 1, 2, \ldots, n_1\}$ be the matrix of true image intensities, $\mathbf{X} = \{X_{ij}; i, j = 1, 2, \ldots, n_1\}$ be the matrix of observed image intensities, $\mathcal{S} = \{(i, j); i, j = 1, 2, \ldots, n_1\}$ be the set of all pixel sites, and $\mathcal{N} = \{N_{ij}; i, j = 1, 2, \ldots, n_1\}$ be a neighborhood system of $\mathcal{S}$. Besag assumed the following: the components of $\mathbf{X}$ are independent of each other conditional on $\mathbf{F}$, each X_{ij} has a same and known conditional pdf $p(X_{ij}|F_{ij})$, and $\mathbf{F}$ is a MRF.

The ICM algorithm is iterative. Besag suggested using the conventional maximum likelihood classifier, which maximizes $p(X_{ij}|F_{ij})$ at each pixel site (i,j), as the initial estimator of F_{ij}. The updating is most conveniently implemented as a raster scan (i.e., the updating pixel site moves from left/right to right/left and from top/bottom to bottom/top in the updating process), but it is helpful to vary the direction of the raster scan from cycle to cycle to reduce small directional effects. Then, the iterations are updated as follows. Let $\widehat{\mathbf{F}} = \{\widehat{F}_{ij}; i,j = 1,2,\ldots,n_1\}$ be the estimator of $\mathbf{F}$ from the previous iteration. Besag suggested updating $\widehat{F}_{ij}$ using all available information up to the current iteration. More specifically, the updated estimator of F_{ij} is defined by the maximizer of the conditional probability $P(F_{ij}|\mathbf{X}, \widehat{F}_s, s \in \mathcal{S}\backslash\{(i,j)\})$ with respect to F_{ij}. By the Bayes Theorem and some algebraic manipulations, it can be shown that

$$P(F_{ij}|\mathbf{X}, \widehat{F}_s, s \in \mathcal{S}\backslash\{(i,j)\}) \propto p(X_{ij}|F_{ij})p_{ij}(F_{ij}|\widehat{F}_s, s \in N_{ij}), \qquad (7.14)$$

where "$\propto$" denotes "proportional to", implying that a multiplier independent of F_{ij} has been omitted, and $p_{ij}(F_{ij}|\widehat{F}_s, s \in N_{ij})$ is the conditional pdf of F_{ij} conditional on its neighboring pixels $\{\widehat{F}_s, s \in N_{ij}\}$.

In comparison with the MAP procedure introduced in the previous subsection, implementation of the ICM algorithm is relatively simple because the right-hand side of expression (7.14) is *locally* defined, and thus the computation involved is relatively simple.

As an example, suppose that the components of $\mathbf{X} = \{X_{ij}, i,j = 1,2,\ldots,n_1\}$ are independent of each other conditional on $\mathbf{F} = \{F_{ij}, i,j = 1,2,\ldots,n_1\}$, where each component X_{ij} is normally distributed with mean F_{ij} and variance σ^2. Then

$$p(\mathbf{X}|\mathbf{F}) = \Pi_{i=1}^{n_1}\Pi_{j=1}^{n_1}p(X_{ij}|F_{ij}) \propto \exp\left[-\frac{1}{2\sigma^2}\sum_{i=1}^{n_1}\sum_{j=1}^{n_1}(X_{ij}-F_{ij})^2\right].$$

Assume that the true image $\mathbf{F}$ is black-and-white (i.e., the image intensity at each pixel site is either 0 or 1) and that its prior distribution is the following Gibbs distribution:

$$p(\mathbf{F}) \propto \exp\{-[\beta_1 Z_1(\mathbf{F}) + \beta_2 Z_2(\mathbf{F})]\},$$

where $Z_1(\mathbf{F})$ denotes the number of discrepant pairs of pixels (i.e., pairs of pixels with opposite colors) in the true image $\mathbf{F}$ that are horizontally or vertically adjacent to each other, $Z_2(\mathbf{F})$ is the number of discrepant pairs of pixels that are diagonally adjacent to each other, and β_1 and β_2 are two positive constants. Then the posterior distribution of $\mathbf{F}$ given $\mathbf{X}$ is proportional to

$$\exp\left[-\frac{1}{2\sigma^2}\sum_{i=1}^{n_1}\sum_{j=1}^{n_1}(X_{ij}-F_{ij})^2\right]\cdot\exp\{-[\beta_1 Z_1(\mathbf{F}) + \beta_2 Z_2(\mathbf{F})]\}.$$

The MAP estimator is then one of the minimizers of the following objective function (as a function of $\mathbf{F}$):

$$\frac{1}{2\sigma^2}\sum_{i=1}^{n_1}\sum_{j=1}^{n_1}(X_{ij}-F_{ij})^2+[\beta_1 Z_1(\mathbf{F})+\beta_2 Z_2(\mathbf{F})].$$

Obviously, the first term of this objective function measures the fidelity of the estimator to the data, and the second term measures the roughness of the estimator. By Besag's ICM procedure, for this example, the observed image intensity X_{ij} can be used as an initial estimator of F_{ij}, since X_{ij} is the maximum likelihood classifier when the conditional distribution $p(X_{ij}|F_{ij})$ is $N(F_{ij},\sigma^2)$. The image intensities can be updated in the raster scan fashion. Thus, at a given site (i,j), the updated image intensity $\widehat{F}_{ij}$ is defined by the minimizer of the following local objective function, as a function of F_{ij}, with all other image intensities estimated from the previous iteration

$$\frac{1}{2\sigma^2}(X_{ij}-F_{ij})^2+[\beta_1 Z_1'(F_{ij})+\beta_2 Z_2'(F_{ij})],$$

where $Z_1'(F_{ij})$ and $Z_2'(F_{ij})$ are numbers of horizontal/vertical and diagonal neighbors of the site (i,j), respectively, at which the image intensities are different from F_{ij}.

The ICM procedure introduced above does not include the line process representing edges in an image, as the MAP procedure does, but Godtliebsen and Sebastiani (1994) generalized the ICM procedure by accommodating the line process as follows. The true image $\mathbf{F}$ and the line process $\mathbf{L}$ are assumed to have the following joint prior distribution:

$$p(\mathbf{F},\mathbf{L})\propto\exp\left\{-\beta\sum_{i,j}\sum_{(i',j')\in N_{ij}}V_p(F_{ij},F_{i'j'})-\lambda\sum_{i,j}V_d(\mathbf{L}_{ij}^*)\right\},\tag{7.15}$$

where $\beta,\lambda>0$ are two parameters, and $\mathbf{L}_{ij}^*$ is a vector composed of the components of $\mathbf{L}$ describing the discontinuity pattern in $N_{ij}\bigcup\{(i,j)\}$. The function $V_p(F_{ij},F_{i'j'})$ is assumed to be

$$V_p(F_{ij},F_{i'j'})=(F_{ij}-F_{i'j'})^2W_{ij,i'j'},$$

where $W_{ij,i'j'}$ is 1 if the (i,j)-th pixel and the (i',j')-th pixel can be connected by a minimum length path consisting of horizontal and vertical steps without crossing any edges, and 0 otherwise. By using this function, the true image $\mathbf{F}$ is constrained to be continuous over regions separated by edge curves.

We define $V_d(\mathbf{L}_{ij}^*)$ in the following way. In the 3×3 neighborhood of a given pixel, there are 12 line components (cf. Figure 7.6), so there exist total 2^{12} possible line configurations in this neighborhood. For a given line configuration $\mathbf{L}_{ij}^*$, it is easy to see from expression (7.15) that

$$P(\mathbf{L}_{ij}^*)\propto\exp[-V_d(\mathbf{L}_{ij}^*)].$$

Therefore, if $P(\mathbf{L}_{ij}^*)\neq 0$, we can set

$$V_d(\mathbf{L}_{ij}^*)=-\log(P(\mathbf{L}_{ij}^*)).$$

If $P(\mathbf{L}^*_{ij}) = 0$, Godtliebsen and Sebastiani suggested assigning the value of $V_d(\mathbf{L}^*_{ij})$ by

$$V_d(\mathbf{L}^*_{ij}) = V_d^{\max} + \alpha \min_{P(\mathbf{L}^*_{i'j'}) \neq 0} \|\mathbf{L}^*_{ij} - \mathbf{L}^*_{i'j'}\|,$$

where $V_d^{\max}$ is the maximum of all $V_d(\mathbf{L}^*_{ij})$ values, $\alpha > 0$ is a constant, and $\| \cdot \|$ is the Euclidean norm.

Godtliebsen and Sebastiani assumed that $P(\mathbf{L}^*_{ij})$ is known in their procedure. In applications, it can be estimated from a training dataset, which could be a specific portion of the image in question, where the edge structure is believed to be representative of the entire image. The parameters α, β and λ in the procedure need to be specified empirically.

Example 7.4 We apply the above image restoration procedure to the mountain image shown in Figure 7.1(a) which has 480 × 480 pixels. In the original image, the image intensity has 256 possible levels; i.i.d. random errors with distribution $N(0, 50^2)$ are added to its individual image intensities, and the noisy image is shown in Figure 7.7(a). In Godtliebsen and Sebastiani's procedure, the entire noisy image is used for estimating the probabilities of different discontinuity patterns $P(\mathbf{L}^*_{ij})$. The restored images are shown in Figures 7.7(b)–7.7(d), respectively, with (i) $\alpha = 16, \beta = 0.5$, and $\lambda = 9$; (ii) $\alpha = 16, \beta = 0.1$, and $\lambda = 18$; and (iii) $\alpha = 16, \beta = 0.1$, and $\lambda = 9$. From these plots, it can be seen that (1) when β is large relative to λ, the restored image is quite noisy (cf. Figure 7.7(b)), (2) when β is small relative to λ, the restored image tends to be oversmoothed and some small edges are blurred (cf. Figure 7.7(c)), and (3) the parameter values specified in the case of Figure 7.7(d) are about right in balancing noise removal and edge preservation.

In the literature, there are many other generalizations and modifications of the MAP and ICM procedures. Several authors, including Blake and Zisserman (1987), Geiger and Girosi (1991), and Geman and Reynolds (1992), have studied various image restoration models without introducing explicit line processes. Hebert and Leahy (1992) discussed MAP image restoration for Poisson data. Bedini *et al.* (1994, 1995) suggested using continuous-valued line processes in image restoration. Ferrari *et al.* (1995) and Greig *et al.* (1989) discussed MAP image restoration of binary and multicolor images. Li (1995a) presented a systematic study about MRF models involving discontinuities, and defined a general discontinuity adaptive MRF model using the *Euler equation*, which is associated with the energy minimization and constrained by a class of *adaptive interaction functions*. McMillan and Berlinear (1996) suggested a hierarchical image restoration procedure using MRF modeling. Fessler *et al.* (2000) derived the exact distribution of the edge-preserving MAP estimators for linear signal modeling with Gaussian measurement noise. Marroquin *et al.* (2001) presented a method for approximating the posterior marginal distributions which results in fast and reliable procedures that may be used for handling ill-posed problems in image processing, including the image restoration problem. For more discussion on this topic, see Chellapa and Jain (1993), Li (1995b), and the references cited there.

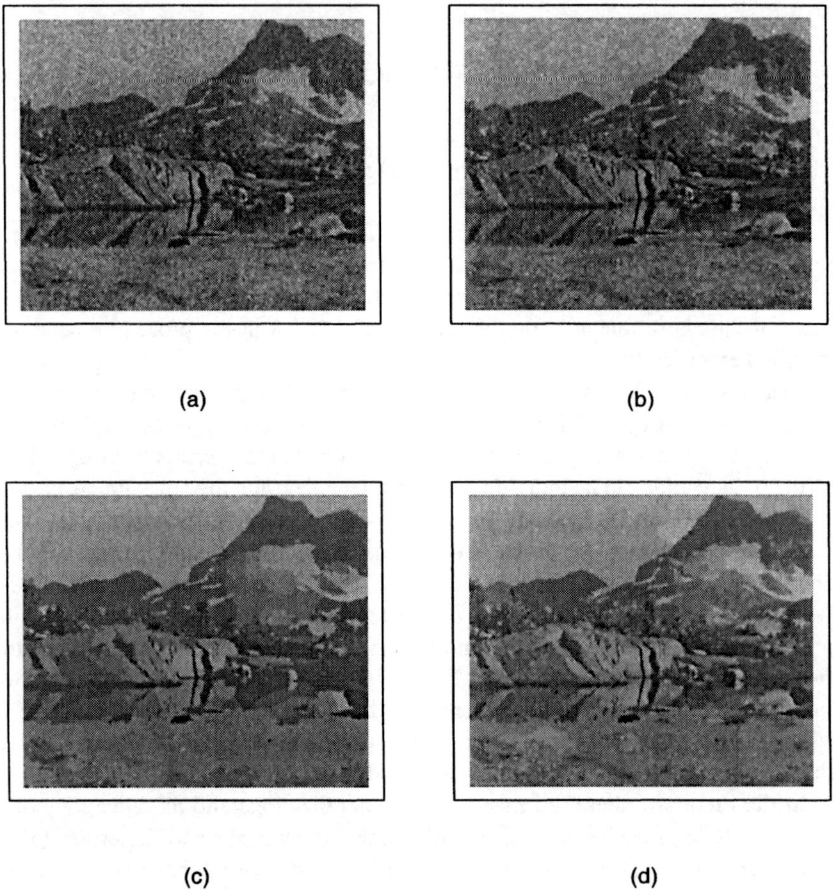

Fig. 7.7 (a): A noisy version of the mountain image contaminated by i.i.d. noise with $N(0, 50^2)$ distribution. (b): Restored image by Godtliebsen and Sebastiani's procedure with $\alpha = 16, \beta = 0.5$, and $\lambda = 9$. (c): Restored image by Godtliebsen and Sebastiani's procedure with $\alpha = 16, \beta = 0.1$, and $\lambda = 18$. (d): Restored image by Godtliebsen and Sebastiani's procedure with $\alpha = 16, \beta = 0.1$, and $\lambda = 9$.

7.3.4 Image restoration by regularization

The image restoration problem can also be formulated as a *regularization* problem, in which the restored image is defined by the minimizer of the following energy function in an appropriate function space:

$$U(g, d) + U(g), \tag{7.16}$$

where the first term $U(g,d)$ measures the *fidelity* of a candidate estimator g of the true image intensity function f to the observed image intensity function d (i.e., the *data*), and the second term $U(g)$ is a *smoothness* measure of g.

In 1-D cases, the first term of expression (7.16) often takes the following form:

$$U(g,d) = \int_a^b \chi(x)[g(x) - d(x)]^2 \, dx,$$

where $\chi(x)$ is a weighting function, and $[a,b]$ is the design interval. In discrete cases, the above integral can be replaced by a summation over all design points. If we also set $\chi(x) \equiv 1$, $U(g,d)$ becomes the more familiar residual sum of squares:

$$U(g,d) = \sum_{i=1}^{n} [g(x_i) - d(x_i)]^2, \tag{7.17}$$

where n is the sample size and $d(x_i)$ is the ith observation obtained at the design point x_i, for $i = 1, 2, \ldots, n$.

The second term is generally defined by

$$U(g) = \sum_{m=1}^{M} \lambda_m \int_a^b \psi\left(g^{(m)}(x)\right) \, dx,$$

where M is the highest order of derivatives of g considered, λ_m are weighting parameters, and ψ is some function. In most applications, ψ is chosen to be the quadratic function. In such cases, if the curvature of the candidate estimator g is used as the smoothness measure, then all λ_m are zero except λ_2, and consequently the second term becomes:

$$U(g) = \lambda_2 \int_a^b \left(g^{(2)}(x)\right)^2 \, dx. \tag{7.18}$$

The objective function defined by expressions (7.16)–(7.18) is the same as the objective function defined by expression (2.18) in Subsection 2.5.1 when $m = 2$, $a = 0$, and $b = 1$. Thus, the minimizer of the former objective function should be a smoothing spline if the function space searched is the Sobolev space $W_2^2([a,b])$. More discussion about the relationship between image restoration by regularization and surface estimation by spline smoothing can be found in Terzopoulos (1986).

The minimization problem of the energy function defined by expression (7.16) is often *ill-posed* if the second term $U(g)$ is omitted because its solution is usually not unique. The major purpose of introducing the second term $U(g)$ in the objective function is to add a smoothness constraint on all candidate estimators such that the minimization problem becomes well-posed. For this reason, that term is often called a *regularizer*; it penalizes the irregularities of the estimator g according to the *a priori* smoothness constraint on g. This regularization method has been used for solving various computer vision problems, including the image restoration problem (cf., e.g., Grimson 1981, Hildreth 1984, Horn and Schunck 1981). Poggio *et al.* (1985) gave a nice review on some early regularization procedures.

It can be seen that the smoothness constraint $U(g)$ is applied globally to g, which can result in an oversmooth estimator of the true image intensity function when the true image intensity function has some irregularities, such as edges, involved. To overcome this problem, a more general smoothness term can be defined as follows:

$$U(g) = \sum_{m=1}^{M} \lambda_m \int_a^b w_m(x) \left(g^{(m)}(x)\right)^2 dx,$$

where $w_m(x)$ are weighting functions.

Ideally, to preserve irregularities in $g^{(m)}$, $w_m(x)$ should be zero at places where $g^{(m)}$ does not exist or is large. However, it is a challenging task to determine weighting functions with such properties in applications. One method for achieving this goal is to use the line process, as described in Subsection 7.3.2; the regularizer can then be defined by both the line process and g. One example of such regularizers is

$$U(\mathbf{g}, \mathbf{l}) = \lambda \sum_{i=2}^{n} [g(x_i) - g(x_{i-1})]^2(1 - l_i) + \lambda\alpha \sum_{i=2}^{n} l_i,$$

where $\mathbf{g} = (g(x_1), g(x_2), \ldots, g(x_n))'$ is a vector of the values of the candidate estimator g at the n design points, $\mathbf{l} = (l_2, l_3, \ldots, l_n)'$ is a line process with each binary (1 if an edge exists between the two design points, and 0 otherwise) line component l_i located between two neighboring design points, and $\lambda, \alpha > 0$ are two parameters. The first term of the above regularizer tends to make the estimator smooth in regions without edges (i.e., to make $|g(x_i) - g(x_{i-1})|$ small when $l_i = 0$), and the second term accounts for a cost $\lambda\alpha$ for each occurrence of $(l_i = 1)$. The energy function with this regularizer and $U(g, d)$ defined in equation (7.17) becomes

$$\sum_{i=1}^{n} [g(x_i) - d(x_i)]^2 + \lambda \left(\sum_{i=2}^{n} [g(x_i) - g(x_{i-1})]^2 (1 - l_i) + \alpha \sum_{i=2}^{n} l_i \right). \quad (7.19)$$

This objective function is often called the *weak string* model, and its 2-D version is called the *weak membrane* model (Blake and Zisserman 1987).

In the literature, there are several different regularization models. For instance, Blake and Zisserman (1987) considered the following regularizer:

$$\sum_{i=1}^{n} [g(x_i) - d(x_i)]^2 + \lambda \sum_{i=2}^{n} \phi_\alpha(g(x_i) - g(x_{i-1})),$$

where $\lambda, \alpha > 0$ are two parameters, and

$$\phi_\alpha(x) = \min(x^2, \alpha)$$

is a truncated quadratic function. The purpose in using the truncated quadratic function in the above regularizer is to allow for discontinuities between neighboring design points, which is a similar idea to robust estimation in statistics (Huber 1981). Several alternative functions have been suggested in the literature in place of ϕ_α. For related discussions, see Bouman and Sauer (1993), Green (1990), Lange (1990), Li (1998), Stevenson *et al.* (1994), and the references cited there. Several authors, including Geiger and Girosi (1991) and Koch *et al.* (1986), introduced continuous line process variables in the range $[0, 1]$ in their regularization models.

Comparing the regularization models described above to the Bayesian image restoration models based on MRF modeling and MAP estimation (cf. expressions (7.15) and (7.19)), it can be seen that the two models look similar. Objective functions of both types of models consist of two terms: one for measuring fidelity of the estimated image to the observed image and the other one for measuring *interaction* among neighboring pixels in the estimated image. The Bayesian image restoration models describe the interaction among neighboring pixels by specifying a prior distribution of the true image, and the regularization models describe the interaction by imposing a smoothness constraint on the true image. In that sense, the former models are more general because they can incorporate not only smoothness interaction but also other types of interactions. However, they are equivalent to each other if the noise involved in the observed image intensities is additive and independent and identically distributed (i.i.d.) with Gaussian distribution, and the prior distribution of the true image imposes a smoothness constraint on the true image (Li 1995a, Marroquin *et al.* 1987).

7.4 IMAGE RESTORATION BY LOCAL SMOOTHING FILTERS

As explained in Section 7.1, noise is often involved in observed images in applications. Various denoising techniques have been suggested in the image processing literature. In this section, we introduce several denoising techniques based on different local smoothing filters. In Subsection 7.4.1, some robust local smoothing filters are introduced, and adaptive local smoothing techniques are discussed in Subsection 7.4.2. Image restoration based on surface estimation is discussed in Subsection 7.4.3.

7.4.1 Robust local smoothing filters

When denoising is our major concern, it is natural to use local smoothing filters. For example, the *neighborhood averaging* mask displayed in Figure 7.8 can be used for the purpose of denoising. By using this mask, the restored image intensity at a given pixel equals the simple average of the image intensities in its 3×3 neighborhood.

However, such filters blur edges at the same time that they remove noise, similarly to the conventional surface estimators discussed in Chapter 5 (cf., e.g., Figure 5.3(c)). Consequently, spatial blurs are generated by such filters even as point blurs are removed by them. To overcome this limitation of the neighborhood averaging filter, various filters based on local median smoothing and other robust local smoothing have been suggested in the literature; some of these are discussed below.

$$\frac{1}{9}\times \begin{array}{|c|c|c|}\hline 1 & 1 & 1 \\ \hline 1 & 1 & 1 \\ \hline 1 & 1 & 1 \\ \hline \end{array}$$

Fig. 7.8 A neighborhood averaging mask.

Local *median filtering* was first suggested in the statistical literature (Tukey 1977), and it became a popular denoising tool in image processing because it can preserve edges to some extent while removing noise (Gallagher and Wise 1981, Huang 1981). In conventional local median filtering, the restored image intensity at a given pixel equals the sample median of the observed image intensities in its neighborhood. For example, suppose that 3×3 neighborhoods are used in the local median filtering, and the observed image intensities in the 3×3 neighborhood of a given pixel are 10, 18, 9, 15, 7, 20, 11, 16, and 18; then the restored image intensity at the given pixel is 15. See Subsection 2.2.2 for a general description about computing the sample median of a dataset.

Example 7.5 Figure 7.9(a) displays a true image that has two step edge segments in the lower-left part of the image. The two step edge segments cross each other at the center of the image with an acute angle. In the top right part of the true image, there are four line edges (cf. Section 6.1 for related discussion): one horizontal line edge, one vertical line edge, and two diagonal line edges. Jump sizes in image intensity levels at all edge segments are 1. We then add i.i.d. random noise with $N(0, 0.25^2)$ distribution to individual image intensity levels, and the noisy image is shown in Figure 7.9(b). The restored image by the local median filter of size 3×3 is shown in Figure 7.9(c), and the restored image by the local median filter of size 5×5 is shown in Figure 7.9(d). From the plots, it can be seen that noise has been removed by the filter to a certain degree, and the step edges are mostly well preserved in both cases. However, it can also be noticed that the line edges are almost removed completely by the filter and that the sharp angle of the step edges is rounded, especially when the window size of the local median filter is large (cf. Figure 7.9(d)).

To preserve horizontal or vertical line edges or to preserve sharp angles formed by horizontal or vertical step edge segments, local median filtering can be performed in cross-shaped neighborhoods, such as the one for a 3×3 neighborhood shown in Figure 7.10(a). By using such median filtering, the restored image intensity at a given pixel is defined by the sample median of the image intensities only from vertically or horizontally neighboring pixels. Similarly, local median filtering can be performed in X-shaped neighborhoods, such as the one for a 3×3 neighborhood shown in Figure 7.10(b), or other types of neighborhoods.

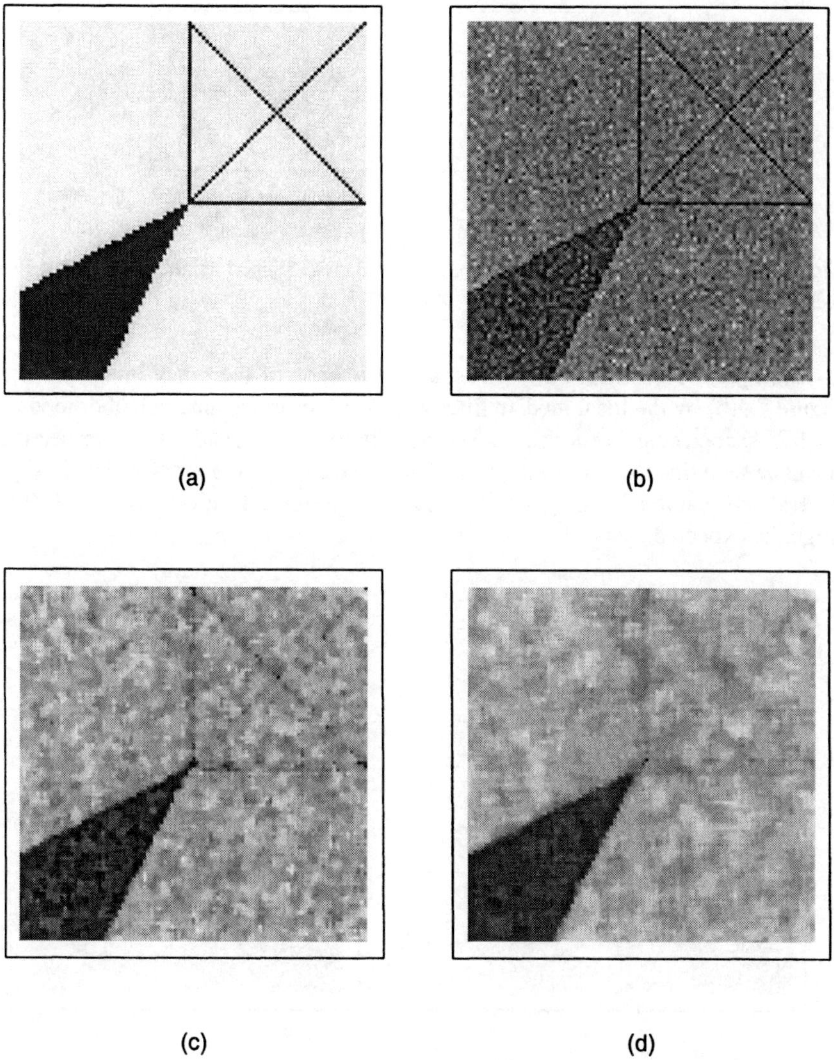

Fig. 7.9 (a): The true image with two step edge segments and four line edge segments. (b): The noisy image contaminated by i.i.d. noise with $N(0, 0.25^2)$ distribution. (c): The restored image by the local median filter of size 3×3. (d): The restored image by the local median filter of size 5×5. In the true image, jump sizes in image intensity levels at all edge pixels are 1.

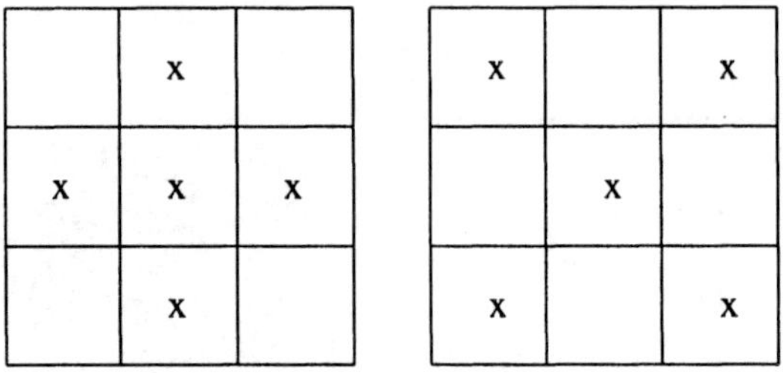

(a) (b)

Fig. 7.10 (a): Pixels marked by "x" form a 3×3 cross-shaped neighborhood. (b): Pixels marked by "x" form a 3×3 X-shaped neighborhood.

Example 7.5 (continued) The restored images of the noisy image shown in Figure 7.9(b) by the local median filter with 5×5 cross-shaped neighborhoods and 5×5 X-shaped neighborhoods are shown in Figures 7.11(a) and 7.11(b), respectively. It can be seen that the two horizontal/vertical line edges are preserved well in Figure 7.11(a) and that the two diagonal line edges are preserved well in Figure 7.11(b), as might be expected.

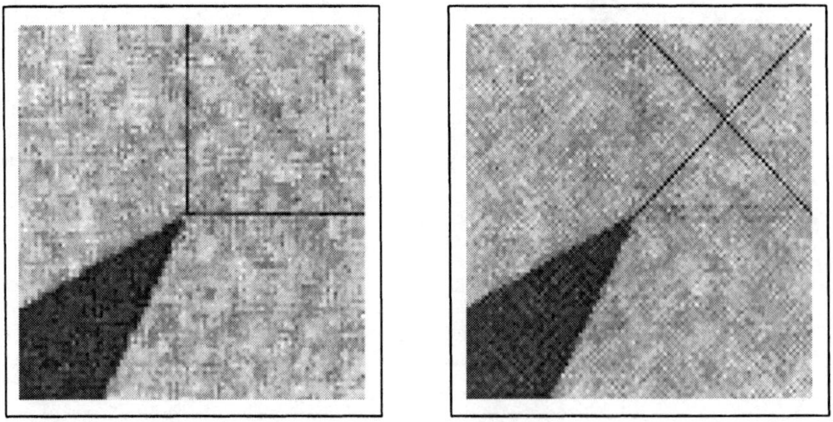

(a) (b)

Fig. 7.11 (a): The restored image of the noisy image shown in Figure 7.9(b) by the local median filter with 5×5 cross-shaped neighborhoods. (b): The restored image of the same noisy image by the local median filter with 5×5 X-shaped neighborhoods.

Local median filtering is often used as a tool for prefiltering the image before further image analysis and processing, such as edge detection, edge enhancement, and image restoration. Its impact on edge detection was studied by several authors, including Bovik *et al.* (1987), Huang and Dunsmuir (1998), Pratt (1991), and Yang and Huang (1981). For instance, Bovik *et al.* (1987) found that local median prefiltering can improve the performance of zero-crossing type edge detectors as well as more conventional gradient-based edge detectors (cf. Section 6.2), when measured by increased noise suppression and insubstantial loss of edge details.

There are some generalizations of the conventional local median filtering. One particularly useful method is *weighted median filtering* (e.g., Arce 1991, Brownrigg 1984, Haavisto *et al.* 1991, Loupas *et al.* 1989), which is introduced below. Let us consider a $(2L+1) \times (2L+1)$ neighborhood of the (i, j)-th pixel, where $L \geq 0$ is an integer. Then the restored image intensity at that pixel is defined by the sample median of the set

$$\{W_{s,t} \diamond X_{i+s,j+t} : s, t = -L, -L+1, \ldots, 0, \ldots, L-1, L\},$$

where $X_{i+s,j+t}$ is the observed image intensity at the $(i+s, j+t)$-th pixel, $W_{s,t}$ is a nonnegative integer, and $W_{s,t} \diamond X_{i+s,j+t}$ denotes the operation of duplicating the value $X_{i+s,j+t}$, $W_{s,t}$ times. Obviously, the observed image intensity $X_{i+s,j+t}$ is more likely to be the sample median of the above set if $W_{s,t}$ is larger; that is, $W_{s,t}$ works as a weight.

In particular, when $W_{0,0} > 1$ and all other $W_{s,t}$ values are 1, the resulting weighted median filters are called *center weighted median filters*, giving more weight only to the central pixel of each neighborhood. When the weight $W_{0,0}$ increases, finer and finer image details are preserved, but the filter's ability to remove noise is decreased at the same time. For more discussion about center weighted median filters, read Ko and Lee (1991), Sun *et al.* (1994), and the references cited there.

7.4.2 Adaptive smoothing and bilateral filtering

The neighborhood averaging mask shown in Figure 7.8 has some ability to remove noise, but it blurs step edges and other image details at the same time. The major reason for such blurring is that pixels located on two different sides of an edge segment would be averaged by such masks.

To overcome this limitation, edge structure of the image should be taken into account when the image is restored. *Adaptive smoothing filters* are those smoothing filters which can adapt themselves to the edge structure of an image (Saint-Marc *et al.* 1991).

Adaptive smoothing filters are implemented in an iterative process. Let $f(x, y)$ and $Z(x, y)$ denote the true and observed image intensities at the location (x, y) in an image, and let $S^{(0)}(x, y) = Z(x, y)$ denote the image intensity before smoothing. Then the smoothed image intensity at (x, y) in the $(t+1)$-th iteration, for any $t \geq 0$, is defined by

$$S^{(t+1)}(x, y) = \frac{1}{N^{(t+1)}(x, y)} \sum_{i=-1}^{1} \sum_{j=-1}^{1} S^{(t)}(x+i, y+j) w^{(t)}(x+i, y+j), \quad (7.20)$$

where $w^{(t)}(x+i, y+j)$ are weights and

$$N^{(t+1)}(x,y) = \sum_{i=-1}^{1} \sum_{j=-1}^{1} w^{(t)}(x+i, y+j).$$

Obviously, $S^{(t+1)}(x,y)$ is defined by a weighted average of the restored image intensities in the 3×3 neighborhood of (x,y) obtained in the previous iteration. Intuitively, the weights should be chosen small for pixels close to edges and relatively large for pixels far away from edges. Unfortunately, for any given pixel, it is often unknown in applications whether it is close to edges or not. Thus, this judgment should be made based on data. A reasonable quantitative measure of the edge structure around the pixel position (x,y) is the magnitude of the image gradient $G_f(x,y) = (f_x(x,y), f_y(x,y))'$, which is denoted by

$$d^{(t)}(x,y) = \sqrt{f_x^2(x,y) + f_y^2(x,y)},$$

where $f_x(x,y)$ and $f_y(x,y)$ can be estimated from the restored image intensities obtained in the previous iteration by methods such as those described in Subsection 6.2.1. If the value of $d^{(t)}(x,y)$ is larger, then it is more likely that (x,y) is close to an edge, and vice versa. Then the weight function used in equation (7.20) can be defined by $w^{(t)}(x,y) = g(d^{(t)}(x,y))$, where g is a nonnegative function satisfying $g(0) = 1$ and $g(d^{(t)}(x,y))$ tends to zero as $d^{(t)}(x,y)$ increases to infinity. In applications,

$$w^{(t)}(x,y) = \exp\left(-\frac{\left(d^{(t)}(x,y)\right)^2}{2\sigma_D^2}\right)$$

is often used, where $\sigma_D > 0$ is a scale parameter.

A generalization of the above adaptive smoothing filter is the *bilateral filtering* procedure suggested by Tomasi and Manduchi (1998). In this procedure, the restored image intensity at pixel position $\mathbf{s} = (x,y)$ is defined by

$$\widehat{f}(\mathbf{s}) = \frac{1}{N(\mathbf{s})} \int_{-\infty}^{\infty} \int_{-\infty}^{\infty} Z(\mathbf{s}^*) K_D(\mathbf{s}^*, \mathbf{s}) K_R(Z(\mathbf{s}^*), Z(\mathbf{s}))\, d\mathbf{s}^*, \qquad (7.21)$$

where $\mathbf{s}^* = (x^*, y^*)$ is a nearby point of $\mathbf{s}$, K_D is a nonnegative bivariate kernel function, K_R is a nonnegative univariate kernel function, and

$$N(\mathbf{s}) = \int_{-\infty}^{\infty} \int_{-\infty}^{\infty} K_D(\mathbf{s}^*, \mathbf{s}) K_R(Z(\mathbf{s}^*), Z(\mathbf{s}))\, d\mathbf{s}^*.$$

By using the bilateral filter (7.21), the restored image intensity $\widehat{f}(\mathbf{s})$ at location $\mathbf{s}$ equals a weighted average of the image intensities at nearby pixels. The weights are determined by the two kernel functions: K_D is for pixel locations, and K_R is for image intensities. The kernel function K_D is usually chosen to be the following bivariate Gaussian density function:

$$K_D(\mathbf{s}^*, \mathbf{s}) = \exp\left(-\frac{\|\mathbf{s}^* - \mathbf{s}\|^2}{2\sigma_d^2}\right),$$

which is radially symmetric and location invariant, and the kernel function K_R is usually chosen to be the following univariate Gaussian density function:

$$K_R(Z(\mathbf{s}^*), Z(\mathbf{s})) = \exp\left(-\frac{(Z(\mathbf{s}^*) - Z(\mathbf{s}))^2}{2\sigma_r^2}\right),$$

where $\|\cdot\|$ is the Euclidean distance, and σ_d and σ_r are two positive scale parameters.

Therefore, the weights used in equation (7.21) are determined by two factors: the Euclidean distance from individual pixel site $\mathbf{s}^*$ to the given pixel site $\mathbf{s}$, and the difference between the observed image intensity at $\mathbf{s}^*$ and the observed image intensity at $\mathbf{s}$. Weights are smaller if the individual pixel sites are farther away from the given pixel site, which is the same idea as that used in kernel smoothing discussed in Chapters 2–5. In addition, weights are smaller for those pixels whose intensity levels are much different from the intensity level of the given pixel. By the second weighting scheme, if the given pixel site $\mathbf{s}$ is close (in Euclidean distance) to pixels on the other side of an edge, these pixels will receive less weights than the pixels located on the same side of the edge as $\mathbf{s}$. In that way, edges are preserved to some degree.

Barash (2002) pointed out that a discrete version of the above bilateral filtering procedure can be written as follows:

$$S^{(t+1)}(x,y) = \frac{1}{N^{(t+1)}(x,y)} \sum_{i=-k}^{k} \sum_{j=-k}^{k} S^{(t)}(x+i, y+j) w^{(t)}(x+i, y+j), \quad (7.22)$$

where $k > 0$ is the window size of the filter,

$$w^{(t)}(x+i, y+j) = \exp\left(-\frac{i^2 + j^2}{2\sigma_d^2}\right) \exp\left(-\frac{(S^{(t)}(x+i, y+j) - S^{(t)}(x,y))^2}{2\sigma_r^2}\right),$$

and

$$N^{(t+1)}(x,y) = \sum_{i=-k}^{k} \sum_{j=-k}^{k} w^{(t)}(x+i, y+j).$$

From the above equations, the bilateral filtering procedure (7.22) is clearly a generalization of the adaptive smoothing procedure (7.20).

7.4.3 Image restoration by surface estimation

In some observed images, extremely large or small image intensities can be observed occasionally; such observations are called *outliers* in statistics (cf. Section 2.2 for a related discussion).

Sinha and Schunck (1992) suggested a two-stage algorithm for edge-preserving image restoration that is robust to outliers. The first stage of the algorithm consists of a *moving local least median squares* (MLMS) approximation. After this stage, an approximation to the true image intensity function is available for the second stage of the algorithm, which eliminates possible outliers, preserves edges, and removes some noise. The second stage further removes noise by a *weighted bicubic spline* (WBS) procedure, in which the edges are preserved by using an appropriate weighting function. These two stages are introduced below in some details.

At the first stage, for a given point (x, y), let $N(x, y)$ be its neighborhood (e.g., 3×3 or other types of neighborhood), in which a linear function

$$g(x, y) = a_0 + a_1 x + a_2 y$$

is used for approximating the image intensity function f. This is done by the following MLMS procedure:

$$\min_{\mathbf{a}} \left\{ \text{median}_{(x_i, y_i) \in N(x,y)} \left(Z_i - g(x_i, y_i) \right)^2 \right\}, \tag{7.23}$$

where $\mathbf{a} = (a_0, a_1, a_2)'$, (x_i, y_i) is the ith pixel in the neighborhood $N(x, y)$, and Z_i is the observed image intensity at (x_i, y_i). Obviously, procedure (7.23) is the same as the conventional local least squares procedure (cf. Subsection 2.2.4), except that in the latter procedure summation is used instead of $\text{median}_{(x_i, y_i) \in N(x,y)}$. By using the median operator in expression (7.23), possible outliers are eliminated, and most edges are preserved; at the same time, noise is removed to a certain degree. Obviously, higher-order polynomials or other parametric functions can also be used in expression (7.23), as long as the number of parameters is less than the number of observations in the neighborhood $N(x, y)$. On the other hand, if the zero-th-order polynomial (i.e., a constant) is used in expression (7.23), then the resulting procedure is equivalent to the local median filtering procedure discussed in Subsection 7.4.1.

Because it is difficult to provide an explicit expression for the solution of the minimization problem (7.23), Sinha and Schunck (1992) instead provided a numerical algorithm for finding such a solution. Suppose that the neighborhood $N(x, y)$ includes M pixels and $3 \leq n^* \ll M$ is an integer. Then we can randomly select n^* pixels from $N(x, y)$ and fit a plane to these pixels. The fitted plane is denoted by f^*, and its median residual squares in $N(x, y)$:

$$\text{median}_{(x_i, y_i) \in N(x,y)} \left(Z_i - f^*(x_i, y_i) \right)^2$$

is then computed. Next, we select another set of n^* pixels from $N(x, y)$, fit a plane to these pixels, and compute its median residual squares in $N(x, y)$. This process continues until all possible sets of n^* pixels from $N(x, y)$ have been selected. Among all possible sets of n^* pixels in $N(x, y)$, the one giving the smallest median residual squares in $N(x, y)$ is then selected, and the corresponding fitted plane is used as an estimator of the solution of (7.23). Sinha and Schunck mentioned that operations of $\binom{M}{n^*}$ samples are parallelizable, and in practice only a limited number of samples need to be selected because adjacent windows share data and could share results as well.

In the second stage, Sinha and Schunck suggested using the WBS procedure for restoring the true image intensity function as follows. For simplicity, Z_i values are still used to denote the new data, which are generated from the approximated image intensity function obtained by the MLMS procedure, at the original pixel locations. Assume that the true image intensity function f can be expressed as a linear combination of the tensor products of the 1-D cubic B-splines:

$$f(x,y) = \sum_{i=1}^{g+3}\sum_{j=1}^{h+3} c_{ij} B_i^3(x) B_j^3(y),$$

where $B_i^3(x)$ and $B_j^3(y)$ are 1-D cubic B-splines (cf. expression (2.24) in Subsection 2.5.4 for their definitions), and g and h are the numbers of related knots in the x and y directions, respectively. The coefficients c_{ij} can be estimated by minimizing the following objective function:

$$\sum_{i=1}^{N}(Z_i - f(x_i,y_i))^2 + \int\int_\Omega w(x,y)\left(f_{xx}^2(x,y) + 2f_{xy}^2(x,y) + f_{yy}^2(x,y)\right)\, dxdy, \tag{7.24}$$

where $w(x,y)$ is a weighting function and Ω is the design space of the image. The weighting function is chosen such that it is large when the data are flat and small when the data are steep. This allows the energy function $f_{xx}^2(x,y)+2f_{xy}^2(x,y)+f_{yy}^2(x,y)$ to be small in flat regions, in which case the function f is smooth, and large in steep regions, so that discontinuities are allowed. Sinha and Schunck suggested choosing the weighting function to be

$$w(x,y) = \frac{\delta}{(1 + f_x^2(x,y) + f_y^2(x,y))^P},$$

where $\delta > 0$ is a scale parameter, and P is either 1 or 2.

7.5 IMAGE RESTORATION BY NONLINEAR DIFFUSION FILTERING

Edge-preserving image restoration is formulated as a nonlinear diffusion problem by Perona and Malik (1990). Some basic terminologies and ideas of this methodology are introduced in Subsection 7.5.1. Its relationship with bilateral filtering and adaptive smoothing is discussed in Subsection 7.5.2. Some modifications and generalizations are discussed in Subsection 7.5.3.

7.5.1 Diffusion filtering

In the computer science literature, a popular denoising technique involves local smoothing with the Gaussian operator. Let $G(x,y;t)$ be a 2-D Gaussian pdf with scale parameter t (cf. its definition in Subsection 5.2.2), and let $I_0(x,y)$ be an observed image. Then the smoothed image by the Gaussian operator is defined by

$$I(x,y,t) = I_0(x,y) \otimes G(x,y;t), \tag{7.25}$$

where $\otimes$ denotes the convolution operator. When the scale parameter t is larger, the smoothed image $I(x, y, t)$ has coarser resolution. More noise is removed in such cases, but many details are lost at the same time. This type of filtering is called *scale-space* filtering in the literature, since the degree of smoothing is controlled by a scale parameter. For more discussions about scale-space filtering, read Babaud *et al.* (1986), Witkin (1983), Yuille and Poggio (1986), and the references cited there.

Hummel (1987) and Koenderink (1984) pointed out that the smoothed image $I(x, y, t)$ defined in equation (7.25) is equivalent to the solution of the following partial differential equation:

$$\begin{cases} \frac{\partial I(x,y,t)}{\partial t} = \nabla^2 I(x, y, t) \\ \text{subject to: } I(x, y, 0) = I_0(x, y), \end{cases} \tag{7.26}$$

where $\nabla^2 = \frac{\partial^2}{\partial^2 x} + \frac{\partial^2}{\partial^2 y}$ is the Laplacian operator. Equation (7.26) can also be expressed as

$$\begin{cases} \frac{\partial I(x,y,t)}{\partial t} = \text{div}\,(c\, G_I(x, y, t)) \\ \text{subject to: } I(x, y, 0) = I_0(x, y), \end{cases} \tag{7.27}$$

where "div" is the *divergence* operator (with respect to the space variables x and y) defined by

$$\text{div}\,((f_1(x, y), f_2(x, y))') = \frac{\partial f_1(x, y)}{\partial x} + \frac{\partial f_2(x, y)}{\partial y},$$

$G_I(x, y, t) = (\partial I(x, y, t)/\partial x, \partial I(x, y, t)/\partial y)'$ is the gradient of $I(x, y, t)$, and $c = 1$.

Equation (7.27) is often used in physics for describing diffusion processes that equilibrate concentration differences without creating or destroying mass. For this reason, it is well-known as a *diffusion equation*. In the context of heat transfer, it is also called the *heat equation*. The constant c is called the *diffusion conductance* or *diffusivity*. Because c is independent of x, y, and t, the diffusion equation is *linear* in the sense that

$$\begin{aligned} &\frac{\partial\,(a_1 I_1(x, y, t) + a_2 I_2(x, y, t))}{\partial t} \\ = \;& a_1 \text{div}\,(c\, G_{I_1}(x, y, t)) + a_2 \text{div}\,(c\, G_{I_2}(x, y, t)), \end{aligned}$$

where I_1 and I_2 are two arbitrary diffusion processes. Diffusion equation (7.27) is also *homogeneous* in the sense that it has homogeneous diffusivity (i.e., the same value of c for all x, y, and t). In such cases, all locations in the image, including the edge pixels, are smoothed equally in the diffusion process. Alternatively, if c depends on x, y, and t, which is denoted as $c(x, y, t)$ for clarity in such cases, then the diffusion equation becomes *nonlinear* and *nonhomogeneous*. Perona and Malik (1990) and some other authors also called such diffusion equations *anisotropic*, and they called the equations with constant diffusivities *isotropic*.

To preserve edges during smoothing, or to smooth *within* regions separated by edges instead of *between* such regions, the diffusivity $c(x, y, t)$ should ideally equal 0 at the edge positions and equal 1 within separated regions. In applications, however, the edge positions are unknown, so they should be estimated from the observed image intensities. From discussions in Section 6.2, we know that the image gradient magnitude $|G_I(x, y, t)|$ provides us information about the edge locations. If its value is large, then the chance that the position (x, y) is on or close to an edge segment is also large, and vice versa. For this reason, we can define

$$c(x, y, t) = g\left(|G_I(x, y, t)|\right), \tag{7.28}$$

where g is a decreasing function with $g(0) = 1$. In such cases with a nonlinear diffusion equation, much smoothing is performed at places where $|G_I(x, y, t)|$ is small (such places are most probably within separated regions), and little smoothing is performed at places where $|G_I(x, y, t)|$ is large (such places are most probably close to edges). Therefore, the diffusion process smoothes the observed image with most edges preserved. As a matter of fact, edges can even be enhanced in the process of diffusion if the function g is chosen properly (Perona and Malik 1990, Section IV-B). Two commonly used g functions suggested by Perona and Malik (1990) are as follows:

$$g_1(s) = \exp\left[-\left(\frac{s}{K}\right)^2\right]$$

and

$$g_2(s) = \frac{1}{1 + \left(\frac{s}{K}\right)^{1+\alpha}},$$

where $K, \alpha > 0$ are two parameters. Obviously, $g_1(s)$ decreases to 0 with a faster rate than does $g_2(s)$ as s tends to infinity.

In applications, the nonlinear diffusion equation (i.e., expression (7.27) with c replaced by $c(x, y, t)$) can be replaced by one of its discrete versions, used in an iterative manner. The $(t + 1)$-th iteration of the discrete version used by Perona and Malik (1990) is defined as follows:

$$\begin{aligned} I_{i,j}^{t+1} &= I_{i,j}^t + \\ &\lambda\left[c_{N_{i,j}}^t \triangle_N I_{i,j}^t + c_{S_{i,j}}^t \triangle_S I_{i,j}^t + c_{E_{i,j}}^t \triangle_E I_{i,j}^t + c_{W_{i,j}}^t \triangle_W I_{i,j}^t\right], \end{aligned}$$

where $\lambda \in [0, 1/4]$ is a parameter and

$$\begin{aligned} &\triangle_N I_{i,j}^t = I_{i-1,j}^t - I_{i,j}^t, \quad \triangle_S I_{i,j}^t = I_{i+1,j}^t - I_{i,j}^t, \\ &\triangle_E I_{i,j}^t = I_{i,j+1}^t - I_{i,j}^t, \quad \triangle_W I_{i,j}^t = I_{i,j-1}^t - I_{i,j}^t, \\ &c_{N_{i,j}}^t = g\left(|\triangle_N I_{i,j}^t|\right), \quad c_{S_{i,j}}^t = g\left(|\triangle_S I_{i,j}^t|\right), \\ &c_{E_{i,j}}^t = g\left(|\triangle_E I_{i,j}^t|\right), \quad c_{W_{i,j}}^t = g\left(|\triangle_W I_{i,j}^t|\right). \end{aligned}$$

In each iteration, the pixels can be updated in the raster scan fashion (i.e., from left to right and from top to bottom in the image).

7.5.2 Relationship with adaptive smoothing and bilateral filtering

Barash (2002) pointed out that both the adaptive smoothing and the bilateral filtering procedures, which we discussed in Subsection 7.4.2, can be regarded as special cases of nonlinear diffusion filtering, as explained below.

For simplicity, let us consider 1-D adaptive smoothing (cf. equation (7.20) for a 2-D version), which can be expressed by

$$I^{t+1}(x) = c_1 I^t(x-1) + c_2 I^t(x) + c_3 I^t(x+1), \tag{7.29}$$

where c_1, c_2, and c_3 are three weights satisfying

$$c_1 + c_2 + c_3 = 1.$$

Clearly, equation (7.29) can be rewritten as

$$I^{t+1}(x) - I^t(x) = c_1 \left[I^t(x-1) - I^t(x)\right] + c_3 \left[I^t(x+1) - I^t(x)\right].$$

If we let $c_1 = c_3 = c$, then the above expression becomes

$$I^{t+1}(x) - I^t(x) = c\left[I^t(x-1) - 2I^t(x) + I^t(x+1)\right],$$

which is a discrete version of the 1-D linear diffusion equation

$$\frac{\partial I(x,t)}{\partial t} = c\,\frac{\partial^2}{\partial x^2} I(x,t).$$

Therefore, the adaptive smoothing scheme can be regarded as a special case of linear diffusion if the weights used are space independent.

If the weights used in adaptive smoothing are space dependent, then the 1-D adaptive smoothing scheme can be written as

$$\begin{aligned} I^{t+1}(x) \quad = \quad & \frac{c^t(x-1) + c^t(x)}{2} I^t(x-1) + c^t(x) I^t(x) + \\ & \frac{c^t(x+1) + c^t(x)}{2} I^t(x+1) \end{aligned} \tag{7.30}$$

with

$$\frac{c^t(x-1) + c^t(x)}{2} + c^t(x) + \frac{c^t(x+1) + c^t(x)}{2} = 1.$$

Expression (7.30) can be written as

$$\begin{aligned} I^{t+1}(x) - I^t(x) \quad = \quad & \frac{c^t(x+1) + c^t(x)}{2} \left[I^t(x+1) - I^t(x)\right] - \\ & \frac{c^t(x-1) + c^t(x)}{2} \left[I^t(x) - I^t(x-1)\right], \end{aligned}$$

which can be regarded as a discrete version of the 1-D nonlinear diffusion equation

$$\frac{\partial I(x,t)}{\partial t} = \frac{\partial}{\partial x}\left(c(x)\frac{\partial I(x,t)}{\partial x}\right).$$

Therefore, the adaptive smoothing scheme can be regarded as a spacial case of nonlinear diffusion if the weights used are space dependent.

As pointed out at the end of Subsection 7.4.2, bilateral smoothing scheme (7.21) can be regarded as a generalized adaptive smoothing scheme (cf. equation (7.22)). Therefore, in conjunction with the above explanation, bilateral smoothing can also be regarded as a special case of diffusion filtering.

7.5.3 Some generalizations and modifications

Since Perona and Malik (1990) first suggested using nonlinear diffusion filtering for edge-preserving image restoration, there have been several generalizations and modifications proposed in the literature. Some of them are described below.

One type of modifications focuses mainly on the function g involved in the diffusivity $c(x, y, t)$ (cf. equation (7.28)); several such possible choices for g are as follows. Charbonnier *et al.* (1994) suggested the following g function:

$$g_3(s) = \frac{1}{\sqrt{1 + \left(\frac{s}{K}\right)^2}},$$

which is similar to Perona and Malik's second g function g_2 when α is small. Weickert (1998) suggested the following g function:

$$g_4(s) = \begin{cases} 1, & \text{if } s = 0 \\ 1 - \exp\left(-\frac{C_m}{(s/K)^{2m}}\right), & \text{if } s > 0, \end{cases}$$

where m is a positive integer and C_m satisfies

$$(1 + 2mC_m)\exp(-C_m) = 1.$$

When $m = 2, 3$, and 4, it can be checked that C_m equals 2.337, 2.918, and 3.315, respectively. Other g functions include Andreu *et al.*'s (2001) function

$$g_5(s) = \frac{1}{s},$$

and Keeling and Stollberger's (2002) function

$$g_6(s) = \frac{1}{s^2}.$$

For $K = 1, \alpha = 1$, and $m = 3$, the six g functions g_1–g_6 are presented in Figure 7.12. It can be seen that the functions g_1–g_4 are all bounded above by 1. Their major difference lies in how quickly they each converge to 0 when s tends to ∞. The two functions g_5 and g_6 do not have upper bounds, and their values could be very large when s is small, which imply that the degree of smoothing in flat regions of the image could be very large compared to the degree of smoothing around edges.

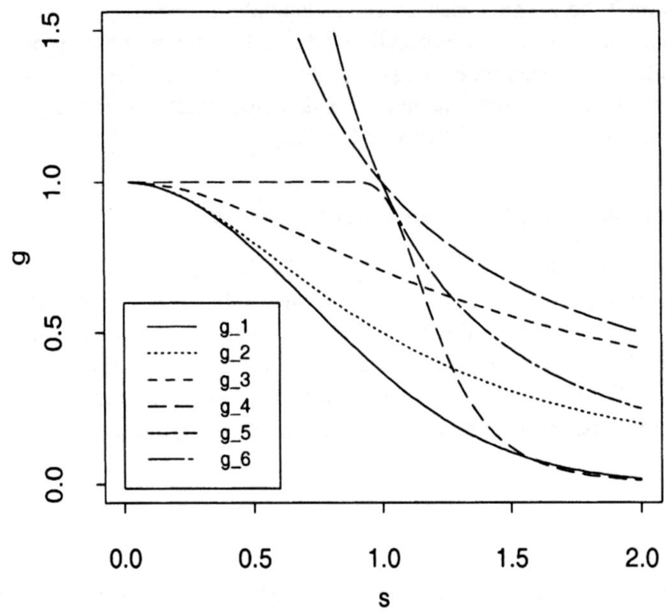

Fig. 7.12 The six g functions g_1–g_6 discussed in Section 7.5.

Catté *et al.* (1992) modified the nonlinear diffusion equation described in the previous subsection as follows:

$$\begin{cases} \frac{\partial I(x,y,t)}{\partial t} = \text{div}\left(g(|G_{I_\sigma}(x,y,t)|)\, G_I(x,y,t)\right) \\ \text{subject to: } I(x,y,0) = I_0(x,y) \text{ and } \frac{\partial I(x,y,t)}{\partial \mathbf{n}}|_{\partial\Omega} = 0, \end{cases} \tag{7.31}$$

where $I_\sigma(x,y,t)$ denotes the convolution between $I(x,y,t)$ and the 2-D Gaussian pdf with scale parameter σ, and $\frac{\partial I(x,y,t)}{\partial \mathbf{n}}|_{\partial\Omega}$ denotes the directional partial derivative of $I(x,y,t)$ along the normal direction $\mathbf{n}$ at the border $\partial\Omega$ of the image domain Ω. In a comparison to the nonlinear diffusion equation described in the previous subsection, the major difference in expression (7.31) is that the diffusivity $g(|G_{I_\sigma}(x,y,t)|)$ depends on $I_\sigma(x,y,t)$ instead of $I(x,y,t)$. Namely, the degree of smoothing in the diffusion process is determined by the Gaussian-smoothed image $I_\sigma(x,y,t)$, which should be more robust to noise, especially around the edges. Weickert *et al.* (1998) modified this procedure further to make it more reliable and efficient for image restoration.

Several authors, including Black *et al.* (1998) and You *et al.* (1996), found that nonlinear diffusion filtering is a kind of robust smoothing. The relationship between nonlinear diffusion filtering and wavelet shrinkage (cf. Section 2.5) has been studied by Mrázek *et al.* (2003). An overview of other image restoration methods based on nonlinear diffusion filtering can be found in Romeny (1994) and Weickert (1998).

7.6 SOME DISCUSSIONS

In this chapter, we have discussed some techniques for restoring true images from their degraded versions. Two types of degradations are considered: point degradation (i.e., noise) and spatial degradation. Other types of degradations involving chromatic or temporal effects are possible in applications, but they have little relation with the monochrome digital images considered in this book; thus, they are omitted.

Image restoration based on Fourier transformations is discussed in Section 7.2. This method results mainly from the following two properties of Fourier transformations: (i) the Fourier transformation of a convolution of two functions equals the product of the Fourier transformations of the two individual functions, and (ii) the Fourier transformation of a linear combination of two functions equals the linear combination of the Fourier transformations of the two individual functions. Therefore, if the observed image follows degradation model (7.1)–(7.2), that is, if it involves position invariant spatial degradation and additive noise, then the Fourier transformation of the true image can be estimated by using the above two properties of the Fourier transformation, after the Fourier transformations of the PSF h and the noise are both estimated. However, since both the PSF h and noise are unobservable in applications, most image restoration methods in this area deal mainly with estimation of h or its Fourier transformation, after finding a way to get rid of the effect of noise. Several approaches are introduced for estimating h or its Fourier transformation in some special cases, although almost all of them are *ad hoc* in nature. It is still an open and challenging problem to estimate h properly in a general situation.

On the other hand, all image restoration procedures discussed in Sections 7.3–7.5 assume that spatial degradation is either absent or known in its mechanism (i.e., h is known). Thus, essentially none of them deal with spatial degradation.

Image restoration by MRF modeling is popular in the computer science literature. Some representative procedures in this direction are discussed in Section 7.3. Geman and Geman (1984) formulated this approach in a systematic and mathematically rigorous way. By this approach, the true image is assumed to be a MRF or, equivalently, to have a Gibbs distribution. The conditional distribution of the observed image given the true image is assumed known, and, in most cases, this conditional distribution is assumed to be a Gaussian distribution. Then the true image is estimated by the MAP procedure, which searches for a maximizer of the posterior distribution of the true image. Edges are preserved by using a line process in MRF modeling. Although Geman and Geman (1984) applied the Gibbs sampler scheme and an annealing algorithm in the MAP step to make the computing task possible, computation involved in the MAP procedure is still enormous. Thus, many modifications and simplifications of Geman and Geman's procedure have been proposed in the literature. Most of them try to simplify the computation in the MAP step (e.g., Besag's (1986) ICM procedure), or to propose different correlation structures among neighboring pixels which lead to different energy functions (e.g., Godtliebsen and Sebastiani's (1994) procedure). Image restoration can also be formulated as a regularization problem, and the close relationship between the MRF modeling approach and the regularization approach is described in Subsection 7.3.4.

The image restoration problem discussed in this chapter is similar to the jump-preserving surface estimation problem discussed in Chapter 5, although point degradation is the only concern in the latter problem. Comparing the image restoration methods based on MRF modeling, which are discussed in Section 7.3, to some jump-preserving surface estimation methods based on kernel smoothing, which are described in Sections 5.2–5.4, each of the two groups of methods has its own strengths and limitations.

First, each group has a set of potentially restrictive assumptions. The MRF methods impose a number of assumptions on the model, some of which are difficult to verify in applications. It is still an open problem regarding how to check whether these MRF methods are robust to their model assumptions, such as the one that the true image is a MRF. With regard to jump-preserving surface estimation, some early methods in this area estimate the true surface in two steps: (i) estimation of the jump location curves, and (ii) estimation of the surface in design subspaces separated by the estimated jump location curves. In order to estimate the true jump location curves properly by curves in the design space, these methods (e.g., those discussed in Section 4.2) also impose some restrictive assumptions on their models. For example, some of them assume that there is a single jump location curve in the entire design space and that the curve is smooth with some other good properties, which are often difficult to satisfy in applications. Comparatively, the jump-preserving surface estimation methods described in Sections 5.2–5.4 impose only mild assumptions on their models, so they should be able to find more applications. However, although these methods are flexible, they have not taken into account the possible probabilistic relationship among neighboring pixels yet, as the MRF methods do.

Second, from a computational viewpoint, most MRF methods are complicated to compute because of their iterative nature, especially when the neighborhoods used are large in size and the number of possible gray levels is also large. Comparatively, computation involved in the jump-preserving surface estimation methods described in Sections 5.2–5.4 is simple, due to the local smoothing natures and to the fact that some of their estimated surfaces have explicit formulas.

Third, because of the iterative nature of most MRF methods, it is usually difficult to study their theoretical properties. A common practice to evaluate their performance is by numerical experiments based on visual impression, which may not be explicit enough to allow for further improvement. As a comparison, we can prove that the estimated surfaces by the procedures described in Sections 5.2–5.4 perform well in the entire design space except in some small regions around the singular points of the jump location curves; these estimated surfaces are also statistically consistent under some regularity conditions.

The median filters discussed in Subsection 7.4.1 have some ability to preserve edges while removing noise. They should be robust to outliers as well, although this issue is not discussed in Subsection 7.4.1. Compared to the conventional neighborhood averaging filter (cf. Figure 7.8), their ability to preserve edges is better, but it is easy to check that their ability to remove noise is weaker if the two types of filters use masks of the same size. Furthermore, edges are still blurred to some degree by them, especially when large masks are used (cf. Figure 7.9(d)). In addition, to preserve line edges, masks of some special shapes (e.g., the cross-shaped and X-shaped masks) should be used. Median filters with such special masks can also be regarded as special cases of weighted median filters, which are discussed at the end of Subsection 7.4.1.

The adaptive smoothing and bilateral filtering procedures discussed in Subsection 7.4.2 are closely related to the sigma filter and M smoother discussed in Subsection 3.4.3 and to the adaptive weighting smoothing procedure discussed in Section 5.5. In adaptive smoothing, the restored image intensity at a given pixel is defined by a weighted average of neighboring observed image intensities. The neighboring observations with larger estimated image gradient magnitudes would receive less weights in the weighted average, and thus edges are preserved to some degree while noise is removed. However, by using adaptive smoothing, blurring is still inevitable, although the degree of blurring is greatly reduced by using the weighting function. The major reason for this phenomenon is that pixels located on the other side of the edge segment around the given pixel still receive *some* weight in the weighted average, although these weights are usually small. It is still unclear how this small blurring at each iteration would affect the final restored image, which is obtained after executing many iterations of the adaptive smoothing algorithm.

Bilateral filtering tries to preserve edges by a similar idea. It determines the weights based on the distances between neighboring pixel locations and the location of the given pixel and also on the differences between the observed gray levels of the neighboring pixels and the observed gray level of the given pixel. The blurring problem mentioned above with regard to adaptive smoothing is also true for bilateral filtering if the kernel function applied to observed gray levels is chosen to be a 2-D Gaussian pdf, as in most bilateral filtering procedures.

Image restoration by nonlinear diffusion filtering is discussed in Section 7.5. The diffusion equations are originally used for describing diffusion processes in physics. Motivated by the connection between linear diffusion equations and smoothing by Gaussian operators, Perona and Malik (1990) suggested using nonlinear diffusion filtering for restoring images with edges preserved. By choosing the diffusivity to be a decreasing function of the image gradient magnitude, observed image intensities within regions separated by edges are smoothed during the diffusion process, observed image intensities around edges have little smoothing involved, and consequently edges are mostly preserved. At this time, we do not know much about the theoretical properties of the restored images by this method. For example, it is not known whether the restored image converges to the true image around edges if the resolution of the true image increases. My intuition tells me that the answer might be a "no", because little smoothing is involved in such areas. As another example, since the diffusivity is chosen to be a decreasing function of the image gradient magnitude, the slope of the true image has a certain effect on the diffusion process and consequently affects the restored image; it is not clear whether this kind of effect can be ignored, as is done in most existing nonlinear diffusion filters. We have not seen much discussion about these kinds of theoretical issues yet in the literature.

Problems

7.1 For the 2-D Fourier transformation discussed in Subsection 7.2.2, do the following problems.

7.1.1. Check that it has the three properties (i)-(iii) stated on page 239.

7.1.2. Check that the discrete Fourier and inverse Fourier transformations are both periodic with periods A and B for u (or x) and v (or y), respectively, where A and B are numbers of rows and columns of pixels in a given image.

7.1.3. Check that the magnitude of the discrete Fourier transformation is symmetric about the origin in the (u, v) domain.

7.2 Suppose that an observed image is a blurred version of the true image and this blurring is caused by a shift of the camera in the vertical direction for a small distance during the exposure. Discuss how to restore the true image from the observed image by using the Fourier transformation approach.

7.3 The blurred image shown in Figure 7.3(b) is obtained from the original image shown in Figure 7.3(a) by applying the neighborhood averaging mask of size 9×9 (cf. Figure 7.8). The related PSF h or its Fourier transformation can be estimated from the blurred version of a point source image blurred by the same neighborhood averaging operator, as described in Subsection 7.2.3.

7.3.1. Calculate a restored image using the inverse filtering procedure (7.8). Try several values of w_1 and comment on the impact of w_1 on the restored image.

7.3.2. Calculate a restored image using the algebraic approach discussed in Subsection 7.2.4.

7.4 For the neighborhood system discussed in Subsection 7.3.1, do the following problems.

7.4.1. For the neighborhood system defined in equation (7.9), find all neighbors of a given site s when $k = 4$ and also find all possible cliques in such cases.

7.4.2. In equation (7.9), if the neighborhood of the (i, j)-th pixel is defined by

$$N_{ij} = \{(i', j') : 0 < |i - i'| + |j - j'| \leq k\},$$

find all neighbors of a given site s when $k = 2$ and also find all possible cliques in such cases.

7.5 In Geman and Geman's MAP procedure discussed in Subsection 7.3.2, verify that equation (7.13) holds.

7.6 Generalize the 1-D regularization model (7.19) to 2-D cases. The resulting 2-D model is the so-called weak membrane model, as mentioned in Subsection 7.3.4.

7.7 Reproduce Figures 7.9 and 7.11.

7.8 Decide whether the local median filters with cross-shaped or X-shaped neighborhoods are special cases of weighted median filters discussed at the end of Subsection 7.4.1; justify your answer. Also, design a weighted median filter which can preserve line edges forming an angle $\pi/6$ with the positive direction of the x-axis.

7.9 The original mountain image shown in Figure 7.1(a) can be downloaded from the ftp site of this book (see front cover for address). A noisy version of this image contaminated by i.i.d. noise with $N(0, 50^2)$ distribution is shown in Figure 7.7(a). With this noisy image, do the following.

7.9.1. Apply the neighborhood averaging operator with mask size 21×21 to the noisy mountain image. Comment on the quality of the restored image.

7.9.2. Apply the conventional local median filter with mask size 21×21 to the same noisy image. Comment on any differences between the current restored image and the restored image obtained in the previous part.

7.9.3. Apply the conventional local median filter with mask size 9×9 to the same noisy image. Comment on any differences between the current restored image and the restored image obtained in the previous part.

7.9.4. Try the local median filter with cross-shaped neighborhoods of the following mask sizes: (i) 3×3, (ii) 7×7, (iii) 15×15. Comment on the results.

7.9.5. Try the local median filter with X-shaped neighborhoods of the following mask sizes: (i) 3×3, (ii) 7×7, (iii) 15×15. Comment on the results.

7.9.6. Try the weighted median filter with mask size 5×5 and the following two sets of weights: (i) $W_{0,0} = 5$, $W_{s,t} = 3$ when $\max(|s|, |t|) = 1$, $W_{s,t} = 1$ when $\max(|s|, |t|) = 2$; and (ii) $W_{0,0} = 24$, $W_{s,t} = 1$ when $\max(|s|, |t|) > 0$. Comment on the results.

7.10 With the noisy mountain image mentioned in the previous problem, do the following.

7.10.1. Apply the adaptive smoothing procedure (7.20) to the image, with the weights determined by

$$w^{(t)}(x, y) = \exp\left[-\frac{\left(d^{(t)}(x, y)\right)^2}{2\sigma_D^2}\right],$$

with the image gradient magnitude $d^{(t)}(x, y)$ determined by the Sobel masks (cf. Figure 4.15), and with σ_D=1 or 5. Comment on the results.

7.10.2. Apply the bilateral filtering procedure (7.22) to the noisy mountain image with (i) $\sigma_d = 1$ and $\sigma_r = 1$, (ii) $\sigma_d = 1$ and $\sigma_r = 10$, (iii) $\sigma_d = 10$ and $\sigma_r = 1$, and (iv) $\sigma_d = 10$ and $\sigma_r = 10$. Comment on these results and the differences between these results and those obtained in the previous part.

7.11 Apply the discrete version of the nonlinear diffusion filter suggested by Perona and Malik (1990) (cf. equations given at the end of Subsection 7.5.1) to the noisy mountain image considered in the previous two problems. In the filter, set $\lambda = 1/4$, $g = g_1$, and K equal to the 0.9 quantile of all absolute values of image gradient magnitudes, updated at every iteration.

References

1. Abdou, I. (1973), "Quantitative methods of edge detection," *USCIPI Report 830*, Image Processing Institute, University of Southern California.

2. Abdou, I., and Pratt, W.K. (1979), "Quantitative design and evaluation of enhancement/thresholding edge detectors," *Proceedings of IEEE*, **67**, 753–763.

3. Allen, D. (1974), "The relationship between variable selection and data augmentation and a method for prediction," *Technometrics*, **16**, 125–127.

4. Aly, E.A., Abd-Rabou, A.S., and Al-Kandari, N.M. (2003), "Tests for multiple change points under ordered alternatives," *Metrika*, **57**, 209–221.

5. Andreu, F., Ballester, C., Caselles, V., and Mazón, J.M. (2001), "Minimizing total variation flow," *Differential and Integral Equations*, **14**, 321–360.

6. Andrews, H.C., and Hunt, B.R. (1977), *Digital Image Restoration,* Englewood Cliffs, NJ: Prentice-Hall, Inc.

7. Anselone, P., and Laurent, P. (1968), "A general method for the construction of interpolating or smoothing spline functions," *Numerical Mathematics*, **12**, 66–82.

8. Antoniadis, A., and Gijbels, I. (2002), "Detecting abrupt changes by wavelet methods," *Journal of Nonparametric Statistics*, **14**, 7–29.

9. Antoniadis, A., Gijbels, I., and Macgibbon, B. (2000), "Non-parametric estimation for the location of a change-point in an otherwise smooth hazard function under random censoring," *Scandinavian Journal of Statistics*, **27**, 501–519.

10. Arce, G.R. (1991), "Multistage order statistic filters for image sequence," *IEEE Transactions on Signal Processing*, **39**, 1146–1163.

11. Argyle, E. (1971), "Techniques for edge detection," *Proceedings of IEEE*, **59**, 285–287.

12. Ash, R.B. (1972), *Real Analysis and Probability*, New York: Academic Press.

13. Babaud, J., Witkin, A., Baudin, M., and Duda, R. (1986), "Uniqueness of the Gaussian kernel for scale-space filtering," *IEEE Transactions on Pattern Analysis and Machine Intelligence*, **8**, 26–33.

14. Bachman, G., Narici, L., and Beckenstein, E. (2000), *Fourier and Wavelet Analysis*, New York: Springer-Verlag.

15. Balakrishnan, N., and Cohen, A.C. (1991), *Order Statistics and Inference: Estimation Methods*, Boston: Academic Press.

16. Barash, D. (2002), "A fundamental relationship between bilateral filtering, adaptive smoothing, and the nonlinear diffusion equation," *IEEE Transactions on Pattern Analysis and Machine Intelligence*, **24**, 844–847.

17. Bates, R.H.T., and McDonnell, M.J. (1986), *Image Restoration and Reconstruction,* Oxford: Clarendon Press.

18. Bedini, L., Gerace, I., and Tonazzini, A. (1994), "A GNC algorithm for constrained image reconstruction with continuous-valued line processes," *Pattern Recognition Letters*, **15**, 907–918.

19. Bedini, L., Gerace, I., and Tonazzini, A. (1995), "Sigmoidal approximations for self-interacting line processes in edge-preserving image restoration," *Pattern Recognition Letters*, **16**, 1011–1022.

20. Ben-Israel, A., and Greville, T.N.E. (1977), *Generalized Inverses: Theory and Applications*, New York: John Wiley & Sons.

21. Besag, J. (1974), "Spatial interaction and the statistical analysis of lattice systems (with discussions)," *Journal of the Royal Statistical Society (Series B)*, **36**, 192–236.

22. Besag, J. (1986), "On the statistical analysis of dirty pictures (with discussion)," *Journal of the Royal Statistical Society (Series B)*, **48**, 259–302.

23. Bhandarkar, S.M., Zhang, Y., and Potter, W.D. (1994), "An edge detection technique using genetic algorithm-based optimization," *Pattern Recognition*, **27** 1159–1180.

24. Black, M.J., Sapiro, G., Marimont, D.H., and Heeger, D. (1998), "Robust anisotropic diffusion," *IEEE Transactions on Image Processing*, **7**, 421–432.

25. Blake, A., and Zisserman, A. (1987), *Visual Reconstruction*, Cambridge, MA: MIT Press.

26. Bouman, C., and Sauer, K. (1993), "A generalized Gaussian image model for edge preserving MAP estimation," *IEEE Transactions on Image Processing*, **2**, 296–310.

27. Bovik, A.C., Huang, T.S., and Munson, D.C. (1987), "The effect of median filtering on edge estimation and detection," *IEEE Transactions on Pattern Analysis and Machine Intelligence*, **9**, 181–194.

28. Breiman, L. (1993), "Fitting additive models to regression data: diagnostics and alternative views," *Computational Statistics and Data Analysis*, **15**, 13–46.

29. Brown, R.L., Durbin, J., and Evans, J.M. (1975), "Techniques for testing the constancy of regression relationships over time (with discussion)," *Journal of the Royal Statistical Society (Series B)*, **37**, 149–192.

30. Brownrigg, D.R.K. (1984), "The weighted median filtering," *Communications of the ACM*, **27**, 807–818.

31. Buja, A., Hastie, T., and Tibshirani, R. (1989), "Linear smoothers and additive models," *The Annals of Statistics*, **17**, 453–555.

32. Burt, D.A., and Coakley, C.W. (2000), "Automatic bandwidth selection for modified M-smoother," *Journal of Statistical Computation and Simulation*, **66**, 295–319.

33. Canny, J. (1986), "A computational approach to edge detection," *IEEE Transactions on Pattern Analysis and Machine Intelligence*, **8**, 679–698.

34. Carlstein, E., and Krishnamoorthy, C. (1992), "Boundary estimation," *Journal of the American Statistical Association*, **87**, 430–438.

35. Casella, G., and Berger, R.L. (1990), *Statistical Inference*, Belmont, CA: Wadsworth.

36. Catté, F., Lions, P.L., Morel, J.M., and Coll, T. (1992), "Image selective smoothing and edge detection by nonlinear diffusion," *SIAM Journal on Numerical Analysis*, **29**, 182–193.

37. Chang, G.S., Yu, B., and Vetterli, M. (2000), "Spatially adaptive wavelet thresholding with context modeling for image denoising," *IEEE Transactions on Image Processing*, **9**, 1522–1531.

38. Charbonnier, P., Blanc-Féraud, L., Aubert, G., and Barlaud, M. (1994), "Two deterministic half-quadratic regularization algorithms for computed imaging," *Proceedings of IEEE International Conference on Image Processing*, **2**, 168–172.

39. Chellapa, R., and Jain, A. (eds.) (1993), *Markov Random Fields: Theory and Applications*, San Diego, CA: Academic Press.

40. Chen, J.S., Huertas, A., and Medioni, G. (1987), "Fast convolution with Laplacian-of-Gaussian masks," *IEEE Transactions on Pattern Analysis and Machine Intelligence*, **9**, 584–590.

41. Chen, M.H., Lee, D., and Pavlidis, T. (1991), "Residual analysis for feature detection," *IEEE Transactions on Pattern Analysis and Machine Intelligence*, **13**, 30–40.

42. Cheney, W., and Kincaid, D. (1994), *Numerical Mathematics and Computing (3rd ed.)*, Pacific Grove, CA: Brooks/Cole Publishing Company.

43. Cheng, K.F., and Lin, P.E. (1981), "Nonparametric estimation of a regression function," *Z. Wahrscheinlichkeitstheorie verw. Gebiete*, **57**, 223–233.

44. Cho, K., Meer, P., and Cabrera, J. (1997), "Performance assessment through bootstrap," *IEEE Transactions on Pattern Analysis and Machine Intelligence*, **19**, 1185–1198.

45. Chu, C.K., Glad, I.K., Godtliebsen, F., and Marron, J.S. (1998), "Edge-preserving smoothers for image processing (with discussion)", *Journal of the American Statistical Association*, **93**, 526–556.

46. Chui, C.K. (1992), *An Introduction to Wavelets*, New York: Academic Press.

47. Chung, K.L. (1974), *A Course In Probability Theory*, New York: Academic Press.

48. Cochran, W.G. (1977), *Sampling Techniques: Probability and Mathematical Statistics*, New York: John Wiley & Sons.

49. Cook, R.D., and Weisberg, S. (1999), *Applied Regression Including Computing and Graphics*, New York: John Wiley & Sons.

50. Cox, D. (1983), "Asymptotics for M-type smoothing spline," *The Annals of Statistics*, **11**, 530–550.

51. Craven, P., and Wahba, G. (1979), "Smoothing noisy data with spline functions," *Numerical Mathematics*, **31**, 377–403.

52. Daubechies, I. (1992), *Ten Lectures on Wavelets*, NSF-CBMS Regional Conference Series in Applied Mathematics, **61**, SIAM Publications, Philadelphia.

53. Davis, L.S. (1975), "A survey of edge detection techniques," *Computer Graphics and Image Processing*, **4**, 248–270.

54. Debnath, L. (2002), *Wavelet Transformations & Their Applications*, Berlin: Birkhäuser.

55. de Boor, C. (1978), *A Practical Guide to Splines*, New York: Springer-Verlag.

56. Demigny, D., and Kamlé, T. (1997), "A discrete expression of Canny's criteria for step edge detector performances evaluation," *IEEE Transactions on Pattern Analysis and Machine Intelligence*, **19**, 1199–1211.

57. Demmler, A., and Reinsch, C. (1975), "Oscillation matrices with spline smoothing," *Numerical Mathematics*, **24**, 375–385.

58. Diestel, R. (2000), *Graph Theory (2nd ed.)*, New York: Springer-Verlag.

59. Donoho, D.L., and Johnstone, I.M. (1994), "Ideal spatial adaptation by wavelet shrinkage," *Biometrika*, **81**, 425–455.

60. Duchon, J. (1977), "Splines minimizing rotation-invariant semi-norms in Sobolev spaces," in *Constructive Theory of Functions of Several Variables*, Berlin: Springer-Verlag, 85–100.

61. Duda, R.O., and Hart, P.E. (1972), "Use of the Hough transformation to detect lines and curves in pictures," *Communications of the ACM*, **15**, 11–15.

62. Duda, R.O., and Hart, P.E. (1973), *Pattern Classification and Scene Analysis*, New York: John Wiley & Sons.

63. Elder, J.H., and Zucker, S.W. (1998), "Local scale control for edge detection and blur estimation," *IEEE Transactions on Pattern Analysis and Machine Intelligence*, **20**, 699–716.

64. Eubank, R.L. (1988), *Spline Smoothing and Nonparametric Regression*, New York: Marcel Dekker, Inc.

65. Eubank, R.L., and Speckman, P. (1991), "A bias reduction theorem with applications in nonparametric regression," *Scandinavian Journal of Statistics*, **18**, 211–222.

66. Eubank, R.L., and Speckman, P. (1994), "Nonparametric estimation of functions with jump discontinuities," in *Change-Point Problems* (edited by Carlstein, E., Müller, H.G., and Siegmund, D.), IMS Lecture Notes, **23**, 130–144.

67. Fan, J., and Gijbels, I. (1996), *Local Polynomial Modelling and Its Applications*, London: Chapman & Hall.

68. Ferrari, P.A., Frigessi, A., and Goncaga de Sá, P. (1995), "Fast approximation maximum *a posterior* restoration of multicolour images," *Journal of the Royal Statistical Society (Series B)*, **57**, 485–500.

69. Fessler, J.A., Erdogan, H., and Wu, W.B. (2000), "Exact distribution of edge-preserving MAP estimators for linear signal models with Gaussian measurement noise," *IEEE Transactions on Image Processing*, **9**, 1049–1055.

70. Frei, W., and Chen, C.C. (1977), "Fast boundary detection," *IEEE Transactions on Computers*, **26**, 988–998.

71. Friedman, J.H. (1991), "Multivariate adaptive regression splines (with discussion)," *The Annals of Statistics*, **19**, 1–141.

72. Friedman, J.H., and Silverman, B.W. (1989), "Flexible parsimonious smoothing and additive modeling (with discussion)," *Technometrics*, **31**, 3–39.

73. Fu, Y.X., and Curnow, R.N. (1990), "Maximum likelihood estimation of multiple change points," *Biometrika*, **77**, 563–573.

74. Fukunaka, K. (1972), *Introduction to Statistical Pattern Recognition*, New York: Academic Press.

75. Gallagher, N.C., Jr., and Wise, G.L. (1981), "A theoretical analysis of the properties of median filtering," *IEEE Transactions on Acoustics, Speech, and Signal Processing*, **29**, 1136–1141.

76. Gan, F.F. (1993), "An optimal design of CUSUM control charts for binomial counts," *Journal of Applied Statistics*, **20**, 445–460.

77. Gan, F.F. (1994), "Design of optimal exponential CUSUM control charts," *Journal of Quality Technology*, **26**, 109–124.

78. Gan, F.F. (1998), "Designs of one- and two-sided exponential EWMA charts," *Journal of Quality Technology*, **30**, 55–69.

79. Gasser, T., Kneip, A., and Köhler, W. (1991), "A flexible and fast method for automatic smoothing," *Journal of the American Statistical Association*, **86**, 643–652.

80. Gasser, T., and Müller, H.-G. (1979), "Kernel estimation of regression functions," in *Smoothing Techniques for Curve estimation*, (edited by Gasser, T., and Rosenblatt, M.), Berlin: Springer-Verlag, 23–68.

81. Gasser, T., and Müller, H.-G. (1984), "Estimating regression functions and their derivatives by the kernel method," *Scandinavian Journal of Statistics*, **11**, 171–185.

82. Geiger, D., and Girosi, F. (1991), "Parallel and deterministic algorithms for MRFs: surface reconstruction," *IEEE Transactions on Pattern Analysis and Machine Intelligence*, **13**, 401–412.

83. Geman, S., and Geman, D. (1984), "Stochastic relaxation, Gibbs distributions and the Bayesian restoration of images," *IEEE Transactions on Pattern Analysis and Machine Intelligence*, **6**, 721–741.

84. Geman, D., and Reynolds, G. (1992), "Constrained restoration and the recovery of distributions," *IEEE Transactions on Pattern Analysis and Machine Intelligence*, **14**, 367–383.

85. Gijbels, I., and Goderniaux, A.C. (2004), "Bandwidth selection for changepoint estimation in nonparametric regression," *Technometrics*, **46**, 76–86.

86. Gijbels, I., Hall, P., and Kneip, A. (1999), "On the estimation of jump points in smooth curves," *Annals of the Institute of Statistical Mathematics*, **51**, 231–251.

87. Gijbels, I., Hall, P., and Kneip, A. (2004), "Interval and band estimation for curves with jumps," *Journal of Applied Probability*, **41**, 65–79.

88. Glasbey, C., and Horgan, G. (1995), *Image Analysis for the Biological Sciences*, New York: John Wiley & Sons.

89. Göb, R., Del Castillo, E., and Drager, K. (1999), "Run length analysis of Shewhart charts applied to drifting processes under an integrative SPC/EPC model," *Metrika*, **50**, 137–161.

90. Godtliebsen, F., and Sebastiani, G. (1994), "Statistical methods for noisy images with discontinuities," *Journal of Applied Statistics*, **21**, 459–477.

91. Gombay, E. (2003), "Sequential change-point detection and estimation," *Sequential Analysis*, **22**, 203–222.

92. Gonzalez, R.C., and Woods, R.E. (1992), *Digital Image Processing*, New York: Addison-Wesley Publishing Company.

93. Gonzalez, R.C., and Woods, R.E. (2002), *Digital Image Processing (2nd ed.)*, New York: Prentice Hall.

94. Goodman, J.W. (1968), *Introduction to Fourier Optics*, New York: McGraw-Hill Book Co.

95. Green, P. (1990), "Bayesian reconstructions from emission tomography data using a modified EM algorithm," *IEEE Transactions on Medical Imaging*, **9**, 84–93.

96. Greig, D.M., Porteous, B.T., and Seheult, A.H. (1989), "Exact maximum *a posteriori* estimation for binary images," *Journal of the Royal Statistical Society (Series B)*, **51**, 271–279.

97. Greville, T. (ed.) (1968), *Theory and Application of Spline Functions*, Madison, WI: University of Wisconsin Press.

98. Grimson, W.E.L. (1981), *From Images to Surfaces: A Computational Study of the Human Early Visual System*, Cambridge, MA: MIT Press.

99. Gu, C. (2002), *Smoothing Spline ANOVA Models*, New York: Springer-Verlag.

100. Gu, C., and Wahba, G. (1993), "Semiparametric analysis of variance with tensor product thin plate splines," *Journal of the Royal Statistical Society (Series B)*, **55**, 353–368.

101. Haavisto, P., Gabbouj, M., and Neuvo, Y. (1991), "Median based idempotent filters," *Journal of Circuits, Systems, and Computers*, **1**, 125–148.

102. Hajek, B. (1988), "Cooling schedules for optimal annealing," *Mathematics of Operations Research*, **13**, 311–329.

103. Hall, P., Peng, L., and Rau, C. (2001), "Local likelihood tracking of fault lines and boundaries," *Journal of the Royal Statistical Society (Series B)*, **63**, 569–582.

104. Hall, P., and Raimondo, M. (1997), "Approximating a line thrown at random onto a grid," *The Annals of Applied Probability*, **7**, 648–665.

105. Hall, P., and Raimondo, M. (1998), "On global performance of approximations to smooth curves using gridded data," *The Annals of Statistics*, **26**, 2206–2217.

106. Hall, P., and Rau, C. (2000), "Tracking a smooth fault line in a response surface," *The Annals of Statistics*, **28**, 713–733.

107. Hall, P., and Rau, C. (2002), "Likelihood-based confidence bands for fault lines in response surfaces," *Probability Theory and Related Fields*, **124**, 26–49.

108. Hall, P., and Titterington, M. (1992), "Edge-preserving and peak-preserving smoothing," *Technometrics*, **34**, 429–440.

109. Hammersley, J.M., and Clifford, P. (1971), "Markov field on finite graphs and lattices," unpublished manuscript.

110. Hansen, F.R., and Elliot, H. (1982), "Image segmentation using simple Markov field models," *Computer Graphics and Image Processing*, **20**, 101–132.

111. Haralick, R.M. (1984), "Digital step edges from zero crossing of second directional derivatives," *IEEE Transactions on Pattern Analysis and Machine Intelligence*, **6**, 58–68.

112. Härdle, W. (1990), *Smoothing Techniques: With Implementation in S*, New York: Springer-Verlag.

113. Härdle, W., Hall, P., and Marron, J.S. (1992), "Regression smoothing parameters that are not far from their optimal," *Journal of the American Statistical Association*, **87**, 227–233.

114. Hastie, T., and Loader, C. (1993), "Local regression: automatic kernel carpentry (with discussion)," *Statistical Science*, **8**, 120–143.

115. Hastie, T., and Tibshirani, R. (1990), *Generalized Additive Models*, New York: Chapman and Hall.

116. Hawkins, D.M. (1987), "Self-starting cusum charts for location and scale," *The Statistician*, **36**, 299–316.

117. Hawkins, D.M. (2001), "Fitting multiple change-point models to data," *Computational Statistics and Data Analysis*, **37**, 323–341.

118. Hawkins, D.M., and Olwell, D.H. (1998), *Cumulative Sum Charts and Charting for Quality Improvement*, New York: Springer-Verlag.

119. Heath, M., Sarkar, S., Sanocki, T., and Bowyer, K.W. (1997), "A robust visual method for assessing the relative performance of edge detection algorithms," *IEEE Transactions on Pattern Analysis and Machine Intelligence*, **19**, 1338–1359.

120. Heath, M., Sarkar, S., Sanocki, T., and Bowyer, K.W. (1998), "Comparison of edge detectors: a methodology and initial study," *Computer Vision and Image Understanding*, **69**, 38–54.

121. Hebert, T., and Leahy, R. (1992), "Statistic-based MAP image restoration from Poisson data using Gibbs priors," *IEEE Transactions on Signal Processing*, **40**, 2290–2303.

122. Heckman, N. (1986), "Spline smoothing in a partly linear model," *Journal of the Royal Statistical Society (Series B)*, **48**, 244–248.

123. Hernández, E., and Weiss, G. (1996), *A First Course on Wavelets*, New York: CRC Press.

124. Hildreth, E.C. (1984), *The Measurement of Visual Motion*, Cambridge, MA: MIT Press.

125. Hinkley, D.V. (1969), "Inference about the intersection in two phase regression," *Biometrika*, **56**, 495–504.

126. Hinkley, D.V. (1970), "Inference about the change-point in a sequence of random variables," *Biometrika*, **57**, 1–17.

127. Hinkley, D.V. (1971), "Inference in two-phase regression," *Journal of the American Statistical Association*, **66**, 736–743.

128. Horn, B.K.P., and Schunck, B.G. (1981), "Determining optical flow," *Artificial Intelligence*, **17**, 185–203.

129. Hough, P.V.C. (1962), "Method and means for recognizing complex patterns," *US Patent 3069654*, December 18, 1962.

130. Huang, D., and Dunsmuir, T.M. (1998), "Computing joint distributions of 2D moving median filters with applications to detection of edges," *IEEE Transactions on Pattern Analysis and Machine Intelligence*, **20**, 340–343.

131. Huang, H.C., and Cressie, N. (2000), "Deterministic/stochastic wavelet decomposition for recovery of signal from noisy data," *Technometrics*, **42**, 262–276.

132. Huang, J.S., and Tseng, D.H. (1988), "Statistical theory of edge detection," *Computer Vision, Graphics and Image Processing*, **43**, 337–346.

133. Huang, T.S. (ed.) (1981), *Two-Dimensional Digital Signal Processing*, New York: Springer-Verlag.

134. Huber, P.J. (1981), *Robust Statistics*, New York: John Wiley & Sons.

135. Huertas, A., and Medioni, G. (1986), "Detection of intensity changes using Laplacian-Gaussian masks," *IEEE Transactions on Pattern Analysis and Machine Intelligence*, **8**, 651–664.

136. Hummel, A. (1987), "Representations based on zero-crossings in scale-space," in *Proceedings of IEEE Computer Vision and Pattern Recognition Conference*, 204–209.

137. Hunter, J.S. (1989), "A one-point plot equivalent to the Shewhart chart with Western Electric rules," *Quality Engineering*, **2**, 13–19.

138. Hunter, J.S. (2002), "Shewhart charts and serially recorded data," *Quality Engineering*, **35**, 64–66.

139. Illingworth, J., and Kittler, J. (1988), "A survey of the Hough transform," *Computer Vision, Graphics, and Image Processing*, **44**, 87–116.

140. Johnson, R.A., and Bagshaw, M. (1974), "The effect of serial correlation on the performance of CUSUM tests," *Technometrics*, **16**, 103–112.

141. Jones, L.A. (2002), "The statistical design of EWMA control charts with estimated parameters," *Journal of Quality Technology*, **34**, 277–288.

142. Kang, K.H., Koo, J.Y., and Park, C.W. (2000), "Kernel estimation of discontinuous regression functions," *Statistics and Probability Letters*, **47**, 277–285.

143. Karlsson, G., and Vetterli, M. (1990), "Theory of two-dimensional multirate filter banks," *IEEE Transactions on Signal Processing*, **38**, 925–937.

144. Keeling, S.L., and Stollberger, R. (2002), "Nonlinear anisotropic diffusion filters for multiscale edge enhancement," *Inverse Problems*, **18**, 175–190.

145. Kim, H.J. (1993), "Robustness of the likelihood ratio test for a change in simple linear regression," *Journal of the American Statistical Association*, **88**, 864–871.

146. Kim, H.J., and Siegmund, D. (1989), "The likelihood ratio test for a change-point in simple linear regression," *Biometrika*, **76**, 409–423.

147. Kimeldorf, G., and Wahba, G. (1970), "A correspondence between Bayesian estimation of stochastic processes and smoothing by splines," *Annals of Mathematical Statistics*, **41**, 495–502.

148. Kimeldorf, G., and Wahba, G. (1971), "Some results on Tchebycheffian spline functions," *Journal of Mathematical Analysis and Applications*, **33**, 82–95.

149. Kiryati, N., Eldar, Y., and Bruckstein, A.M. (1991), "A probabilistic Hough transform," *Pattern Recognition*, **24**, 303–316.

150. Ko, S.J., and Lee, Y.H. (1991), "Center weighted median filters and their applications to image enhancement," *IEEE Transactions on Circuits and Systems*, **38**, 984–993.

151. Koch, C., Marroquin, J., and Yuille, A. (1986), "Analog 'neuronal' networks in early vision," *Proceedings of National Academic Science of USA*, **83**, 4263–4267.

152. Koenderink, J. (1984), "The structure of images," *Biological Cybernetics*, **50**, 363–370.

153. Koo, J.Y. (1997), "Spline estimation of discontinuous regression functions," *Journal of Computational and Graphical Statistics*, **6**, 266–284.

154. Kooperberg, C., and Stone, C.J. (1991), "A study of logspline density estimation," *Computational Statistics and Data Analysis*, **12**, 327–347.

155. Korostelev, A.P., and Tsybakov, A.B. (1993), *Minimax Theory of Image Reconstruction*, Lecture Notes in Statistics, **82**, New York: Springer.

156. Kumar, K., and Wu, B. (2001), "Detection of change points in time series analysis with fuzzy statistics," *International Journal of Systems Science*, **32**, 1185–1192.

157. Lange, K. (1990), "Convergence of EM image reconstruction algorithm with Gibbs smoothing," *IEEE Transactions on Medical Imaging*, **9**, 439–446.

158. Lehmann, E.L. (1983), *Theory of Point Estimation*, New York: John Wiley & Sons.

159. Lehmann, E.L. (1986), *Testing Statistical Hypotheses (2nd ed.)*, New York: John Wiley & Sons.

160. Li, K.C. (1985), "From Stein's unbiased risk estimates to the method of generalized cross-validation," *The Annals of Statistics*, **13**, 1352–1377.

161. Li, K.C. (1986), "Asymptotic optimality of C_L and generalized cross validation in ridge regression with application to spline smoothing," *The Annals of Statistics*, **14**, 1101–1112.

162. Li, K.C. (1987), "Asymptotic optimality of C_p, C_L, cross-validation and generalized cross validation: discrete index set," *The Annals of Statistics*, **15**, 958–975.

163. Li, S.Z. (1995a), "On discontinuity-adaptive smoothness priors in computer vision," *IEEE Transactions on Pattern Analysis and Machine Intelligence*, **17**, 576–586.

164. Li, S.Z. (1995b), *Markov Random Field Modeling in Computer Vision,* New York: Springer-Verlag.

165. Li, S.Z. (1998), "Close-form solution and parameter selection for convex minimization-based edge-preserving smoothing," *IEEE Transactions on Pattern Analysis and Machine Intelligence*, **20**, 916–932.

166. Lindgren, B.W. (1993), *Statistical Theory (4th ed.)*, New York: Chapman & Hall/CRC.

167. Loader, C.R. (1996), "Change point estimation using nonparametric regression," *The Annals of Statistics*, **24**, 1667–1678.

168. Loader, C.R. (1999), "Bandwidth selection: classical or plug-in?," *The Annals of Statistics*, **27**, 415–438.

169. Loupas, T., McDicken, W.N., and Allan, P.L. (1989), "An adaptive weighted median filter for speckle suppression in medical ultrasonic images," *IEEE Transactions on Circuits and Systems*, **36**, 129–135.

170. Macleod, I. (1970), "On finding structure in pictures," in *Picture Processing and Psychopictorics* (edited by Lipkin, B.S., and Rosenfeld, A.), New York: Academic Press.

171. Macleod, I. (1972), "Comments on techniques for edge detection," *Proceedings of IEEE*, **60**, 344.

172. Mallat, S.G. (1989), "A theory for multiresolution signal decomposition: the wavelet representation," *IEEE Transactions on Pattern Analysis and Machine Intelligence*, **11**, 674–693.

173. Mallows, C.L. (1973), "Some comments on C_p," *Technometrics*, **15**, 661–675.

174. Marr, D., and Hildreth, E. (1980), "Theory of edge detection," *Proceedings of the Royal Society in London*, **B207**, 187–217.

175. Marroquin, J.L., Mitter, S., and Poggio, T. (1987), "Probabilistic solution of ill-posed problems in computational vision," *Journal of the American Statistical Association*, **82**, 76–89.

176. Marroquin, J.L., Velasco, F.A., Rivera, M., and Nakamura, M. (2001), "Gauss-Markov measure field models for low-level vision," *IEEE Transactions on Pattern Analysis and Machine Intelligence*, **23**, 337–347.

177. McDonald, J.A., and Owen, A.B. (1986), "Smoothing with split linear fits," *Technometrics*, **28**, 195–208.

178. McMillan, N.J., and Berlinear, L.M. (1996), "Hierarchical image reconstruction using Markov random fields," *Bayesian Statistics*, **5**, 675–683.

179. Meer, P., and Georgescu, B. (2001), "Edge detection with embedded confidence," *IEEE Transactions on Pattern Analysis and Machine Intelligence*, **23**, 1351–1365.

180. Montgomery, D.C. (1996), *Introduction to Statistical Quality Control (3rd ed.)*, New York: John Wiley & Sons.

181. Morrone, M.C., Navangione, A., and Burr, D. (1995), "An adaptive approach to scale selection for line and edge detection," *Pattern Recognition Letters*, **16**, 667–677.

182. Moussouris, J. (1974), "Gibbs and Markov systems with constraints," *Journal of Statistical Physics*, **10**, 11–33.

183. Mrázek, P., Weickert, J., and Steidl, G. (2003), "Correspondences between wavelet shrinkage and nonlinear diffusion," in *Scale-Space 2003* (edited by Griffin, L.D., and Lillholm, M.), 101–116.

184. Müller, H.-G. (1988), *Nonparametric Regression Analysis of Longitudinal Data*, New York: Springer-Verlag.

185. Müller, H.G. (1992), "Change-points in nonparametric regression analysis," *The Annals of Statistics*, **20**, 737–761.

186. Müller, H.G., and Song, K.S. (1994a), "Maximin estimation of multidimensional boundaries," *Journal of Multivariate Analysis*, **50**, 265–281.

187. Müller, H.G., and Song, K.S. (1994b), "Cube splitting in multidimensional edge estimation," in *Change-Point Problems* (edited by Carlstein, E., Müller, H.G., and Siegmund, D.), IMS Lecture Notes, **23**, 210–223.

188. Müller, H.G., and Wang, J.L. (1990), "Nonparametric analysis of changes in hazard rates for censored survival data: an alternative to change-point models," *Biometrika*, **77**, 305–314.

189. Nadaraya, E.A. (1964), "On estimating regression," *Theory of Probability and Its Applications*, **9**, 141–142.

190. Nalwa, V.S., and Binford, T.O. (1986), "On detecting edges," *IEEE Transactions on Pattern Analysis and Machine Intelligence*, **8**, 699–714.

191. O'Gorman, F., and Clowes, M.B. (1976), "Finding picture edges through collinearity of feature points," *IEEE Transactions on Computers*, **C-25**, 449–456.

192. O'Sullivan, F. (1983), "The analysis of some penalized likelihood estimation schemes," *Technical Report 726*, Department of Statistics, University of Wisconsin-Madison.

193. O'Sullivan, F., and Qian, M. (1994), "A regularized contrast statistic for object boundary estimation - implementation and statistical evaluation," *IEEE Transactions on Pattern Analysis and Machine Intelligence*, **16**, 561–570.

194. Page, E.S. (1954), "Continuous inspection schemes," *Biometrika*, **41**, 100–115.

195. Page, E.S. (1961), "Cumulative sum charts," *Technometrics*, **3**, 1–9.

196. Page, E.S. (1963), "Controlling the standard deviation by cusums and warning lines," *Technometrics*, **5**, 307–315.

197. Palmer, P., Kittler, J., and Petrou, M. (1994), "Using focus of attention with the Hough transform for accurate line parameter estimation," *Pattern Recognition*, **27**, 1127–1134.

198. Peli, T., and Malah, D. (1982), "A study of edge detection algorithms," *Computer Graphics and Image Processing*, **20**, 1–21.

199. Perona, P., and Malik, J. (1990), "Scale space and edge detection using anisotropic diffusion," *IEEE Transactions on Pattern Analysis and Machine Intelligence*, **12**, 629–639.

200. Petrou, M., and Kittler, J. (1991), "Optimal edge detectors for ramp edges," *Transactions on Pattern Analysis and Machine Intelligence*, **13**, 483–491.

201. Phillips, G.M., and Taylor, P.J. (1996), *Theory and Applications of Numerical Analysis (2nd ed.)*, New York: Academic Press.

202. Picard, D. (1985), "Testing and estimating change-points in time series," *Advances in Applied Probability*, **17**, 841–867.

203. Poggio, T., Torre, V., and Koch, C. (1985), "Computational vision and regularization theory," *Nature*, **317**, 314–319.

204. Polzehl, J., and Spokoiny, V.G. (2000), "Adaptive weights smoothing with applications to image restoration," *Journal of the Royal Statistical Society (Series B)*, **62**, 335–354.

205. Polzehl, J., and Spokoiny, V.G. (2003a), "Varying coefficient regression modeling," Preprint 818, WIAS 2003. *http://www.wias-berlin.de/publications/preprints/818*.

206. Polzehl, J., and Spokoiny, V.G. (2003b), "Local likelihood modeling by adaptive weights smoothing," Preprint 27, Preprint Series DFG-SPP 1114. *http://www.math.uni-breman.de/zetem/DFG-Schwerpunkt/preprints/prep27.pdf*.

207. Polzehl, J., and Spokoiny, V.G. (2003c), "Image denoising: pointwise adaptive approach," *The Annals of Statistics*, **31**, 30–57.

208. Pratt, W.K. (1991), *Digital Image Processing (2nd ed.)*, New York: John Wiley & Sons.

209. Prewitt, J.M.S. (1970), "Object enhancement and extraction," in *Picture Processing and Psychopictorics* (edited by Lipkin, B.S., and Rosenfeld, A.), New York: Academic Press.

210. Priestley, M.B., and Chao, M.T. (1972), "Non-parametric function fitting," *Journal of the Royal Statistical Society (Series B)*, **34**, 385–392.

211. Qiu, P. (1991), "Estimation of a kind of jump regression function," *Systems Science and Mathematical Sciences*, **4**, 1–13.

212. Qiu, P. (1994), "Estimation of the number of jumps of the jump regression functions," *Communications in Statistics-Theory and Methods*, **23**, 2141–2155.

213. Qiu, P. (1997), "Nonparametric estimation of the jump regression surface," *Sankhya (Series A)*, **59**, 268–294.

214. Qiu, P. (1998), "Discontinuous regression surfaces fitting," *The Annals of Statistics*, **26**, 2218–2245.

215. Qiu, P. (1999), "Comparisons of several local smoothing jump detectors in one-dimensional nonparametric regression," *The 1999 ASA Proceedings of the Statistical Computing Section*, 150–155.

216. Qiu, P. (2002), "A nonparametric procedure to detect jumps in regression surfaces," *Journal of Computational and Graphical Statistics*, **11**, 799–822.

217. Qiu, P. (2003a), "A jump-preserving curve fitting procedure based on local piecewise-linear kernel estimation," *Journal of Nonparametric Statistics*, **15**, 437–453.

218. Qiu, P. (2003b), "Jump-preserving surface reconstruction from noisy data," Manuscript, School of Statistics, University of Minnesota, Minneapolis, MN 55455.

219. Qiu, P. (2004), "The local piecewisely linear kernel smoothing procedure for fitting jump regression surfaces," *Technometrics*, **46**, 87–98.

220. Qiu, P., Asano, Chi., and Li, X. (1991), "Estimation of jump regression functions," *Bulletin of Informatics and Cybernetics*, **24**, 197–212.

221. Qiu, P., and Bhandarkar, S.M. (1996), "An edge detection technique using local smoothing and statistical hypothesis testing," *Pattern Recognition Letters*, **17**, 849–872.

222. Qiu, P., Chappell, R., Obermeyer, W., and Benca, R. (1999), "Modelling daily and subdaily cycles in rat sleep data," *Biometrics*, **55**, 930–935.

223. Qiu, P., and Hawkins, D.M. (2001), "A rank-based multivariate CUSUM procedure," *Technometrics*, **43**, 120–132.

224. Qiu, P., and Hawkins, D.M. (2003), "A nonparametric multivariate CUSUM procedure for detecting shifts in all directions," *Journal of the Royal Statistical Society (Series D)*, **52**, 151–164.

225. Qiu, P., and Yandell, B. (1997), "Jump detection in regression surfaces," *Journal of Computational and Graphical Statistics*, **6**, 332–354.

226. Qiu, P., and Yandell, B. (1998), "A local polynomial jump detection algorithm in nonparametric regression," *Technometrics*, **40**, 141–152.

227. Quandt, R.E. (1958), "The estimation of the parameters of a linear regression system obeys two separate regimes," *Journal of the American Statistical Association*, **53**, 873–880.

228. Quandt, R.E. (1960), "Tests of the hypothesis that a linear regression system obeys two separate regimes," *Journal of the American Statistical Association*, **55**, 324–330.

229. Reinsch, C.H. (1967), "Smoothing by spline functions," *Numerical Mathematics*, **10**, 177–183.

230. Reinsch, C.H. (1971), "Smoothing by spline functions II," *Numerical Mathematics*, **16**, 451–454.

231. Reynolds, M.R., Jr, Amin, R.W., and Arnold, J.C. (1990), "CUSUM charts with variable sampling intervals," *Technometrics*, **32**, 371–384.

232. Rice, J. (1984), "Boundary modification for kernel regression," *Communications in Statistics – Theory and Methods*, **13**, 893–900.

233. Roberts, L.G. (1965), "Machine perception of three-dimensional solids," in *Optical and Electro-Optical Information Processing* (edited by Tippett, J.T., *et al.*), Cambridge, MA: MIT Press.

234. ter Haar Romeny, B.M. (eds.) (1994), *Geometry-Driven Diffusion in Computer Vision*, Boston: Kluwer.

235. Rosenfeld, A., and Kak, A.C. (1976), *Digital Picture Processing,* New York: Academic Press.

236. Rosenfeld, A., and Kak, A.C. (1982), *Digital Picture Processing (2nd ed.),* New York: Academic Press.

237. Rudemo, M., and Stryhn, H. (1994), "Approximating the distribution of maximum likelihood contour estimators in two-region images," *Scandinavian Journal of Statistics*, **21**, 41–55.

238. Rue, H., Chu, C.K., Godtliebsen, F., and Marron, J.S. (2002), "M-smoother with local linear fit," *Journal of Nonparametric Statistics*, **14**, 155–168.

239. Ruppert, D., Sheather, S.J., and Wand, M.P. (1995), "An efficient bandwidth selector for local least squares regression," *Journal of the American Statistical Association*, **90**, 1257–1270.

240. Ruppert, D., and Wand, M.P. (1994), "Multivariate locally weighted least squares regression," *The Annals of Statistics*, **22**, 1346–1370.

241. Ruskai, M.B. (1992), *Wavelets and Their Applications*, Boston: Jones and Bartlett Publishers, Inc.

242. Saint-Marc, P., Chen, J., and Medioni, G. (1991), "Adaptive smoothing: a general tool for early vision," *IEEE Transactions on Pattern Analysis and Machine Intelligence*, **13**, 514–529.

243. Sarkar, S., and Boyer, K.L. (1991), "On optimal infinite impulse response edge detection filters," *IEEE Transactions on Pattern Analysis and Machine Intelligence*, **13**, 1154–1171.

244. Schoenberg, I.J. (1964), "Spline functions and the problem of graduation," *Proceedings of the National Academy of Sciences of the USA*, **52**, 947–950.

245. Seber, G.A.F. (1977), *Linear Regression Analysis*, New York: John Wiley & Sons.

246. Sen, P.K. (1994), "Some change-point problems in survival analysis: relevance of nonparametrics in applications," *Journal of Applied Statistical Science*, **1**, 425–444.

247. Shea, D.J., Worley, S.J., Stern, I.R., and Hoar, T.J. (1994), "An introduction to atmospheric and oceanographic data," *NCAR/TN-404+IA*, Climate and Global Dynamics Division, National Center For Atmospheric Research, Boulder, Colorado.

248. Shiau, J. (1985), "Smoothing spline estimation of functions with discontinuities," *Technical Report # 768*, Department of Statistics, University of Wisconsin.

249. Shiau, J. (1987), "A note on MSE coverage intervals in a partial Spline model," *Communications in Statistics – Theory and Methods*, **16**, 1851–1866.

250. Shiau, J., and Wahba, G. (1988), "Rates of convergence of some estimators for a semiparametric model," *Communications in Statistics – Simulation and Computation*, **17**, 1117–1133.

251. Shin, M.C., Goldgof, D., and Bowyer, K.W. (2001), "Comparison of edge detector performance through use in an object recognition task," *Computer Vision and Image Understanding*, **84**, 160–178.

252. Shiohama, T., Taniguchi, M., and Puri, M.L. (2003), "Asymptotic estimation theory of change-point problems for time series regression models and its applications," *Probability, Statistics and Their Applications: Papers in Honor of Rabi Bhattacharya*, 257–284, IMS Lecture Notes Monograph Series, **41**, Beachwood, OH.

253. Silverman, B. (1984), "Spline smoothing: the equivalent variable kernel method," *The Annals of Statistics*, **12**, 898–916.

254. Sinha, S.S., and Schunck, B.G. (1992), "A two-stage algorithm for discontinuity-preserving surface reconstruction," *IEEE Transactions on Pattern Analysis and Machine Intelligence*, **14**, 36–55.

255. Smith, A.F.M. (1975), "A Bayesian approach to inference about a change-point in a sequence of random variables," *Biometrika*, **62**, 407–416.

256. Speckman, P. (1985), "Spline smoothing and optimal rates of convergence in nonparametric regression," *The Annals of Statistics*, **13**, 970–983.

257. Srivastava, M.S., and Wu, Y. (1993), "Comparison of EWMA, CUSUM and Shiryayev-Roberts procedures for detecting a shift in the mean," *The Annals of Statistics*, **21**, 645–670.

258. Stevenson, R.L., Schmitz, B.E., and Delp, E.J. (1994), "Discontinuity preserving regularization of inverse visual problems," *IEEE Transactions on Systems, Man, and Cybernetics*, **24**, 455–469.

259. Stoll, M. (2001), *Real Analysis (2nd ed.)*, New York: Addison-Wesley Longman, Inc.

260. Stone, C. (1982), "Optimal global rates of convergence for nonparametric regression," *The Annals of Statistics*, **10**, 1040–1053.

261. Stone, C. (1985), "Additive regression and other nonparametric models," *The Annals of Statistics*, **15**, 885–941.

262. Stoumbos, Z.G., and Sullivan, J.H. (2002), "Robustness to non-normality of the multivariate EWMA control chart," *Journal of Quality Technology*, **34**, 260–276.

263. Sullivan, J.H. (2002), "Estimating the locations of multiple change points in the mean," *Computational Statistics*, **17**, 289–296.

264. Sun, T., Gabbouj, M., and Neuvo, Y. (1994), "Center weighted median filters: some properties and their applications in image processing," *Signal Processing*, **35**, 213–229.

265. Tagare, H.D., and deFigueiredo, R.J.P. (1990), "On the localization performance measure and optimal edge detection," *IEEE Transactions on Pattern Analysis and Machine Intelligence*, **12**, 1186–1190.

266. Tan, H.L., Gelfand, S.B., and Delp, E.J. (1989), "A comparative cost function approach to edge detection," *IEEE Transaction on Systems, Man, and Cybernetics*, **19**, 1337–1349.

267. Tan, H.L., Gelfand, S.B., and Delp, E.J. (1992), "A cost minimization approach to edge detection using simulated annealing," *IEEE Transactions on Pattern Analysis and Machine Intelligence*, **14**, 3–18.

268. Terzopoulos, D.T. (1986), "Regularization of inverse visual problems involving discontinuities," *IEEE Transactions on Pattern Analysis and Machine Intelligence*, **8**, 413–424.

269. Tomasi, C., and Manduchi, R. (1998), "Bilateral filtering for gray and color images," *Proceedings of the 1998 IEEE International Conference on Computer Vision*, 839–846, Bombay, India.

270. Torre, V., and Poggio, T. (1986), "On edge detection," *IEEE Transactions on Pattern Analysis and Machine Intelligence*, **8**, 147–163.

271. Tukey, J.W. (1977), *Exploratory Data Analysis,* Reading, MA: Addison-Wesley.

272. van der Heijden, F. (1995), "Edge and line feature extraction based on covariance models," *IEEE Transactions on Pattern Analysis and Machine Intelligence*, **17**, 16–33.

273. Wahba, G. (1979), "Convergence rates of 'thin plate' smoothing splines when the data are noisy," in *Smoothing Techniques for Curve Estimation*, Berlin: Springer-Verlag, 223–245.

274. Wahba, G. (1984), "Partial spline models for the semiparametric estimation of functions of several variables," in *Statistical Analysis of Time Series*, Proceedings of the Japan-US Joint Seminar, Tokyo, 319–329.

275. Wahba, G. (1990), *Spline Models for Observational Data,* CBMS-NSF Regional Conference Series in Applied Mathematics, Philadelphia, PA: SIAM.

276. Wahba, G., and Wold, S. (1975), "A completely automatic French curve," *Communications in Statistics*, **4**, 1–17.

277. Wang, Y. (1995), "Jump and sharp cusp detection by wavelets," *Biometrika*, **82**, 385–397.

278. Wang, Y. (1998), "Change curve estimation via wavelets," *Journal of the American Statistical Association*, **93**, 163–172.

279. Watson, G.S. (1964), "Smooth regression analysis," *Sankhya (Series A)*, **26**, 359–372.

280. Weickert, J. (1998), *Anisotropic Diffusion in Imaging Processing*, Stuttgart, Germany: Teubner.

281. Weickert, J., ter Haar Romeny, B.M., and Viergever, M.A. (1998), "Efficient and reliable schemes for nonlinear diffusion filtering," *IEEE Transactions on Image Processing*, **7**, 398–410.

282. Witkin, A. (1983), "Scale-space filtering," in *International Joint Conference in Artificial Intelligence*, Karlsruhe, Germany, 1019–1021.

283. Woodall, W.H. (1984), "On the Markov chain approach to the two-sided CUSUM procedure," *Technometrics*, **26**, 41–46.

284. Woodall, W.H. (1986), "The design of CUSUM quality control charts," *Journal of Quality Technology*, **18**, 99–102.

285. Woodall, W.H. (2000), "Controversies and contradictions in statistical process control," *Journal of Quality Technology*, **32**, 341–350.

286. Worsley, K.J. (1983a), "Testing for a two-phase multiple regression," *Technometrics*, **25**, 35–42.

287. Worsley, K.J. (1983b), "The power of the likelihood ratio and cumulative sum tests for a change in a binomial probability," *Biometrika*, **70**, 455–464.

288. Worsley, K.J. (1986), "Confidence regions and test for a change-point in a sequence of exponential family random variables," *Biometrika*, **73**, 91–104.

289. Wu, J.S., and Chu, C.K. (1993a), "Kernel type estimators of jump points and values of a regression function," *The Annals of Statistics*, **21**, 1545–1566.

290. Wu, J.S., and Chu, C.K. (1993b), "Nonparametric function estimation and bandwidth selection for discontinuous regression functions," *Statistica Sinica*, **3**, 557–576.

291. Yang, G.J., and Huang, T.S. (1981), "The effect of median filtering on edge location estimation," *Computer Graphics and Image Processing*, **15**, 224–245.

292. Yao, Y.C. (1987), "Approximating the distribution of the maximum likelihood estimate of the change-point in a sequence of independent random variables," *The Annals of Statistics*, **15**, 1321–1328.

293. Yashchin, E. (1992), "Analysis of CUSUM and other Markov-type control schemes by using empirical distributions," *Technometrics*, **34**, 54–63.

294. Yashchin, E. (1993), "Performance of CUSUM control schemes for serially correlated observations," *Technometrics*, **35**, 37–52.

295. Yitzhaky, Y., and Peli, E. (2003), "A method for objective edge detection evaluation and detector parameter selection," *IEEE Transactions on Pattern Analysis and Machine Intelligence*, **25**, 1027–1033.

296. You, Y.L., Xu, W., Tannenbaum, A., and Kaveh, M. (1996), "Behavioral analysis of anisotropic diffusion in image processing," *IEEE Transactions on Image Processing*, **5**, 1539–1553.

297. Yuille, A., and Poggio, T. (1986), "Scaling theorems for zero crossings," *IEEE Transactions on Pattern Analysis and Machine Intelligence*, **8**, 15–25.

298. Zhang, H. (1994), "Maximal correlation and adaptive splines," *Technometrics*, **36**, 196–201.

Index

WILEY SERIES IN PROBABILITY AND STATISTICS

ESTABLISHED BY WALTER A. SHEWHART AND SAMUEL S. WILKS

The ***Wiley Series in Probability and Statistics*** is well established and authoritative. It covers many topics of current research interest in both pure and applied statistics and probability theory. Written by leading statisticians and institutions, the titles span both state-of-the-art developments in the field and classical methods.

Reflecting the wide range of current research in statistics, the series encompasses applied, methodological and theoretical statistics, ranging from applications and new techniques made possible by advances in computerized practice to rigorous treatment of theoretical approaches.

This series provides essential and invaluable reading for all statisticians, whether in academia, industry, government, or research.

ABRAHAM and LEDOLTER · Statistical Methods for Forecasting
AGRESTI · Analysis of Ordinal Categorical Data
AGRESTI · An Introduction to Categorical Data Analysis
AGRESTI · Categorical Data Analysis, *Second Edition*
ALTMAN, GILL, and McDONALD · Numerical Issues in Statistical Computing for the Social Scientist
AMARATUNGA and CABRERA · Exploration and Analysis of DNA Microarray and Protein Array Data
ANDĚL · Mathematics of Chance
ANDERSON · An Introduction to Multivariate Statistical Analysis, *Third Edition*
* ANDERSON · The Statistical Analysis of Time Series
ANDERSON, AUQUIER, HAUCK, OAKES, VANDAELE, and WEISBERG · Statistical Methods for Comparative Studies
ANDERSON and LOYNES · The Teaching of Practical Statistics
ARMITAGE and DAVID (editors) · Advances in Biometry
ARNOLD, BALAKRISHNAN, and NAGARAJA · Records
* ARTHANARI and DODGE · Mathematical Programming in Statistics
* BAILEY · The Elements of Stochastic Processes with Applications to the Natural Sciences
BALAKRISHNAN and KOUTRAS · Runs and Scans with Applications
BARNETT · Comparative Statistical Inference, *Third Edition*
BARNETT and LEWIS · Outliers in Statistical Data, *Third Edition*
BARTOSZYNSKI and NIEWIADOMSKA-BUGAJ · Probability and Statistical Inference
BASILEVSKY · Statistical Factor Analysis and Related Methods: Theory and Applications
BASU and RIGDON · Statistical Methods for the Reliability of Repairable Systems
BATES and WATTS · Nonlinear Regression Analysis and Its Applications
BECHHOFER, SANTNER, and GOLDSMAN · Design and Analysis of Experiments for Statistical Selection, Screening, and Multiple Comparisons
BELSLEY · Conditioning Diagnostics: Collinearity and Weak Data in Regression

*Now available in a lower priced paperback edition in the Wiley Classics Library.
†Now available in a lower priced paperback edition in the Wiley–Interscience Paperback Series.

† BELSLEY, KUH, and WELSCH · Regression Diagnostics: Identifying Influential Data and Sources of Collinearity
BENDAT and PIERSOL · Random Data: Analysis and Measurement Procedures, *Third Edition*
BERRY, CHALONER, and GEWEKE · Bayesian Analysis in Statistics and Econometrics: Essays in Honor of Arnold Zellner
BERNARDO and SMITH · Bayesian Theory
BHAT and MILLER · Elements of Applied Stochastic Processes, *Third Edition*
BHATTACHARYA and WAYMIRE · Stochastic Processes with Applications
† BIEMER, GROVES, LYBERG, MATHIOWETZ, and SUDMAN · Measurement Errors in Surveys
BILLINGSLEY · Convergence of Probability Measures, *Second Edition*
BILLINGSLEY · Probability and Measure, *Third Edition*
BIRKES and DODGE · Alternative Methods of Regression
BLISCHKE AND MURTHY (editors) · Case Studies in Reliability and Maintenance
BLISCHKE AND MURTHY · Reliability: Modeling, Prediction, and Optimization
BLOOMFIELD · Fourier Analysis of Time Series: An Introduction, *Second Edition*
BOLLEN · Structural Equations with Latent Variables
BOROVKOV · Ergodicity and Stability of Stochastic Processes
BOULEAU · Numerical Methods for Stochastic Processes
BOX · Bayesian Inference in Statistical Analysis
BOX · R. A. Fisher, the Life of a Scientist
BOX and DRAPER · Empirical Model-Building and Response Surfaces
* BOX and DRAPER · Evolutionary Operation: A Statistical Method for Process Improvement
BOX, HUNTER, and HUNTER · Statistics for Experimenters: An Introduction to Design, Data Analysis, and Model Building
BOX and LUCEÑO · Statistical Control by Monitoring and Feedback Adjustment
BRANDIMARTE · Numerical Methods in Finance: A MATLAB-Based Introduction
BROWN and HOLLANDER · Statistics: A Biomedical Introduction
BRUNNER, DOMHOF, and LANGER · Nonparametric Analysis of Longitudinal Data in Factorial Experiments
BUCKLEW · Large Deviation Techniques in Decision, Simulation, and Estimation
CAIROLI and DALANG · Sequential Stochastic Optimization
CASTILLO, HADI, BALAKRISHNAN, and SARABIA · Extreme Value and Related Models with Applications in Engineering and Science
CHAN · Time Series: Applications to Finance
CHATTERJEE and HADI · Sensitivity Analysis in Linear Regression
CHATTERJEE and PRICE · Regression Analysis by Example, *Third Edition*
CHERNICK · Bootstrap Methods: A Practitioner's Guide
CHERNICK and FRIIS · Introductory Biostatistics for the Health Sciences
CHILÈS and DELFINER · Geostatistics: Modeling Spatial Uncertainty
CHOW and LIU · Design and Analysis of Clinical Trials: Concepts and Methodologies, *Second Edition*
CLARKE and DISNEY · Probability and Random Processes: A First Course with Applications, *Second Edition*
* COCHRAN and COX · Experimental Designs, *Second Edition*
CONGDON · Applied Bayesian Modelling
CONGDON · Bayesian Statistical Modelling
CONOVER · Practical Nonparametric Statistics, *Third Edition*
COOK · Regression Graphics
COOK and WEISBERG · Applied Regression Including Computing and Graphics
COOK and WEISBERG · An Introduction to Regression Graphics

*Now available in a lower priced paperback edition in the Wiley Classics Library.
†Now available in a lower priced paperback edition in the Wiley–Interscience Paperback Series.

CORNELL · Experiments with Mixtures, Designs, Models, and the Analysis of Mixture Data, *Third Edition*
COVER and THOMAS · Elements of Information Theory
COX · A Handbook of Introductory Statistical Methods
* COX · Planning of Experiments
CRESSIE · Statistics for Spatial Data, *Revised Edition*
CSÖRGŐ and HORVÁTH · Limit Theorems in Change Point Analysis
DANIEL · Applications of Statistics to Industrial Experimentation
DANIEL · Biostatistics: A Foundation for Analysis in the Health Sciences, *Eighth Edition*
* DANIEL · Fitting Equations to Data: Computer Analysis of Multifactor Data, *Second Edition*
DASU and JOHNSON · Exploratory Data Mining and Data Cleaning
DAVID and NAGARAJA · Order Statistics, *Third Edition*
* DEGROOT, FIENBERG, and KADANE · Statistics and the Law
DEL CASTILLO · Statistical Process Adjustment for Quality Control
DEMARIS · Regression with Social Data: Modeling Continuous and Limited Response Variables
DEMIDENKO · Mixed Models: Theory and Applications
DENISON, HOLMES, MALLICK and SMITH · Bayesian Methods for Nonlinear Classification and Regression
DETTE and STUDDEN · The Theory of Canonical Moments with Applications in Statistics, Probability, and Analysis
DEY and MUKERJEE · Fractional Factorial Plans
DILLON and GOLDSTEIN · Multivariate Analysis: Methods and Applications
DODGE · Alternative Methods of Regression
* DODGE and ROMIG · Sampling Inspection Tables, *Second Edition*
* DOOB · Stochastic Processes
DOWDY, WEARDEN, and CHILKO · Statistics for Research, *Third Edition*
DRAPER and SMITH · Applied Regression Analysis, *Third Edition*
DRYDEN and MARDIA · Statistical Shape Analysis
DUDEWICZ and MISHRA · Modern Mathematical Statistics
DUNN and CLARK · Basic Statistics: A Primer for the Biomedical Sciences, *Third Edition*
DUPUIS and ELLIS · A Weak Convergence Approach to the Theory of Large Deviations
* ELANDT-JOHNSON and JOHNSON · Survival Models and Data Analysis
ENDERS · Applied Econometric Time Series
ETHIER and KURTZ · Markov Processes: Characterization and Convergence
EVANS, HASTINGS, and PEACOCK · Statistical Distributions, *Third Edition*
FELLER · An Introduction to Probability Theory and Its Applications, Volume I, *Third Edition,* Revised; Volume II, *Second Edition*
FISHER and VAN BELLE · Biostatistics: A Methodology for the Health Sciences
FITZMAURICE, LAIRD, and WARE · Applied Longitudinal Analysis
* FLEISS · The Design and Analysis of Clinical Experiments
FLEISS · Statistical Methods for Rates and Proportions, *Third Edition*
FLEMING and HARRINGTON · Counting Processes and Survival Analysis
FULLER · Introduction to Statistical Time Series, *Second Edition*
FULLER · Measurement Error Models
GALLANT · Nonlinear Statistical Models
GHOSH, MUKHOPADHYAY, and SEN · Sequential Estimation
GIESBRECHT and GUMPERTZ · Planning, Construction, and Statistical Analysis of Comparative Experiments
GIFI · Nonlinear Multivariate Analysis
GIVENS and HOETING · Computational Statistics

*Now available in a lower priced paperback edition in the Wiley Classics Library.
†Now available in a lower priced paperback edition in the Wiley–Interscience Paperback Series.

GLASSERMAN and YAO · Monotone Structure in Discrete-Event Systems
GNANADESIKAN · Methods for Statistical Data Analysis of Multivariate Observations, *Second Edition*
GOLDSTEIN and LEWIS · Assessment: Problems, Development, and Statistical Issues
GREENWOOD and NIKULIN · A Guide to Chi-Squared Testing
GROSS and HARRIS · Fundamentals of Queueing Theory, *Third Edition*
† GROVES · Survey Errors and Survey Costs
* HAHN and SHAPIRO · Statistical Models in Engineering
HAHN and MEEKER · Statistical Intervals: A Guide for Practitioners
HALD · A History of Probability and Statistics and their Applications Before 1750
HALD · A History of Mathematical Statistics from 1750 to 1930
HAMPEL · Robust Statistics: The Approach Based on Influence Functions
HANNAN and DEISTLER · The Statistical Theory of Linear Systems
HEIBERGER · Computation for the Analysis of Designed Experiments
HEDAYAT and SINHA · Design and Inference in Finite Population Sampling
HELLER · MACSYMA for Statisticians
HINKELMAN and KEMPTHORNE: · Design and Analysis of Experiments, Volume 1: Introduction to Experimental Design
HOAGLIN, MOSTELLER, and TUKEY · Exploratory Approach to Analysis of Variance
HOAGLIN, MOSTELLER, and TUKEY · Exploring Data Tables, Trends and Shapes
* HOAGLIN, MOSTELLER, and TUKEY · Understanding Robust and Exploratory Data Analysis
HOCHBERG and TAMHANE · Multiple Comparison Procedures
HOCKING · Methods and Applications of Linear Models: Regression and the Analysis of Variance, *Second Edition*
HOEL · Introduction to Mathematical Statistics, *Fifth Edition*
HOGG and KLUGMAN · Loss Distributions
HOLLANDER and WOLFE · Nonparametric Statistical Methods, *Second Edition*
HOSMER and LEMESHOW · Applied Logistic Regression, *Second Edition*
HOSMER and LEMESHOW · Applied Survival Analysis: Regression Modeling of Time to Event Data
† HUBER · Robust Statistics
HUBERTY · Applied Discriminant Analysis
HUNT and KENNEDY · Financial Derivatives in Theory and Practice
HUSKOVA, BERAN, and DUPAC · Collected Works of Jaroslav Hajek—with Commentary
HUZURBAZAR · Flowgraph Models for Multistate Time-to-Event Data
IMAN and CONOVER · A Modern Approach to Statistics
† JACKSON · A User's Guide to Principle Components
JOHN · Statistical Methods in Engineering and Quality Assurance
JOHNSON · Multivariate Statistical Simulation
JOHNSON and BALAKRISHNAN · Advances in the Theory and Practice of Statistics: A Volume in Honor of Samuel Kotz
JOHNSON and BHATTACHARYYA · Statistics: Principles and Methods, *Fifth Edition*
JOHNSON and KOTZ · Distributions in Statistics
JOHNSON and KOTZ (editors) · Leading Personalities in Statistical Sciences: From the Seventeenth Century to the Present
JOHNSON, KOTZ, and BALAKRISHNAN · Continuous Univariate Distributions, Volume 1, *Second Edition*
JOHNSON, KOTZ, and BALAKRISHNAN · Continuous Univariate Distributions, Volume 2, *Second Edition*
JOHNSON, KOTZ, and BALAKRISHNAN · Discrete Multivariate Distributions

*Now available in a lower priced paperback edition in the Wiley Classics Library.
†Now available in a lower priced paperback edition in the Wiley–Interscience Paperback Series.

JOHNSON, KOTZ, and KEMP · Univariate Discrete Distributions, *Second Edition*
JUDGE, GRIFFITHS, HILL, LÜTKEPOHL, and LEE · The Theory and Practice of Econometrics, *Second Edition*
JUREČKOVÁ and SEN · Robust Statistical Procedures: Aymptotics and Interrelations
JUREK and MASON · Operator-Limit Distributions in Probability Theory
KADANE · Bayesian Methods and Ethics in a Clinical Trial Design
KADANE AND SCHUM · A Probabilistic Analysis of the Sacco and Vanzetti Evidence
KALBFLEISCH and PRENTICE · The Statistical Analysis of Failure Time Data, *Second Edition*
KASS and VOS · Geometrical Foundations of Asymptotic Inference
KAUFMAN and ROUSSEEUW · Finding Groups in Data: An Introduction to Cluster Analysis
KEDEM and FOKIANOS · Regression Models for Time Series Analysis
KENDALL, BARDEN, CARNE, and LE · Shape and Shape Theory
KHURI · Advanced Calculus with Applications in Statistics, *Second Edition*
KHURI, MATHEW, and SINHA · Statistical Tests for Mixed Linear Models
* KISH · Statistical Design for Research
KLEIBER and KOTZ · Statistical Size Distributions in Economics and Actuarial Sciences
KLUGMAN, PANJER, and WILLMOT · Loss Models: From Data to Decisions, *Second Edition*
KLUGMAN, PANJER, and WILLMOT · Solutions Manual to Accompany Loss Models: From Data to Decisions, *Second Edition*
KOTZ, BALAKRISHNAN, and JOHNSON · Continuous Multivariate Distributions, Volume 1, *Second Edition*
KOTZ and JOHNSON (editors) · Encyclopedia of Statistical Sciences: Volumes 1 to 9 with Index
KOTZ and JOHNSON (editors) · Encyclopedia of Statistical Sciences: Supplement Volume
KOTZ, READ, and BANKS (editors) · Encyclopedia of Statistical Sciences: Update Volume 1
KOTZ, READ, and BANKS (editors) · Encyclopedia of Statistical Sciences: Update Volume 2
KOVALENKO, KUZNETZOV, and PEGG · Mathematical Theory of Reliability of Time-Dependent Systems with Practical Applications
LACHIN · Biostatistical Methods: The Assessment of Relative Risks
LAD · Operational Subjective Statistical Methods: A Mathematical, Philosophical, and Historical Introduction
LAMPERTI · Probability: A Survey of the Mathematical Theory, *Second Edition*
LANGE, RYAN, BILLARD, BRILLINGER, CONQUEST, and GREENHOUSE · Case Studies in Biometry
LARSON · Introduction to Probability Theory and Statistical Inference, *Third Edition*
LAWLESS · Statistical Models and Methods for Lifetime Data, *Second Edition*
LAWSON · Statistical Methods in Spatial Epidemiology
LE · Applied Categorical Data Analysis
LE · Applied Survival Analysis
LEE and WANG · Statistical Methods for Survival Data Analysis, *Third Edition*
LEPAGE and BILLARD · Exploring the Limits of Bootstrap
LEYLAND and GOLDSTEIN (editors) · Multilevel Modelling of Health Statistics
LIAO · Statistical Group Comparison
LINDVALL · Lectures on the Coupling Method
LINHART and ZUCCHINI · Model Selection
LITTLE and RUBIN · Statistical Analysis with Missing Data, *Second Edition*
LLOYD · The Statistical Analysis of Categorical Data

*Now available in a lower priced paperback edition in the Wiley Classics Library.
†Now available in a lower priced paperback edition in the Wiley–Interscience Paperback Series.

MAGNUS and NEUDECKER · Matrix Differential Calculus with Applications in Statistics and Econometrics, *Revised Edition*
MALLER and ZHOU · Survival Analysis with Long Term Survivors
MALLOWS · Design, Data, and Analysis by Some Friends of Cuthbert Daniel
MANN, SCHAFER, and SINGPURWALLA · Methods for Statistical Analysis of Reliability and Life Data
MANTON, WOODBURY, and TOLLEY · Statistical Applications Using Fuzzy Sets
MARCHETTE · Random Graphs for Statistical Pattern Recognition
MARDIA and JUPP · Directional Statistics
MASON, GUNST, and HESS · Statistical Design and Analysis of Experiments with Applications to Engineering and Science, *Second Edition*
McCULLOCH and SEARLE · Generalized, Linear, and Mixed Models
McFADDEN · Management of Data in Clinical Trials
* McLACHLAN · Discriminant Analysis and Statistical Pattern Recognition
McLACHLAN, DO, and AMBROISE · Analyzing Microarray Gene Expression Data
McLACHLAN and KRISHNAN · The EM Algorithm and Extensions
McLACHLAN and PEEL · Finite Mixture Models
McNEIL · Epidemiological Research Methods
MEEKER and ESCOBAR · Statistical Methods for Reliability Data
MEERSCHAERT and SCHEFFLER · Limit Distributions for Sums of Independent Random Vectors: Heavy Tails in Theory and Practice
MICKEY, DUNN, and CLARK · Applied Statistics: Analysis of Variance and Regression, *Third Edition*
* MILLER · Survival Analysis, *Second Edition*
MONTGOMERY, PECK, and VINING · Introduction to Linear Regression Analysis, *Third Edition*
MORGENTHALER and TUKEY · Configural Polysampling: A Route to Practical Robustness
MUIRHEAD · Aspects of Multivariate Statistical Theory
MULLER and STOYAN · Comparison Methods for Stochastic Models and Risks
MURRAY · X-STAT 2.0 Statistical Experimentation, Design Data Analysis, and Nonlinear Optimization
MURTHY, XIE, and JIANG · Weibull Models
MYERS and MONTGOMERY · Response Surface Methodology: Process and Product Optimization Using Designed Experiments, *Second Edition*
MYERS, MONTGOMERY, and VINING · Generalized Linear Models. With Applications in Engineering and the Sciences
† NELSON · Accelerated Testing, Statistical Models, Test Plans, and Data Analyses
† NELSON · Applied Life Data Analysis
NEWMAN · Biostatistical Methods in Epidemiology
OCHI · Applied Probability and Stochastic Processes in Engineering and Physical Sciences
OKABE, BOOTS, SUGIHARA, and CHIU · Spatial Tesselations: Concepts and Applications of Voronoi Diagrams, *Second Edition*
OLIVER and SMITH · Influence Diagrams, Belief Nets and Decision Analysis
PALTA · Quantitative Methods in Population Health: Extensions of Ordinary Regressions
PANKRATZ · Forecasting with Dynamic Regression Models
PANKRATZ · Forecasting with Univariate Box-Jenkins Models: Concepts and Cases
* PARZEN · Modern Probability Theory and Its Applications
PEÑA, TIAO, and TSAY · A Course in Time Series Analysis
PIANTADOSI · Clinical Trials: A Methodologic Perspective
PORT · Theoretical Probability for Applications
POURAHMADI · Foundations of Time Series Analysis and Prediction Theory
PRESS · Bayesian Statistics: Principles, Models, and Applications

*Now available in a lower priced paperback edition in the Wiley Classics Library.
†Now available in a lower priced paperback edition in the Wiley–Interscience Paperback Series.

PRESS · Subjective and Objective Bayesian Statistics, *Second Edition*
PRESS and TANUR · The Subjectivity of Scientists and the Bayesian Approach
PUKELSHEIM · Optimal Experimental Design
PURI, VILAPLANA, and WERTZ · New Perspectives in Theoretical and Applied Statistics
† PUTERMAN · Markov Decision Processes: Discrete Stochastic Dynamic Programming
QIU · Image Processing and Jump Regression Analysis
* RAO · Linear Statistical Inference and Its Applications, *Second Edition*
RAUSAND and HØYLAND · System Reliability Theory: Models, Statistical Methods, and Applications, *Second Edition*
RENCHER · Linear Models in Statistics
RENCHER · Methods of Multivariate Analysis, *Second Edition*
RENCHER · Multivariate Statistical Inference with Applications
* RIPLEY · Spatial Statistics
RIPLEY · Stochastic Simulation
ROBINSON · Practical Strategies for Experimenting
ROHATGI and SALEH · An Introduction to Probability and Statistics, *Second Edition*
ROLSKI, SCHMIDLI, SCHMIDT, and TEUGELS · Stochastic Processes for Insurance and Finance
ROSENBERGER and LACHIN · Randomization in Clinical Trials: Theory and Practice
ROSS · Introduction to Probability and Statistics for Engineers and Scientists
† ROUSSEEUW and LEROY · Robust Regression and Outlier Detection
* RUBIN · Multiple Imputation for Nonresponse in Surveys
RUBINSTEIN · Simulation and the Monte Carlo Method
RUBINSTEIN and MELAMED · Modern Simulation and Modeling
RYAN · Modern Regression Methods
RYAN · Statistical Methods for Quality Improvement, *Second Edition*
SALTELLI, CHAN, and SCOTT (editors) · Sensitivity Analysis
* SCHEFFE · The Analysis of Variance
SCHIMEK · Smoothing and Regression: Approaches, Computation, and Application
SCHOTT · Matrix Analysis for Statistics, *Second Edition*
SCHOUTENS · Levy Processes in Finance: Pricing Financial Derivatives
SCHUSS · Theory and Applications of Stochastic Differential Equations
SCOTT · Multivariate Density Estimation: Theory, Practice, and Visualization
* SEARLE · Linear Models
SEARLE · Linear Models for Unbalanced Data
SEARLE · Matrix Algebra Useful for Statistics
SEARLE, CASELLA, and McCULLOCH · Variance Components
SEARLE and WILLETT · Matrix Algebra for Applied Economics
SEBER and LEE · Linear Regression Analysis, *Second Edition*
† SEBER · Multivariate Observations
† SEBER and WILD · Nonlinear Regression
SENNOTT · Stochastic Dynamic Programming and the Control of Queueing Systems
* SERFLING · Approximation Theorems of Mathematical Statistics
SHAFER and VOVK · Probability and Finance: It's Only a Game!
SILVAPULLE and SEN · Constrained Statistical Inference: Inequality, Order, and Shape Restrictions
SMALL and McLEISH · Hilbert Space Methods in Probability and Statistical Inference
SRIVASTAVA · Methods of Multivariate Statistics
STAPLETON · Linear Statistical Models
STAUDTE and SHEATHER · Robust Estimation and Testing
STOYAN, KENDALL, and MECKE · Stochastic Geometry and Its Applications, *Second Edition*

*Now available in a lower priced paperback edition in the Wiley Classics Library.
†Now available in a lower priced paperback edition in the Wiley–Interscience Paperback Series.

STOYAN and STOYAN · Fractals, Random Shapes and Point Fields: Methods of Geometrical Statistics
STYAN · The Collected Papers of T. W. Anderson: 1943–1985
SUTTON, ABRAMS, JONES, SHELDON, and SONG · Methods for Meta-Analysis in Medical Research
TANAKA · Time Series Analysis: Nonstationary and Noninvertible Distribution Theory
THOMPSON · Empirical Model Building
THOMPSON · Sampling, *Second Edition*
THOMPSON · Simulation: A Modeler's Approach
THOMPSON and SEBER · Adaptive Sampling
THOMPSON, WILLIAMS, and FINDLAY · Models for Investors in Real World Markets
TIAO, BISGAARD, HILL, PEÑA, and STIGLER (editors) · Box on Quality and Discovery: with Design, Control, and Robustness
TIERNEY · LISP-STAT: An Object-Oriented Environment for Statistical Computing and Dynamic Graphics
TSAY · Analysis of Financial Time Series
UPTON and FINGLETON · Spatial Data Analysis by Example, Volume II: Categorical and Directional Data
VAN BELLE · Statistical Rules of Thumb
VAN BELLE, FISHER, HEAGERTY, and LUMLEY · Biostatistics: A Methodology for the Health Sciences, *Second Edition*
VESTRUP · The Theory of Measures and Integration
VIDAKOVIC · Statistical Modeling by Wavelets
VINOD and REAGLE · Preparing for the Worst: Incorporating Downside Risk in Stock Market Investments
WALLER and GOTWAY · Applied Spatial Statistics for Public Health Data
WEERAHANDI · Generalized Inference in Repeated Measures: Exact Methods in MANOVA and Mixed Models
WEISBERG · Applied Linear Regression, *Third Edition*
WELSH · Aspects of Statistical Inference
WESTFALL and YOUNG · Resampling-Based Multiple Testing: Examples and Methods for p-Value Adjustment
WHITTAKER · Graphical Models in Applied Multivariate Statistics
WINKER · Optimization Heuristics in Economics: Applications of Threshold Accepting
WONNACOTT and WONNACOTT · Econometrics, *Second Edition*
WOODING · Planning Pharmaceutical Clinical Trials: Basic Statistical Principles
WOODWORTH · Biostatistics: A Bayesian Introduction
WOOLSON and CLARKE · Statistical Methods for the Analysis of Biomedical Data, *Second Edition*
WU and HAMADA · Experiments: Planning, Analysis, and Parameter Design Optimization
YANG · The Construction Theory of Denumerable Markov Processes
* ZELLNER · An Introduction to Bayesian Inference in Econometrics
ZHOU, OBUCHOWSKI, and McCLISH · Statistical Methods in Diagnostic Medicine

*Now available in a lower priced paperback edition in the Wiley Classics Library.
†Now available in a lower priced paperback edition in the Wiley–Interscience Paperback Series.

Lightning Source UK Ltd.
Milton Keynes UK
UKOW03n1524050914

238153UK00001B/51/P

9 780471 420996